Synthesis and Chemistry of Agrochemicals V

ACS SYMPOSIUM SERIES **686**

Synthesis and Chemistry of Agrochemicals V

Don R. Baker, EDITOR
Zeneca Agricultural Products

Joseph G. Fenyes, EDITOR
Buckman Laboratories International

Gregory S. Basarab, EDITOR
DuPont Agricultural Products

David A. Hunt, EDITOR
American Cyanamid Company

Developed from a symposium sponsored by the Division of Agrochemicals of the American Chemical Society,

American Chemical Society, Washington, DC

ISBN 0–8412–3546–5

ISSN 0097–6156

This book is printed on acid-free, recycled paper.

Distributed by Oxford University Press

PRINTED IN THE UNITED STATES OF AMERICA

Foreword

THE ACS SYMPOSIUM SERIES was first published in 1974 to provide a mechanism for publishing symposia quickly in book form. The purpose of the series is to publish timely, comprehensive books developed from ACS sponsored symposia based on current scientific research. Occasionally, books are developed from symposia sponsored by other organizations when the topic is of keen interest to the chemistry audience.

Before agreeing to publish a book, the proposed table of contents is reviewed for appropriate and comprehensive coverage and for interest to the audience. Some papers may be excluded in order to better focus the book; others may be added to provide comprehensiveness. When appropriate, overview or introductory chapters are added. Drafts of chapters are peer-reviewed prior to final acceptance or rejection, and manuscripts are prepared in camera-ready format.

As a rule, only original research papers and original review papers are included in the volumes. Verbatim reproductions of previously published papers are not accepted.

ACS BOOKS DEPARTMENT

Contents

Control of Insects and Acids

CONTROL OF PLANT FUNGI

INDEXES

Preface

THE LAST HALF OF THE 20TH CENTURY has seen major changes in the synthesis and development of new agrochemicals. This whole period has been in a state of flux in most of the aspects of the design, synthesis, and development of new active materials. Now as the dawn of the 21st century approaches, we see various agricultural biotechnological efforts supplementing and replacing standard agrochemicals. It appears, however, that biotechnology is not the panacea it was once touted to be in the late 1970s. Similarly, it is also clear that the regulatory field for agrochemicals has been drastically altered over the last few years.

The editors of these volumes have organized synthesis symposia at each ACS National meeting since 1984. These symposia have been sponsored by the Agrochemicals Division of the ACS. The aim of these symposia has been to provide a forum for presenting the synthesis and chemistry of new agrochemical agents. These symposia have provided a focus for agrochemical synthesis chemists to share their best work. The current book has chapters taken from the symposia presented at the Washington, DC ACS National Meeting held in August of 1994 through the San Francisco ACS National Meeting held in the spring of 1997. Each of the chapter authors was requested to present the current status of the work at the time of preparation of the manuscript. In many cases considerably more effort has been done since the work was presented at the meeting. These chapters present a unique look into the discovery process of our current and future agrochemical products. Not all of the chapters represent materials that will eventually be commercialized. It is hoped that these chapters will also provide a learning experience both for those working in this field and for others interested in the biological effects of organic molecules.

As with the previous volumes, our goal is to inform the reader of the current trends in research for safe, efficient, biologically active agrochemicals. The organization of this book is similar to that of the preceding volumes. After the overview chapter there is a section dedicated to Dr. Marinus Los and his important discovery of the Imidazolinone herbicides. The next section describes the discovery of new herbicides and plant control materials. The following section deals with the control of insects, acarids, and nematodes. The final section covers the control of plant fungal diseases.

Acknowledgments

We express our thanks to the authors who have shared with us the details of their interesting research results—first at the symposia and here now in going that extra mile in preparing the chapters ready for publication. We hope that you our readers—whether you are our fellow synthesis chemists, or microbiologists, entomologists, plant physiologists, biotechnologists, or medicinal and pharmaceutical chemists—will find the chapters interesting, useful and above all stimulating!

Last, but not least, we also wish to thank our employers, Zeneca Ag Products, Buckman Laboratories International, DuPont, and American Cyanamid Company, without their generous support of these symposia, this volume and the previous ones could not have been published.

DON R. BAKER
Zeneca Ag Products
Western Research Center
1200 South 47th Street
Richmond, CA 94804–0024

JOSEPH G. FENYES
Buckman Laboratories International
1256 North McLean Boulevard
Memphis, TN 38108

GREGORY S. BASARAB
DuPont Agricultural Products
Stine-Haskell Research Center
Newark, DE 1714

DAVID A. HUNT
American Cyanamid Company
P.O. Box 400
Princeton, NJ 08543

Chapter 1

Synthesis of Agrochemicals and Agricultural Biotechnology Entering the 21st Century

David A. Hunt, Don R. Baker, Joseph G. Fenyes, and Gregory S. Basarab

[1]Cyanamid Agricultural Research Center, American Cyanamid Company, P.O. Box 400, Princeton, NJ 08543–0400

[2]Zeneca Agricultural Products, 1200 South 47th Street, Richmond, CA 94804

[3]Buckman Laboratories International, Inc., 1256 North McLean Boulevard, Memphis, TN 38108

[4]Stine-Haskell Research Center, DuPont Agricultural Products, Newark, DE 19714

The last few years have been witness to dramatic changes in the agrochemical synthesis field. Since many of the approaches of the agrochemical sector to finding new active agents are similar to those in the pharmaceutical field, the newer technologies developed for pharmaceutical research have been embraced by agrochemical research. Among these are combinatorial chemistry and robotics synthesis as well as aggressive acquisition of large compound libraries from academic and commercial sources. This delivers greater inputs into the screening process. Robotics are also increasingly being employed for the screening process itself. The prime goal is to satisfy the ever increasing need for new leads compounds. The better new leads thus generated are then optimized by more conventional means or else by means of automated synthesis of more directed compound libraries. Consequently, the potential for agrochemical scientists to continue to provide safer, lower use-rate, more environmentally benign crop protection agents remains high. Concurrently during the past decade, great strides have been made in the adaptation and integration of biotechnology to agricultural production. Much of the work in this area has focused in developing new biological methods of pest control, crops with engineered resistance to pests and herbicides, and products with higher nutritional content. Overall this new technology promises to reduce the environmental load of chemical pesticides. While the development of molecular biology and its subsequent application to agriculture has been no less than spectacular, it has raised many questions concerning the landscape of pest control methodology in the future of global agriculture, particularly with regard to the place of traditional synthesis chemistry.

As we approach the dawn of the 21st century there is an ever increasing need for agricultural food and fiber. The world population continues to increase creating progressively more production pressure on dwindling arable land resources (*1*). Many fear that as we proceed several decades into the 21st century that the world population will become so large that it can no longer feed itself and that the environment will concurrently be destroyed. Many foresee a growth in agrochemical technology to ensure that such catastrophes can be avoided. However, environmental and health fears associated with agrochemicals has greatly increased the regulatory pressure on their use. This has greatly increased the costs of registration and re-registration of agrochemicals. Research and development costs for new agrochemicals have risen to the point that only major markets can justify the great development time and expense for new products. Minor crops find fewer and fewer materials registered for their use, much to the misfortune of those farmers who are effected. Likewise, resistance to single mode of action agrochemicals continues to grow and be a major problem. Resistance is now a common problem in most of the agrochemical markets. It has become such a problem that major efforts are being made to screen large numbers of compounds looking for new lead materials which have novel modes of action compared to current products in order to provide effective control of resistant and non-resistant pest populations. The net result is that agrochemical technology can be expected to partner with biotechnology to solve the problems identified and support the market demands of the world's farmers and indirectly, the world's population.

The basic understanding of biological systems on the molecular level coupled with the power of synthesis chemistry has created opportunities which were unimaginable just a decade ago. Molecular biology has enabled researchers to know specific enzyme and receptor targets for small molecular inhibition and characterize the genes that encode them before the compounds move from research to development. This allows refinement of the agrochemical design process either to increase compound potency or decrease the binding to off-target proteins. The power of combinatorial chemistry to provide orders of magnitude greater numbers of compounds for lead identification and the concomitant growth in the development of molecular-based high throughput screens have rapidly altered the research environment during the past few years. In the case of combinatorial chemistry, the pharmaceutical industry first embraced the technology resulting in the birth of companies focusing on the generation and sale of compound libraries. However, agricultural applications of this burgeoning technology are now being explored and exploited.

The many and varied ways of discovery of new agrochemicals have been exemplified in this series of ACS Books (*2-5*). The important discovery of the imidazolinone herbicides by Dr. Marinus Los and his American Cyanamid Co. coworkers is exemplified in the first section. The following three sections demonstrate discoveries in the control of weeds and plant growth; the control of insects, acarids and nematodes; and the control of fungi. Throughout the chapters of this book, the relationship between agrochemical discoveries and their biochemical target-sites of action get particular attention. As our ability to manipulate genes and their protein products improve, the importance of agricultural biotechnology and its partnership with chemistry can be expected to be amplified.

Agricultural Biotechnology

In the preface of a 1987 ACS monograph entitled "Biotechnology in Agricultural Chemistry" the authors state "...The only place biotechnology can be sold as a science is on the stock market. Everywhere else, biotechnology is merely a means of developing tools or products that must be applied in some practical way. Biotechnology must compete with the performance, ease of use, effectiveness, and economy of all other methods for doing the same job. It must face the traditions and biases of the user"(*6*). Those criteria today have been met with success in many cases.

Biotechnology appears to be ushering science into a new era, with agricultural applications having seemingly boundless potential to enhance productivity, the environment, food safety, and nutritional quality. Some of the most significant developments over the past five years have resided in the field of transgenic plants which express foreign genes thereby conferring biologically beneficial traits to major crops, ranging from longer shelf life to pest resistance. Transgenic plant technology directed toward herbicide tolerance and insect control attained full approval for agronomic use in key crops in North America in 1995 and 1996. US farmer acceptance of this new tool indicates that this technology in a variety of forms will continue to be a significant force in the future of global crop protection. However, global acceptance is still very much controversial.

In a recent article in the Chronicle of Higher Education (*7*), Nina Fedoroff of the Life Sciences Consortium and Biotechnology Institute at Pennsylvania State University eloquently expresses concern over the need to develop methods to increase agricultural productivity in the face of a rapidly expanding global population. With the current population growth rate, people are consuming more than they are producing. A strong proponent of biotechnology, Ms. Fedoroff indicates in her article that concomitant development and utilization of a variety of technologies will be required in order to meet the increasing needs of the human community.

As agrochemicals were lauded as the ultimate in pest control methods decades ago, so too has biotechnology today in some circles. With all of this promise, could there possibly be a down side to biotechnology? Uncertainties have been raised regarding questions of both long- and short-term impacts on health matters, food safety, the environment, and costs relating to both social and economic issues. For example, researchers have long known that transgenic plants can form hybrids with wild relatives. Scientists in Denmark have recently shown that these hybrids can transmit genetic traits to subsequent generations under field conditions (*8*). Besides speculations of the rapid development of resistance, harmful traits may be transferred between transgenic crops. The incorporation of new genetic material into a host organism gives rise to the production of non-native proteins, which has led to health-related concerns. This type of phenomenon was observed recently when a Brazil nut gene was engineered into soybeans in order to provide enhanced levels of methionine for animal feed. The project was terminated when tests revealed that extracts from the genetically altered soybeans caused allergic reactions (*9,10*).

Based on these factors, does synthesis chemistry still have contributions for crop protection, or has it been made obsolete by the development and commercialization of biotechnologically engineered products? Even with all of this unrealized potential, it appears that biotechnology is not the panacea it was once touted to be in the late 1970's.

Likewise, it is clear that the regulatory landscape for crop protection chemicals has been drastically and irreversibly altered over the past decade. The ability to feed a rapidly growing global population, the increased awareness of the impact that pesticides have on the environment, and the need to minimize cost are clearly the major driving forces for the development of new technologies for agriculture in the 21st century. While synthesis chemistry has established itself as a key factor in conventional farming and biocontrol methods have gained acceptance, synthetic chemicals and genetically engineered biopesticides are not independently the solution for global pest control and enhanced agricultural productivity. Most likely both will be necessary in the farmer's arsenal based on their ability to provide economic value. New products derived from both chemical synthesis and biotechnology with unique modes of action will be required due to the dynamic nature of agriculture. It is most likely that pest management will come to rely on an integration of technologies, combining current methods and technologies with new ones. This approach accurately reflects the state of life sciences research today - the convergence of a variety of scientific disciplines to discover novel applications for both the agricultural and pharmaceutical industries.

This evolution of science as applied to agriculture is evident in that many agricultural chemical companies and universities are incorporating new tools from both molecular biology and chemistry into product discovery efforts. The following are typical of the great variety of these efforts:

1. The development of plant systems capable of being chemically induced to provide multi-target defenses against pathogens - a process similar in nature to animal cellular immunity.
2. On going attempts to discover natural crop protection agents from natural products.
3. The development of genetically-mediated syntheses of natural products.
4. Development of genes responsible for small molecule synthesis which can be stacked and shuffled, thereby producing novel complex structures.
5. Use of anti sense technology to identify potential herbicide targets.
6. The use of antibody technology as a method to develop surrogate receptors in compound screening.
7. Microbial applications of biotechnology for agricultural uses provide natural products as both commercial entities and templates for analog synthesis, as well as systems for the degradation of waste pesticides and other environmentally hazardous materials.

What is the future for synthesis chemistry in light of the rapid development of biotechnology? The advances in the life sciences hinge on increased knowledge in organic chemistry and, by definition, molecular biology has its roots in organic chemistry. Still, "...the science is best described as being between infancy and childhood" (*11*). With the rapid development of molecular biology, the formal boundaries between these disciplines have become vague. Synthesis chemistry will continue to play an important role in crop protection methods in the future and will continue to evolve to meet challenges presented to molecular biology in support of future developments in biotechnology.

Literature Cited

1. Klassen, W.; in *Eighth International Congress of Pesticide Chemistry, Options 2000;* Ragsdale, N.N.; Kearney, P. C.; Plimmer, J. R., Eds; ACS Conference Proceedings Series, Washington, DC, 1995; pp 1-32.
2. Baker, D. R.; Fenyes, J. G.; Basarab, G. S.; Eds., *Synthesis and Chemistry of Agrochemicals IV,* ACS Symposium Series, 584, Washington, DC, 1995.
3. Baker, D. R.; Fenyes, J. G.; Steffens, J. J.; Eds., *Synthesis and Chemistry of Agrochemicals III,* ACS Symposium Series, 504, Washington, DC, 1991.
4. Baker, D. R.; Fenyes, J. G.; Moberg W. K.; Eds., *Synthesis and Chemistry of Agrochemicals II,* ACS Symposium Series, 443, Washington, DC, 1989.
5. Baker, D. R.; Fenyes, J. G.; Moberg W. K.; Cross B.; Eds., *Synthesis and Chemistry of Agrochemicals,* ACS Symposium Series, 355, Washington, DC, 1987.
6. LeBaron, H. M.; Mumma, R. O.; Honeycutt, R. C.; Duesing, J. H. in *Biotechnology in Agricultural Chemistry*; LeBaron, H. M.; Mumma, R. O.; Honeycutt, R. C.; Duesing, J. H., Eds.; ACS Symposium Series 334; American Chemical Society: Washington, DC, 1987; p xi.
7. V. Fedoroff, "Food for a Hungry World: We Must Find Ways to Increase Agricultural Productivity", *The Chronicle of Higher Education* **1997**, *43*, Number 41, pp B4-B5.
8. Mikkelsen, T. R.; Andersen, B.; Joergensen, R. B. *Nature* **1996**, *380 (6569)*, p 31.
9. Beardsley, T. "Advantage: Nature. Could Escaped Genes from Bioengineered Crops Give Weeds a Critical Boost?", *Scientific American*; May, **1996**, p 33.
10. Nordlee, J. A.; Taylor, S. L.; Townsend, J. A.; Thomas, L. A.; Bush, R. K. *The New England Journal of Medicine,* **1996**, *334*, pp 688-692.
11. Trost, B.M. "Sculpting Horizons in Organic Chemistry," *Science* **1985**, *227*, p 908.

Creative Invention and the Imidazolinone Herbicides

Chapter 2

Anomalies and the Discovery of the Imidazolinone Herbicides

Marinus Los

Cyanamid Agricultural Research Center, American Cyanamid Company, P.O. Box 400, Princeton, NJ 08543–0400

The imidazolinone herbicides have become important tools used by farmers around the world. The role that anomalies, both biological and chemical, played in the discovery and development of the imidazolinones is detailed. These anomalies led to new avenues for exploration, to new biologically-active structures, and sustained the project for more than fourteen years.

The imidazolinone story spans some fourteen years of research. It is of interest and instructive to reflect on the factors that both started the project and sustained it for the length of time. During this time there were milestones, better described as anomalies, which fueled the project and it is these anomalies that will be the basis for this paper.

Anomalies are not an unusual happening in any research project and could, in fact, be cited as one of the characteristics of a successful one. In the imidazolinone project these anomalies occurred, for example, in the formation of a minor by-product from a seemingly trival reaction or as an unexpected outcome of what appeared to be a rational alternative synthesis for a known compound. Thomas S. Kuhn of the Massachusetts Institute of Technology stated it very clearly in his book "The Structure of Scientific Revolutions": "Discovery commences with the awareness of anomaly, i.e. with the recognition that nature has somehow violated the paradigm-induced expectations that govern normal science. It then continues with a more or less extended exploration of anomaly. And it closes only when the paradigm theory has been adjusted so that the anomalous has become the expected."[2]. This in essence describes the imidazolinone project.

The project started with the screening for herbicidal activity of a phthalimide **1** which had been prepared in a pharmaceutical project as an analog of an anti-convulsant. This compound was not very active but had biological properties rarely seen in weakly active herbicides - it controlled two very important perennial weeds, *Cyperus rotundus* and *Convolvulus arvensis*.

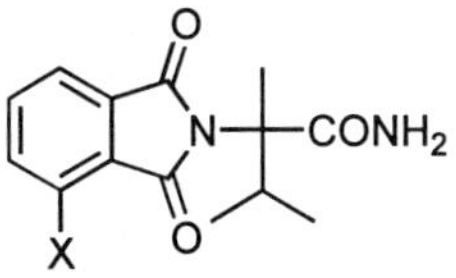

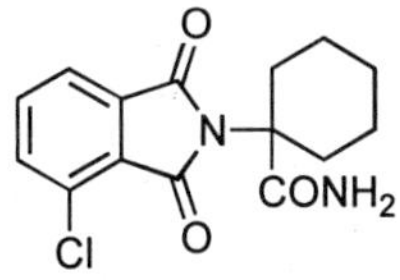

X
1 H
2 Cl

3 (AC 94377)

This then was the impetus for a small synthesis effort. This did not result in a "better" herbicide but instead yielded a compound **2** with essentially no herbicidal activity but rather activity not unlike that of gibberellic acid. This was an anomaly and, as Professor Kuhn stated, led to the "exploration of the area of anomaly". By the usual chemical manipulations of the structure, it was possible to greatly increase the level of activity in compound **3**. It is remarkable that a compound of such simplicity can indeed mimic the rather complex structure of gibberellic acid with its many chiral centers.

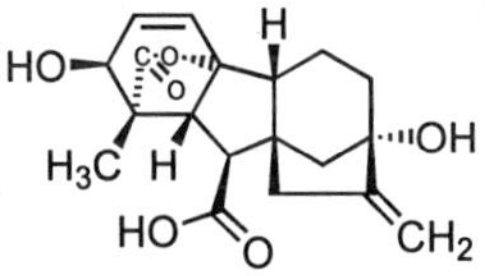

GA_3

AC 94377 was field tested extensively and relatively large quantities were required. This rather simple molecule can be assembled in a number of ways. The 3-chlorophthalic anhydride can be reacted with the approparite α-amino nitrile, amide, or acid. All three routes were examined. The one from the α-amino amide yielded the phthalamic acid **4** as a mixture of isomers which could be cyclized with acetic anhydride to the imide **3** in high yield. As an alternative dehydrating agent, the use of trifluoroacetic anhydride was examined. Although it also afforded **3** as the major product, a small quantity of a by-product was isolated which proved to be the result of the further dehydration of **3** to give **5**.

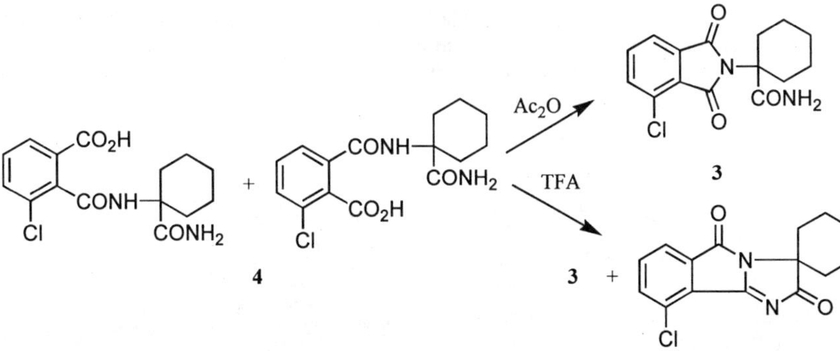

4 **3** **5**

This was another anomaly and when it found that the biological properties of **5** were very similar to those of **3**, this area of chemistry demanded further exploration. A good method was required for the synthesis of compounds such as **5**. In fact, the cyclization of the imides such a **1** with either sodium hydride or sodium hydroxide pels proved to be most efficient.

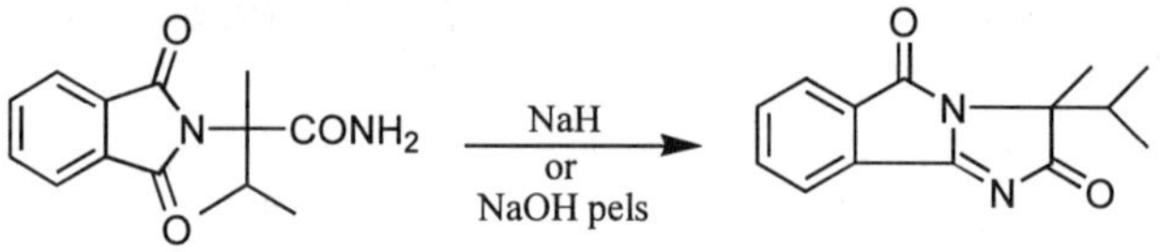

1 6

These tricyclic derivatives such as **6** proved to be pivotal in the discovery and development of the imidazolinones. First, they were novel structures with interesting biological activities and thus deserved attention. Secondly, while they themselves did not reach the levels of activity and selectivity required for commercial development, their structures did lend themselves to further modification and manipulation.

It was quickly discovered that compounds such as **6** could be reduced with sodium borohydride to the corresponding dihydro derivative which retained high herbicidal activity. The mixture of isomers could be readily separated by fractional crystallization. The field testing of **6** required that it be formulated.

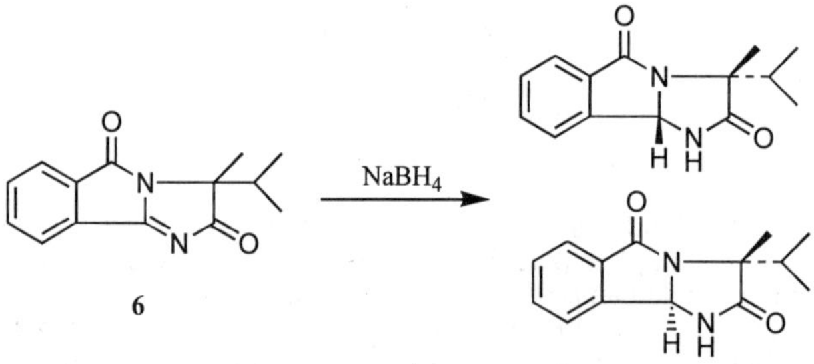

6

On standing, the experimental formulation deposited a nicely crystalline material, analysis of which indicated that this material had a molecular formula consisting of **6** plus the elements of methanol. The structure could not be assigned unambiguously based on nmr data and thus attempts were made to synthesize the compound.

The reaction of **6** with methanol was examined under a variety of conditions. Under acidic conditions, methanol added to **6** to give a product identical to that isolated from the experimental formulation and this proved to have structure **7**. Conditions were established under which a variety of alcohols, thiols, amines and water would add to structures such as **6**.

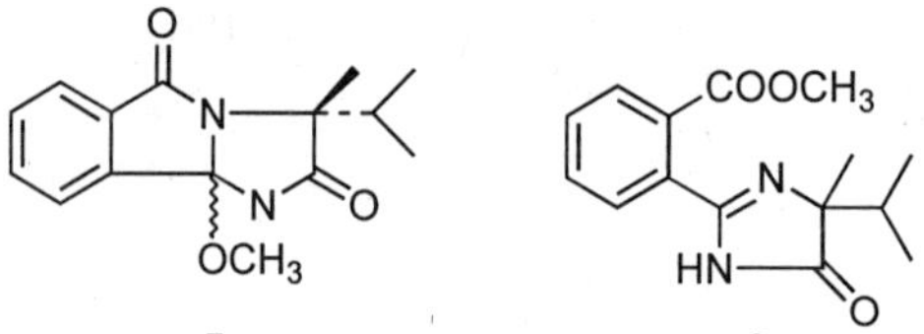

7 8

Another set of conditions employed to try to produce the product formed in the formulation involved the treatment of **6** with methanol in the presence of sodium methoxide. This resulted in an extremely rapid reaction and the formation of yet another unexpected product. This proved to be the imidazolinone **8** which not only was a novel structure but also displayed a high level of herbicidal activity. This then was another anomaly and the beginning of a most successful project.

It was thought at that time that this was our first synthesis of an imidazolinone. In fact, one had been prepared unwittingly during the exploration of alternate routes to a rather simple intermediate for compounds such as **6**. The phthalamic acid **9** had been readily converted to the imide **10** with acetic anhydride. When thionyl chloride was substitute for the acetic anhydride, **10** was again the major product but two other products, **1** and **6**, were also isolated.

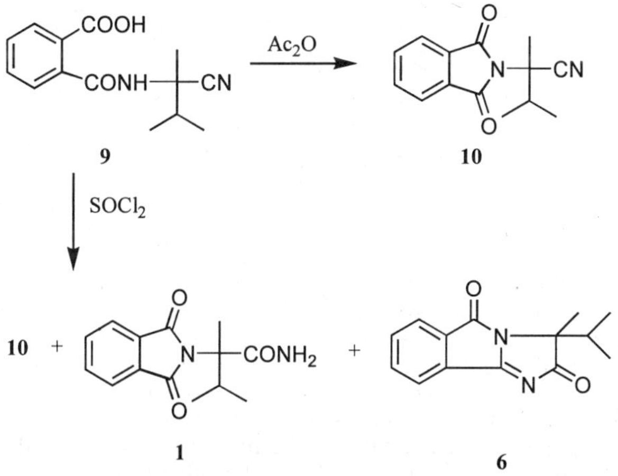

A considerable amount of a water-soluble black tar was also produced.

This reaction was examined further. The hydrogen chloride produced in the reaction was thought to play an important role in the outcome of the reaction and thus the effect of HCl gas on **9** was examined under a variety of conditions. Eventually it was shown that when **9** in dichlorobenzene is treated at room temperature with dry HCl gas, there is produced in over 80% total yield, equal amounts of the imide **1** and the imidazolinone hydrochloride **11**. The black tar prepared in the initial thionyl chloride reaction proved to consist primarily of the hydrochloride **11**.

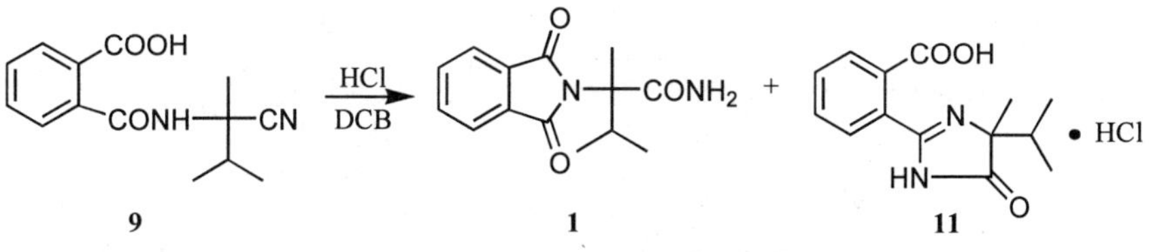

The reaction is believed to proceed as follows:

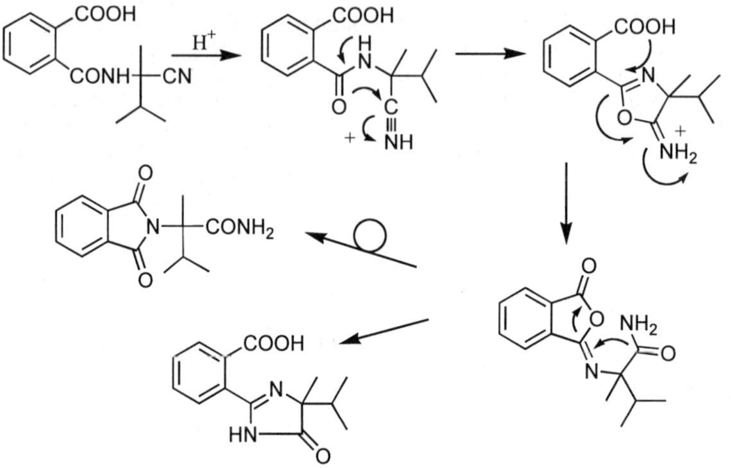

The method of preparation of the tricyclic compounds such as **6** and their conversion to imidazolinones has severe limitations in that the use of an unsymmetrical anhydride results in a mixture of tricyclic compounds and subsequently a mixture of imidazolinones. Usually this is most undesirable. Yet in an early case involving 4-methylphthalic anhydride, the product was indeed the expected mixture of imidazolinones which is sold commercially as ASSERT® herbicide[3]. That this product is a mixture is actually an asset since each isomer is active on different and important weeds in wheat. Thus the m-isomer controls wild oats whereas the p-isomer has activity on wild mustard. It would still be desirable to have an economically competitive method to make either isomer regiospecifically.

Real difficulties arose when the surprisingly active pyridyl-substituted imidazolinones were discovered. At that time, pyridines were not a common constituent of agrochemicals. They were expensive and could involve quite difficult chemitsry. Since all pyridine-2,3-dicarboxylic acids are unsymmetrical, to make economically feasible products in the pyridine and quinoline series required that methodology be developed that did not waste one half of the starting diacid. Efforts to minimize this problem will be discussed later.

It is appropriate to mention another reaction of the tricyclic derivatives. In the benzene series, treatment of **6** with 6N HCl resulted in the formation of the imidazolinone hydrochloride **11** in good yield. This could be interpreted as the simple hydrolysis at the benzoyl carbonyl group.

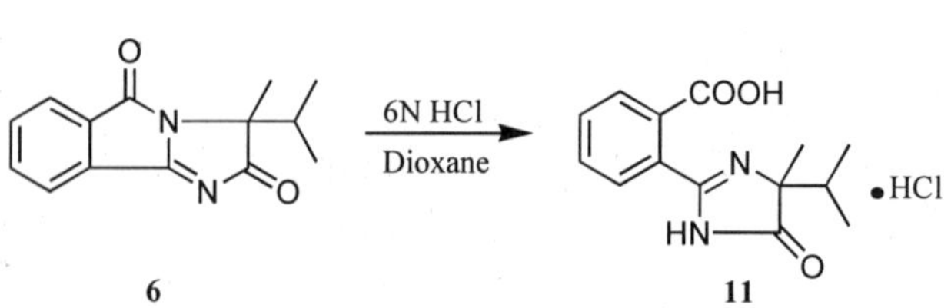

6 **11**

The same reaction applied in the pyridine series resulted in an unexpected and curious result:

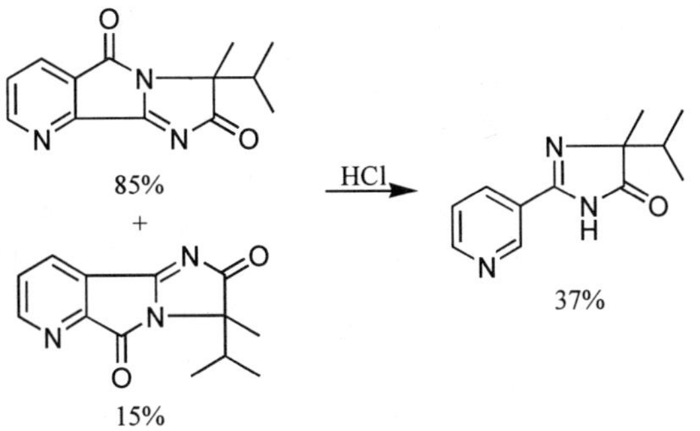

This result can be explained the following way:

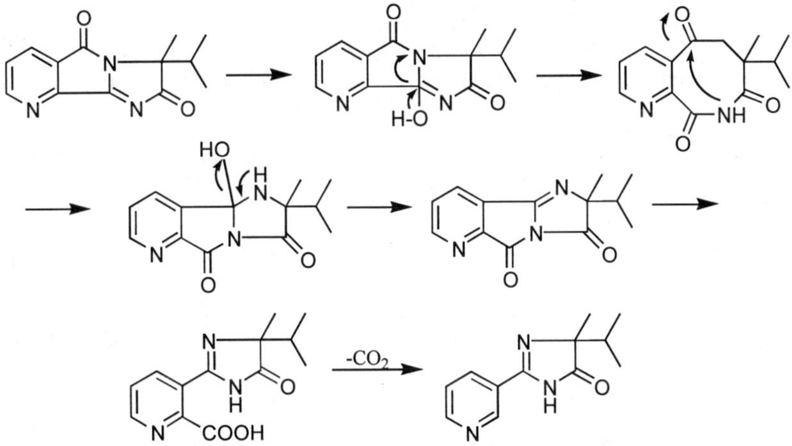

That this is indeed a plausible mechanism was demonstrated by treating a single isomer **12** of the tricyclic derivative from 4-methylphthalic acid with hydrochloric acid and isolating a small amount of the imidazolinone **13** which could only have been formed by the rearrangement described above.

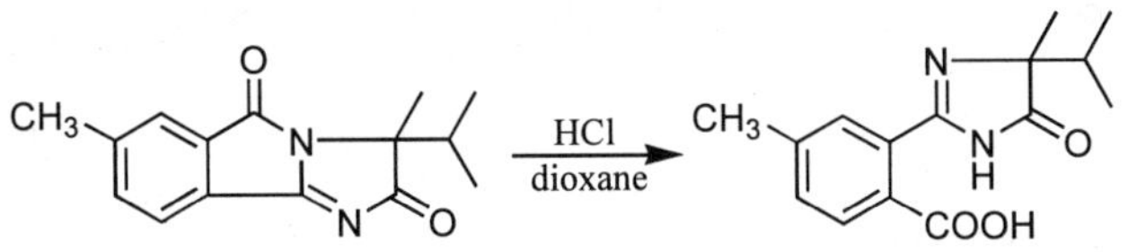

12 **13**

Much chemistry was performed on the tricyclic mixture **15** in the pyridine series produced by the cyclization of **14**.

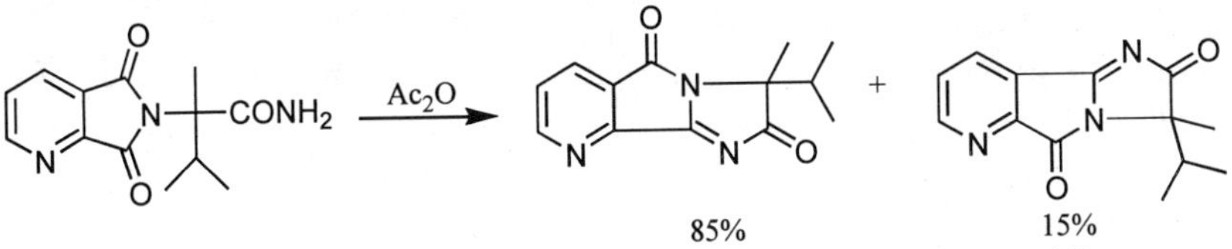

14 **15a** **15b**

With the nicotinic acid **16** readily available, attempts were made to use this as the starting material for an alternate synthesis of pure **15a**. The reaction of **16** with acetic anhydride did yield primarily **15a** but contaminated with 15% of a new material.

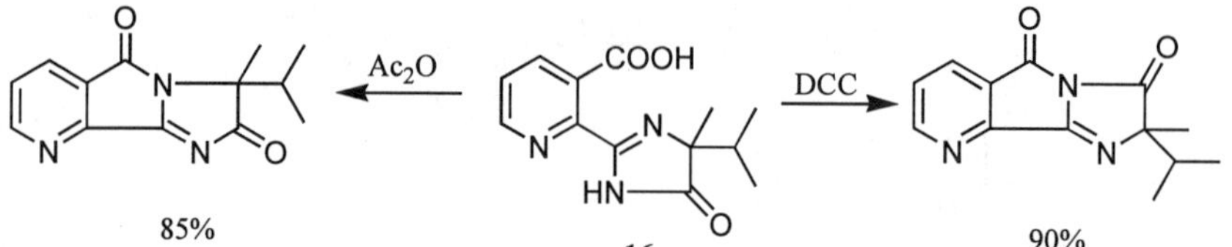

85% **15a** + **17** (15%)

16

90% **17** + **15a** (10%)

In an attempt to increase the purity of the product, the acid **16** was treated with dicyclohexylcarbodiimide. The major product now proved to be a new tricyclic compound **17** formed in 90% yield with properties clearly different from the earlier produced material **15**. This unexpected result provided an intermediate that allowed access to a variety of new substituents at C-3 of the pyridine ring - an opportunity to better define the structural requirements for herbicidal activity.

In order to explore this area of pyridine/quinoline chemistry, it became necessary to develop new synthetic methods, particularly those that would lead to the biologically-active nicotinic acids rather than mixtures. One such process is based on some Russian work which used $PCl_5/POCl_3$ to cyclize **18**.[4]

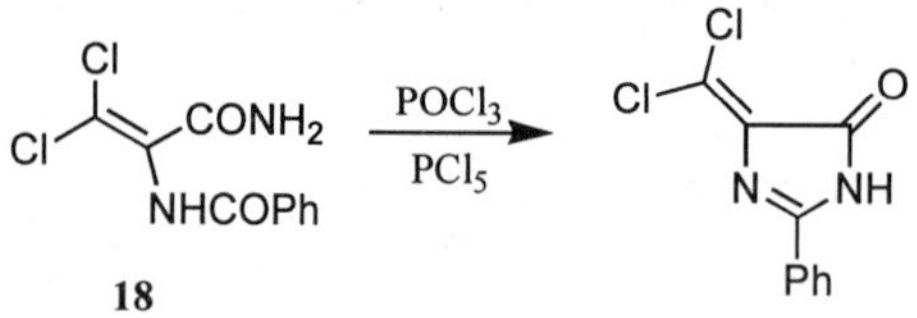

18

This reaction was applied to the pyridine series as shown below. The reaction conditions as described by the Russians are severe and work-up difficult and it was found possible to greatly simplify these by utilizing only PCl_5 in toluene.

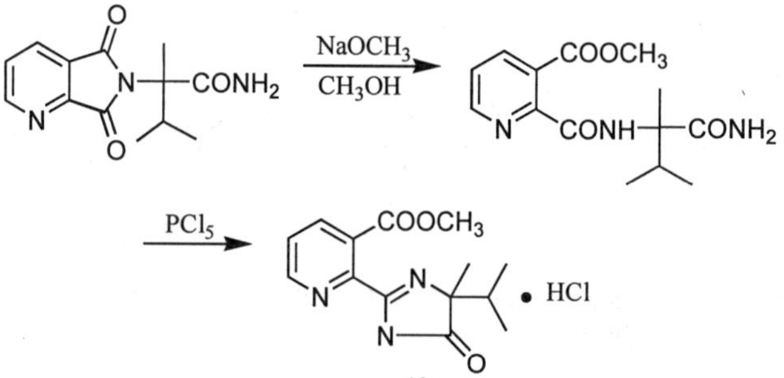

19

The ester **19**, as the hydrochloride salt, could be isolated simply by filtration of the reaction mixture.

It was reasoned that the reaction could be further simplified by substituting the Vilsmeier reagent for the PCl_5. Surprisingly, this reaction took an unexpected, anomalous course to yield the zwitterionic molecule **20** which still retained high herbicidal activity.

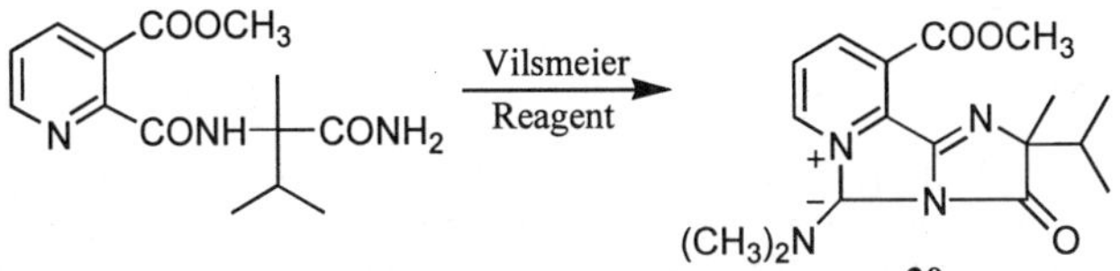

20

This new chemistry did not, however, satisfy the need to prepare the variety of analogs required to explore the extent of the discovery. The solution eventually came from an extension of the work of A.I. Meyers who had demonstrated the directing and stabilizing effect of the oxazoline ring in the metallation of aromatic rings.[5] Could the imidazolinone ring serve the same purpose? In fact, the imidazolinone ring proved to be an excellent substitute.

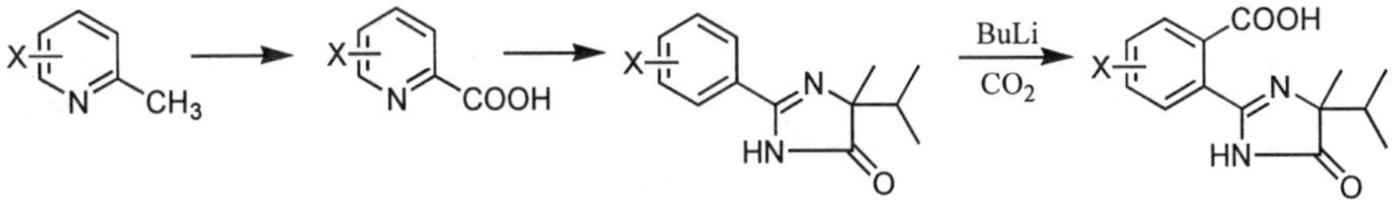

The requirements for the synthesis of an analog was thus reduced to the availability of a substituted α-picoline or picolinic acid. The substituent obviously had to be stable to the conditions for metallation. Examples of this powerful method are:

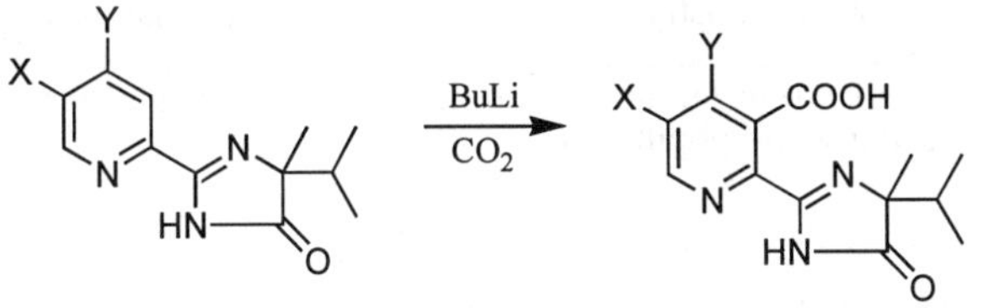

	X	Y
21	CH_3O	H
22	H	Cl
23	C_2H_5	H

It would be highly desirable to be able to control in some measure the stability and thus the persistence of these compounds. One possible means by which this might be achieved might be through the destabilization of the imidazolinone ring. That this might be achieved was based on the known mode of action of the antibacterial monobactams.[6]

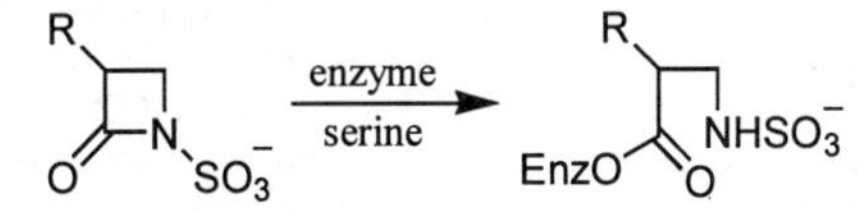

The monobactams, N-sulfonated and N-phosphorylated β-lactams, are excellent suicide substrates for the important bacterial transpeptidases. It was not possible to prepare these N-substituted derivatives of the imidazolinones. Easily prepared, however, was the N-tosyl derivative **24**. It had been established that the free acid was necessary for high activity.

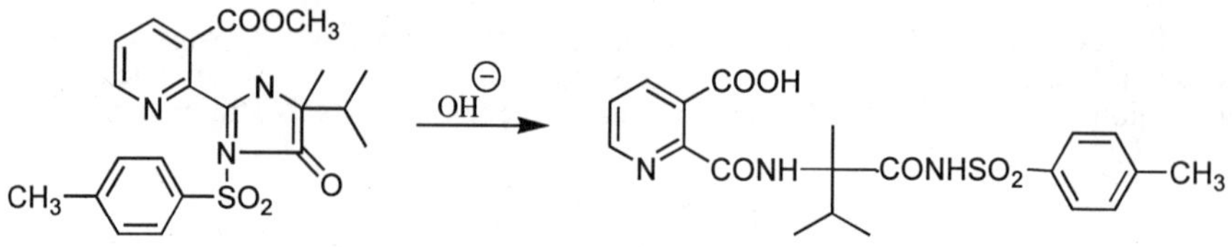

24 25

Surprisingly, mild hydrolysis of the ester in **24** was accompanied by cleavage of the imidazolinone, the hoped for result. Perhaps even more surprising and in some ways disappointing was the fact that the hydrolysis product **25** was also remarkably active as a herbicide and an excellent inhibitor of the same target enzyme as the imidazolinones. This then was yet another anomaly, a novel class of inhibitors, the basis for a new synthesis program.

The imidazolinone project started with a lead, active at 4Kg/ha, generated in a random herbicide screening program. Many leads of this ilk are discovered, yet which of these have the potential to be transformed into a product whose use rate is 25 g/ha? This is a perplexing question. Some few steps in the biosynthesis of some amino acids have proven to be excellent target sties for herbicides. Thus, inhibitors of acetolactate synthase, EPSP synthase, and glutamine synthase fall in this category.[7] Yet commercial products belonging to the sulfonylurea and imidazolinone families as well as glyphosate had their genesis in random screening programs, not a rational synthesis program. It is in fact extremely difficult to even determine what might be a good target site for new herbicides. Molecular biology many eventually be able to provide guidance on this in the future. New products and those in development are primarily ALS or protoporphyrinogen oxidase inhibitors, both excellent target sites and sites for which potent inhibitors can be found. These products will be economical to use because of their incredible potency.

The imidazolinones represent one of the first of these new classes of herbicides which have set new standards for the industry. They have demonstrated that excellent weed control is possible in an effective, safe and environmentally benign way. It is products such as these that will allow farmers to produce the remarkable harvests needed to feed an ever-increasing world population.

Literature Cited

1. For an excellent overview of the chemistry and biology for the imidazolinone area, see *The Imidazolinone Herbicides*; Shaner, D.L,; O'Connor, S.L., Eds.; CRC: Boca Raton, 1991.
2. Kuhn, T.S. *The Structure of Scientific Revolutions*; 3rd ed.; University of Chicago: Chicago, 1996.
3. Los, M. In *The Imidazolinone Herbicides*; Shaner, D.L.; O'Connor, S.L.; Eds,; CRC: Boca Raton, 1991; p 8.
4. Drach, B.S.; Miskevich, G.N. *Zh. Org. Chim.*, **1978**, *14*, 943-947.
5. Reumay, M: Meyers, A.I. *Tetrahedron*, **1985**, *41*, 837-860.
6. Russell, A.d.; Furr, J.R. *J. Antimicrob. Chemother.*, **1982**, *9*, 329-331.
7. Ray, T.B. In *Herbicides as Inhibitors of Amino Acid Biosynthesis*, Böger, P,; Sandmann, G.; Eds.; CRC: Boca Raton, 1989; Chapter 6.

Chapter 3

The Discovery of the Imidazolinone Herbicides from the Perspective of a Discovery Chemistry Manager

Gerald Berkelhammer

Cyanamid Agricultural Research Center, American Cyanamid Company, P.O. Box 400, Princeton, NJ 08543–0400

Marinus Los's discovery of the imidazolinone herbicides is placed in the historical perspective of agricultural research at the American Cyanamid Company. Dr. Los's contributions to the project, as well as two milestones in the work leading to the imidazolinones, are discussed.

Marinus Los's discovery of the imidazolinone herbicides stands with Benjamin Duggar's discovery of the tetracycine antibiotics as one of the two most important scientific achievements in the history of the American Cyanamid Company.

This paper will place Dr. Los's achievement in the historical perspective of agrochemical research at Cyanamid, deal with Dr. Los's contribution to the imidazolinone project, and finally, discuss two pivotal milestones in the imidazolinine story.

Short History of Cyanamid Agricultural R&D

American Cyanamid was created as an agricultural chemical company. It was founded in 1907 as the American licensee for the new German process to make calcium cyanamide by fixing atmospheric nitrogen. The product was sold as the first synthetic fertilizer. Although Cyanamid could be said to have entered the pesticide business in 1920 when it bought at 50% interest in the Owl Fumigating Company of Azusa, California, a company that specialized in cyanide funigation of orange and lemon trees, the first really important moves into pesticides occurred after World War II. A Cyanamid research executive was one of those who examined Gerhard Schrader's laboratory at IG Farben after the war, and he brought home news of parathion, TEPP, and the other Schrader organophosphates. This inaugurated some intensive synthesis and process work in organophosphate chemistry, which led to the discovery and commercialization of several products over the next 15 years or so, including malathion, phorate, and dimethoate.

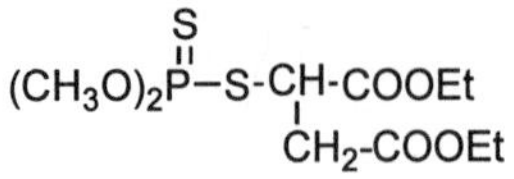

Malathion

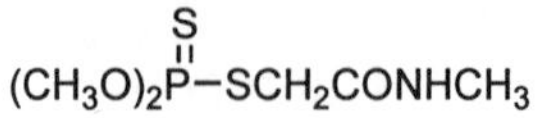

Dimethoate

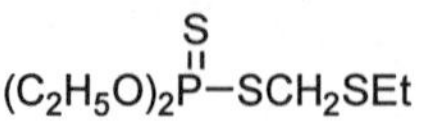

Phorate

Also emerging in the 50's were a few compounds that were not phosphates or insecticides.

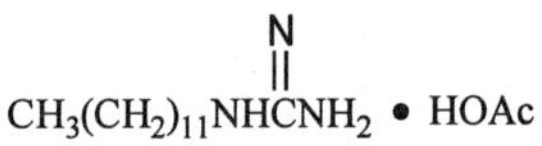

Dodine

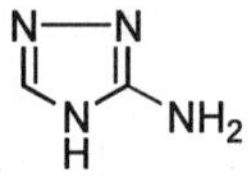

Amitrole

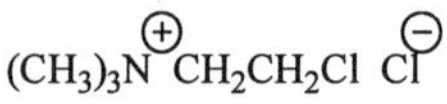

Chlormequat

Of these only dodine, a fungicide, was a Cyanamid discovery. 3-Aminotriazole, a herbicide, was discovered by what was then called the American Chemical Paint Company and co-developed by Cyanamid. Chlormequat was discovered by Tolbert at Michigan State University and licensed to Cyanamid and became a rather important PGR.

The above compounds were discovered in the late 1940's and 1950's. The decade of the 1960's presented a different picture.

Temephos is an interesting compound in that it is so safe that it is allowed to be used in drinking water, the only insecticide that can be so used. It gained some recent renown when Cyanamid donated several million dollars worth for use in Jimmy Carter's Guinea Worm Eradication Program, but its sales have been largely limited to public health and mosquito control. Similarly, although phosfolan had some very good years on Egyptian cotton, it never was registered in the U.S. and it was not a big winner over time.

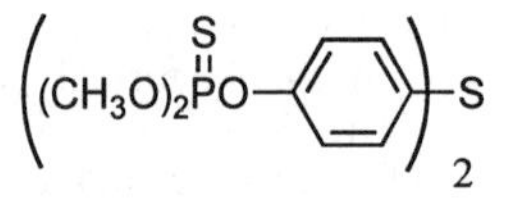

Temephos

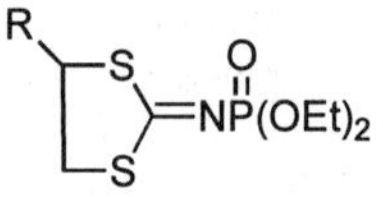

R = H: Phosfolan
R = CH_3: Mephosfolan

A new director of agricultural research, George Sutherland, was appointed in 1970. Sutherland changed discovery approaches at the Princeton Agricultural Center in several ways. One was to make discovery chemistry more goal-oriented and lead-oriented. A second was to breach the walls between discovery synthesis groups in different end-use areas, allowing a concentration of synthesis effort in promising new areas of chemistry as they emerged.

Another important event occurring at about the same time was the acceptance by the corporation of the so-called Affleck prospectus. Dr James Affleck, trained as a Ph.D. organic chemist and later President and Chairman of the Board of American Cyanamid, was CEO of the Agricultural Division at the time, and he persuaded corporate management in 1969 to expand the discovery program in herbicides and PGR's by 15 people in 1970.

Whether the new resources and an accompanying, almost palpable, change to a can-do atmosphere had anything to do with it or not, the fact is that by the end of 1971, Cyanamid's R&D had produced three outstanding development prospects:

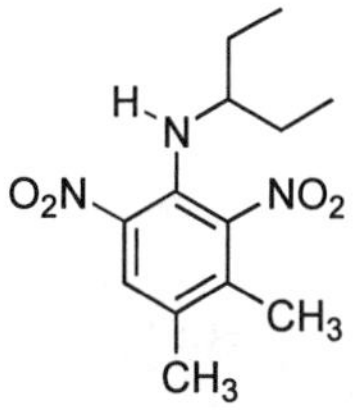

Pendimethalin

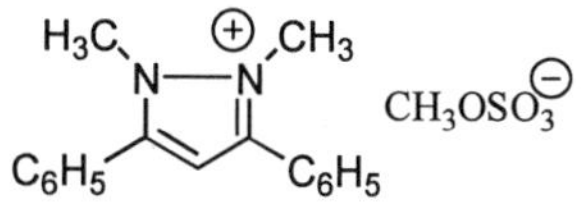

Difenzoquat

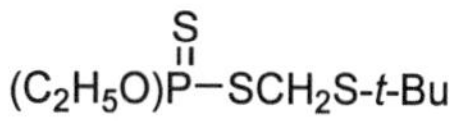

Terbufos

These were the turnaround products for Cyanamid's agricultural business (two of them herbicides), the first from Cyanamid's own research after 25 years of trying.

It was into this background and this atmosphere that Dr. Marinus Los was promoted to group leader of the Plant Growth Regulant Synthesis Group in early 1971.

Background of Marinus Los

Dr. Los was born in the Netherlands, was educated in England and Scotland, did postdoctoral work in Canada, and joined the American Cyanamid Company at the Lederle Laboratories in Pearl River, New York in 1960. From 1960 to 1969, at Pearl River and then Princeton, New Jersey, he worked on the synthesis of steroids as metabolic and estrus regulators. After a year at Edinburgh University in the pharmacology department, he rejoined Cyanamid in 1970 in the Plant Growth Regulant Synthesis Group and became group leader shortly afterwards.

The Imidazolinones

Compounds **1** and **2** afford an idea of the distance Los and his group traveled in the imidazolinone project.

Compound **1** is the plant growth regulator lead that Los picked up in 1971 as the new PGR group leader. Compound **2** is currently Cyanamid's most important crop chemical, imazethapyr, which was synthesized 10 years later. The only characteristic the two molecules obviously have in common is a methyl and an isopropyl group joined to the same carbon. The journey from **1** to **2** involved a great number of chemists and biologists. At one point over twenty chemists, more than one-half Cyanamid's total synthesis effort in agrochemicals at the time, were engaged in the project. Many of these individuals made very important contributions to the project. However, there was no doubt that Marinus Los led the charge and set the direction.

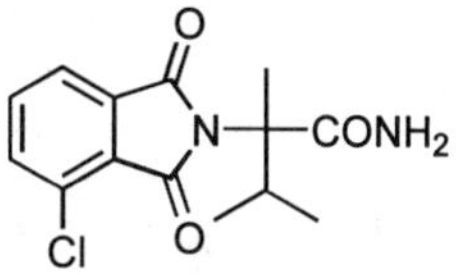

1
Beginning

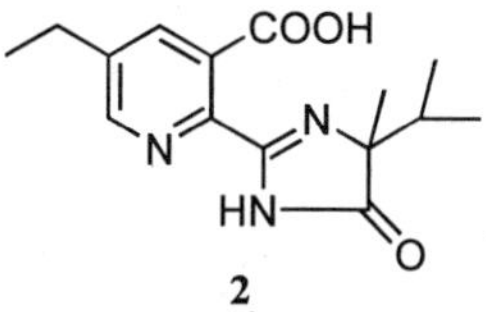

2
Imazethapyr
One of the "Ends"

As early as December, 1973, when his interest was still primarily in plant growth regulators, Dr. Los realized that the tricyclic compounds that he had earlier discovered as ring-closure products of compounds like **1** could likely be ring-opened to afford compounds somewhat similar to DPX 1840, a PGR reported by DuPont (1). The following is some material Los presented at an internal meeting of Cyanamid chemists and biologists engaged in PGR synthesis and testing held in December 1973.

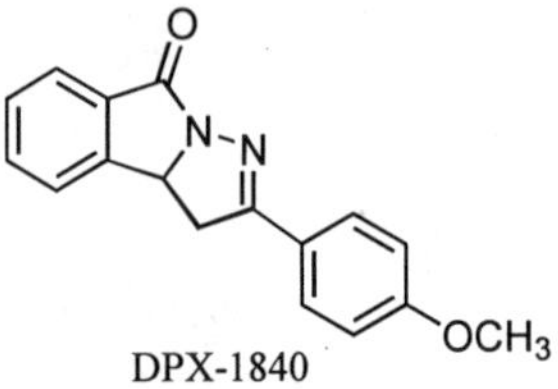

DPX-1840

Auxin transport inhibitor
Breaks apical dominance

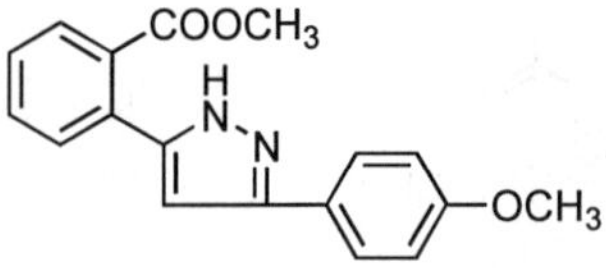

DuPont PGR
Sugar increase, yield increase, etc.

Prepare compounds such as:

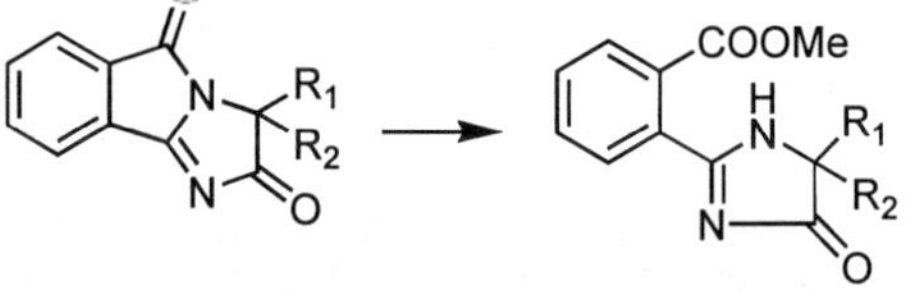

As it happens, it wasn't until over a year later and for totally different reasons that this alcoholysis was actually studied, but it is clear that Los was thinking about imidazolinones long before they showed up in the laboratory. During all this time and until the successful synthesis of imazamethabenz methyl, imazapyr, imazaquin, and imazethapyr herbicides, Dr. Los called the shots and made the major decisions.

Major Imidazolinone Milestones. Two of these decisions became major turning points in the imidazolinone project. The discoveries that led to both these decisions contained serendipitous elements.

PGR's to Herbicides. The first occurred when synthesis of the interesting PGR AC 94377 (2) for a field trial affoarded a small amount of the tricyclic product **3**.

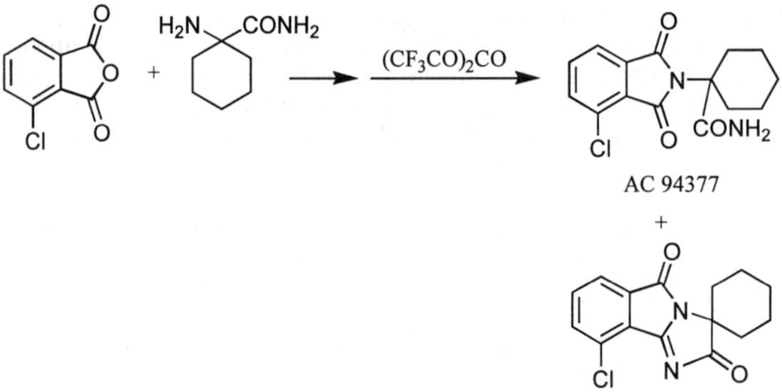

3

This cyclized version of AC 94377 had essentially the same PGR properties of AC 94377 itself but was a bit slower acting, a fact attributed to slow hydrolysis back to AC 94377. The question was whether this was an interesting discovery worthy of followup or merely a distraction which, because of limited resources, would perhaps put off the discovery of a phthalimide more active than AC 94377. The decision was not that difficult to make. Suppose a competitor discovered these tricyclic put off the discovery of a phthalimide more active than AC 94377. The decision was not that difficult to make. Suppose a competitor discovered these tricyclic compounds and found a way to make them that did not proceed through our patented phthalimides. We could not be sure that some of these isoindolediones might not compete favorably as plant regulators with the phthalimdies. There was really very little choice; we had to make a number of isoindolediones and file a patent. It was a pleasant surprise when, just a short way into this little side project, the tricyclic compounds began exhibiting interesting herbicide activity instead of, or in addition to, growth regulant properties.

In particular, compound **4** soon became a field candidate for use against perennial weeds.

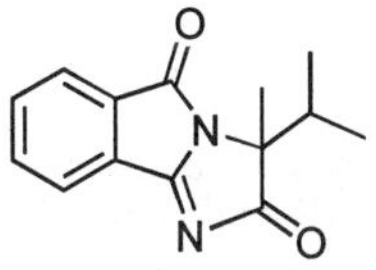

4

Annoying Crystals. The second important discovery occurred when some annoying white crystals separated out of a formulation of **4** that was about to be sent to the field.

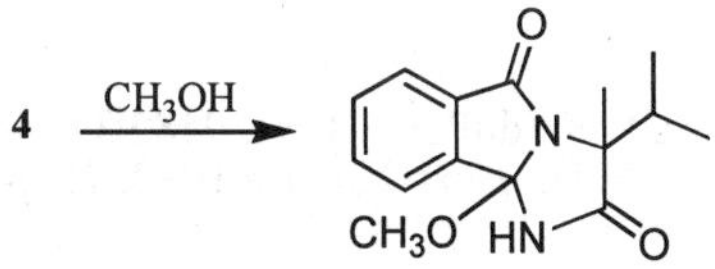

Spectroscopic analysis indicated that methanol, which was a minor constituent of the formulation, had added across the double bond to give the adduct shown. No doubt Los's first reaction was to tell the formulations group to get the bloody methanol out of their surfactant and come up with a new formulation quickly so we could get this compound to the field in time. Upon reflection, the methanol adduct looked interesting and deserved to be tested. Indeed, it turned out to have some herbicidal activity.

The First Imidazolinone. Of much greater significant was that when an attempt was made to make some more of the adduct, this time under basic conditions, the methoxide, instead of adding to the double bond, cleaved the carbonyl-nitrogen bond, thereby forming **5**.

So here was our first imidazolinone. It appeared 15 months after Dr. Los first wrote down its general structure and the equation to produce it. Though it would undoubtedly have been made in due course anyway, the immediate cause of its preparation was the natural curiosity of an organnic chemist about an unexpected precipitate caused by a little methanol in a formulation.

Lessons to be Learned

What are the lessons to be learned from the imidazolinone story? No doubt, there are many. One deserving special emphasis is that the combination of a highly creative and soundly trained individual practicing the art and science of organic chemistry was, and still is, far and away the best means of discovering new biologically active agrochemicals and pharmaceuticals. Organic chemists can discover something

4

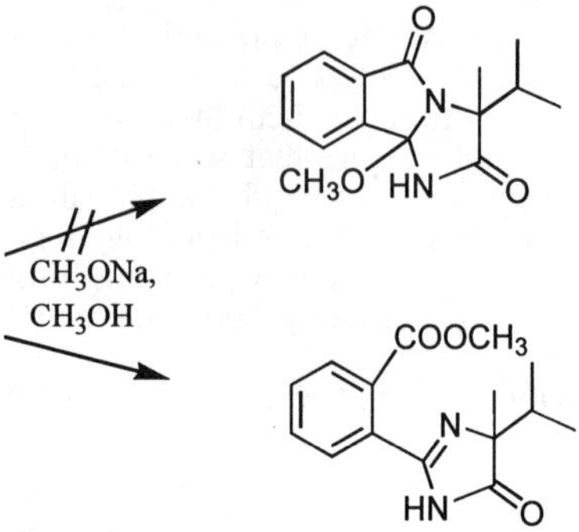

5

because they are intelligent or they are lucky, they can find something because of curiosity, or based on a solid knowledge of biochemistry, or a desire to protect the economic interests of an employer through patents, or a combination of all of these. But whatever the reason for their discoveries, the things that organic chemists find are new compositions of matter, one of which just might be the one they, or somebody else, is looking for. So this story has two heroes. One is Marinus Los, the creative and talented chemist who discovered the imidazolinones, and the other is the grand and glorious science of organic chemisty itself.

Literature Cited

1. Johnson, A.L.; Sweetser, P.B. (duPont), US 3,948,937 (4/6/76).
2. Los, M.; Kust, C.A.; Lamb, G., *Hort. Science* **1980**, *15*, p. 22.

Chapter 4

Why Are Imidazolinones Such Potent Herbicides

Dale L. Shaner and Bijay K. Singh

Cyanamid Agricultural Research Center, American Cyanamid Company, P.O. Box 400, Princeton, NJ 08543–0400

The imidazolinones control weeds at 32 to 125 g/ha, making them some of the most potent herbicides on the market. They kill plants by inhibiting acetohydroxyacid synthase (AHAS), the first common enzyme in the branched chain amino acid biosynthetic pathway. Mode of action studies suggest that factors in addition to the potency at the target site are important in determining herbicidal activity. These factors include the absorption and translocation properties of the imidazolinones, localization of AHAS, the developmental regulation of the branched chain amino acid biosynthetic pathway, and in vivo inhibition of AHAS by imidazolinones. These results illustrate the importance of taking a holistic view of herbicide activity rather than focusing solely on the site of action of the herbicide.

In 1986 American Cyanamid Company introduced imazaquin, the first imidazolinone herbicide, for broad spectrum weed control in soybeans (*1*). The synthesis effort that led to the imidazolinones started with a compound that killed several important perennial weeds at 4 kg/ha (*2, 3*). Analog synthesis work on this lead compound led to a compound which had GA-like activity. From this chemistry, additional synthesis work yielded a promising herbicide, particularly on perennial weeds. Further work in this area of chemistry eventually led to the synthesis of the first imidazolinone (*4*). Currently, six different imidazolinones are used as commercial herbicides in various cropping and non-cropping situations (Figure 1). The use rate of the common herbicides in cropping situation (e.g. corn and soybean) is in the range of 32-125 g/ha which makes them among the most potent herbicides on the market. In this paper, we have attempted to provide the possible reasons for such high potency of the imidazolinone herbicides.

Absorption of Imidazolinones

One of the best features of the imidazolinone herbicides is the flexibility in the way they may be applied in the field. These herbicides are active when applied to the foliage of plants or to the soil because they are rapidly taken up by the shoots and the roots. The imidazolinones are weak acids with a pKa between 3.6 and 5.3, and these compounds are taken up by weak acid ion trapping (*5, 6*). After foliar application, uptake of imidazolinones occurs rapidly and reaches maximum level within a few hours after treatment. Thus, these compounds have good rain fastness provided the rainfall occurs 6 hours or more after application. Addition of non-ionic surfactants and urea ammonium nitrate greatly increase the penetration of the imidazolinone herbicides into leaves. Rapid absorption of these compounds is evident from one study where 61% of the applied ^{14}C-imazaquin was absorbed by *Xanthium strumarium* foliage in 3 days (*7*).

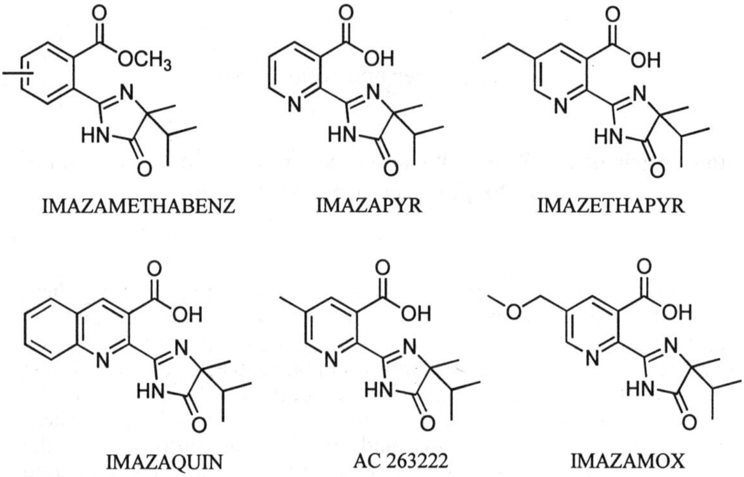

Figure 1. The commercial imidazolinone herbicides.

Imidazolinones are absorbed by roots by a passive process that is dependent on the pH of the external solution and the lipophilicity of the herbicide. Study of the structure-activity relationship between the physico-chemical parameters of 5'-substituted pyridine imidazolinones and absorption by roots of corn and sunflower showed that the lipophilicity of the analogs was the predominant factor affecting root absorption of these imidazolinone (*8*, *9*). Uptake increased as lipophilicity increased. Similar to foliage, roots also readily absorb imidazolinones. For example, 66% of the applied imazaquin was rapidly taken up by sunflower roots (*10*).

Translocation of Imidazolinones

Imidazolinones are rapidly translocated from the leaves and the roots to the other parts of the plants which indicates that these compounds can move through both xylem and phloem. Translocation from the roots to the shoots takes place via xylem and is correlated with the lipophilicity of the compound (*8*, *9*). For example, imazaquin is translocated much more effectively than imazethapyr which, in turn, is translocated better than imazapyr.

Translocation of imidazolinones from the leaves to other parts of the plants occurs through the phloem. Transport via phloem varies with the chemical structure, however multiple physical-chemical factors, which includes the pKa of the imidazolinone as well as its lipophilicity, determine the phloem mobility of the imidazolinones. Imazapyr is highly phloem mobile, with greater than 60% of the absorbed herbicide moving out of the treated leaf of perennial grasses and maize (*10*). The differences in the mobility of the imidazolinones in different species may also be related to the pathway of metabolism in the symplast.

Inhibition of AHAS by imidazolinones

There are four main lines of evidence which prove that the site of action of imidazolinones is acetohydroxyacid synthase (AHAS), the first common enzyme in the branched chain amino acid biosynthetic pathway (Figure 2). First, an imidazolinone treatment caused a simultaneous increase in the total free amino acid levels in the treated tissue and a corresponding decrease in the soluble protein levels (*11*, *12*). Further analysis revealed that the levels of most of the amino acids increased after treatment with the exception of the branched chain amino acids (*12*). Second, growth inhibition by an imidazolinone was prevented when the growth medium was supplemented with the branched chain amino acids (*11*, *12*). Third, imidazolinones were in vitro inhibitors of AHAS extracted from plants (*13*). Fourth and the most definitive result was the demonstration that imidazolinone resistant plants, selected in the presence of herbicidal concentrations of an imidazolinone, contained an altered AHAS that was no longer inhibited by the imidazolinones (*14-16*). Further analysis revealed that a single amino acid change in the AHAS protein was responsible for resistance to the imidazolinones (*17*, *18*).

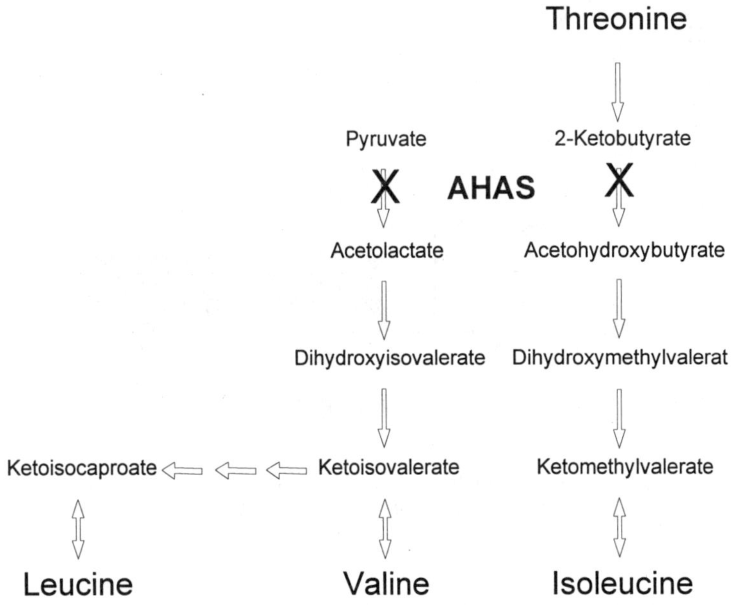

Figure 2. The biosynthetic pathway of branched chain amino acids. Acetohydroxyacid synthase (AHAS) is inhibited by the imidazolinone herbicides as indicated by an X on the reactions catalyzed by this enzyme.

In vitro assays have shown that imidazolinones are potent, uncompetitive inhibitors of AHAS which suggests that they bind to the enzyme-pyruvate complex (*13*). Furthermore, imidazolinones are slow, tight-binding inhibitors of the enzyme (*19*). In vivo studies have revealed an interesting feature of the interaction of imidazolinones with AHAS which may be one of the key reasons for the remarkable herbicidal activity of these compounds. When a plant is treated with a lethal dose of an imidazolinone, the level of AHAS activity extracted from these plants is significantly reduced. This effect of inhibitors can be discerned within an hour after treatment and in 24 hours, the extactable enzyme activity is reduced more than 80 % (Figure 3; *19*, *20*). The loss of extractable AHAS activity in the imidazolinone treated plants may be because the imidazolinones interact with the enzyme in such a way in vivo that the herbicide does not easily separate from the enzyme during the extraction procedure. Alternatively, the herbicide may alter the AHAS protein structure in such a way that it is enzymatically inactive. Whatever the reason may be, imidazolinone treatment of a plant causes a rapid loss of the in vivo AHAS activity.

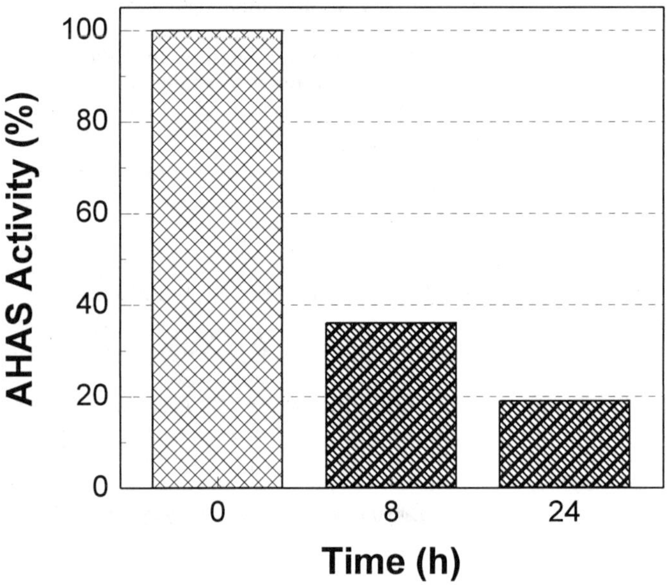

Figure 3. Decrease in extractable AHAS activity in corn shoots treated with 10 μM imazapyr. Data is presented as per cent of AHAS activity extracted from the untreated plant.

Localization and developmental regulation of AHAS

All plant parts contain AHAS because of the ubiquitous presence of AHAS mRNA (*21-23*), the encoded AHAS protein (our unpublished data) and AHAS activity (*24*, *25*). Within each plant part, AHAS seems to be present predominantly in the actively growing tissue. As indicated in Figure 4, AHAS activity was highest in the youngest leaf tissue. The variation in enzyme activity with leaf age corresponded to the amount of AHAS protein (our unpublished data). Similar differences were observed in the levels of AHAS protein and the enzyme activity in different organs of lima bean (*25*, our unpublished data). These observations are also supported by in situ hybridizations of anti-mRNA probes to plant sections (*23*).

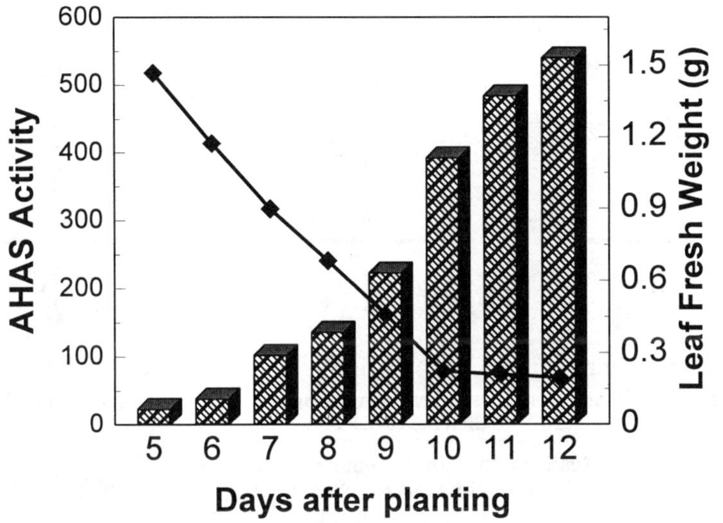

Figure 4. Changes in leaf fresh weight (bar) and AHAS activity (line) with increasing leaf age of the unifoliate leaf of lima bean seedling.

The presence of AHAS at high levels in young, meristematic tissues is consistent with the fact that the flux of carbon through the branched chain amino acid biosynthetic pathway occurs primarily in the young tissue (*26*, *27*). Therefore, high level expression of AHAS and the whole branched chain amino acid biosynthetic pathway in young tissues allows the tissue to meet the demand of high levels of amino acids required for protein synthesis in actively growing tissues.

Existence of this mechanism of regulation of branched chain amino acid biosynthesis in plants is quite significant for the action of imidazolinone herbicides. Since imidazolinones inhibit AHAS, the very first enzyme common to the biosynthesis of all three branched chain amino acids, the availability of these amino acids is diminished (an observation that led to the discovery of the mode of action of imidazolinones; *12*). Therefore, synthesis of new proteins, especially in the growing points, is inhibited, and this inhibition ultimately results in plant death.

Whole Plant Activity of Imidazolinones

In addition to the imidazolinones, three other classes of AHAS inhibitors (sulfonylurea, triazolopyrimidine sulfonamide and pyrimidyl thio-benzoate) are commercial herbicides. These compound classes are much more potent inhibitors of AHAS than the imidazolinones (*27*), yet the field use rates of all of these herbicide classes is similar. The results described earlier and summarized in Figure 5 provide evidence as to why imidazolinones are such potent herbicides.

Imidazolinone herbicides are taken up rapidly by both roots and shoots (Figure 5A). They are rapidly translocated from the site of uptake to other plant parts and accumulate at the growing points (Figure 5B). As described earlier, the growing points contain the highest levels of AHAS activity and the whole branched chain amino acid biosynthetic pathway is most active in these tissues. Therefore, the

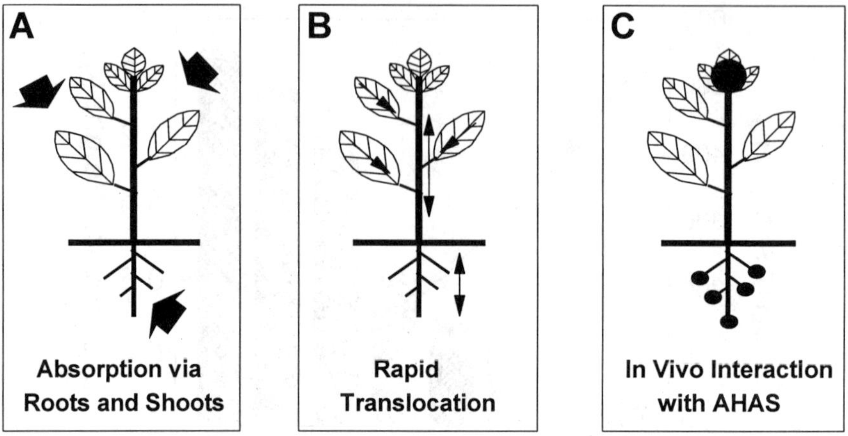

Figure 5. The features of imidazolinones that make them highly potent herbicides. Imidazolinones are rapidly taken up by both shoots and roots (A), translocate rapidly to other plant parts (B), and accumulate in growing points where they inhibit AHAS activity (C).

imidazolinone herbicides are present in the highest concentrations in the tissues where AHAS activity is most needed to provide branched chain amino acids for new protein synthesis. In these tissues, imidazolinones interact with AHAS in such a way that they make the enzyme inactive in vivo (Figure 5C) resulting in an inhibition of branched chain amino acid biosynthesis. Lack of branched chain amino acids causes immediate cessation of plant growth which leads to the death of growing points and eventually the whole plant.

Literature Cited

1. Orwick, P. L.; Marc, P. A.; Umeda, K.; Shaner, D. L., Los, M.; Ciarlante, D.R. *Proc South Weed Sci. Soc.* **1983**, 36, 90.
2. Los, M. In *Pesticide Synthesis Through Rational Approaches;* Magee, P.S., Kohn, G.K. Menn, J.J., Ed.; American Chemical Society: New York, 1984; 255, pp 29-44.
3. Los, M. In *Proc. 6th Internationaal Congr. Pestic Chem (IUPAC)*; Blackwell, London, 1987; pp 35-42.
4. Los, M. In *The Imidazolinone Herbicides*; Shaner, D. L., O'Connor, S. L., Ed; CRC Press: Boca Raton, FL, 1991, pp 1-5.
5. Van Ellis, M. R.; Shaner, D. L. *Pest. Sci.* **1988**, 23, 25-34.
6. Beyer, E. M.; Brown, H. M.; Duffy, M. J. *British Crop Protection ConferenceWeeds* 1987, 2, 531-540.
7. Shaner, D.L.; Robson, P.A. *Weed Sci.* **1985**, 33, 469-471.
8. Little, D. L.; Shaner, D. L.; Ladner, D. W.; Tecle, B.; Ilnicki, R. D. *Pest. Sci.* **1994**, 41, 161-169.
9. Little, D. L.; Ladner, D. W.; Shaner, D. L.; Ilnicki, R. D. *Pest. Sci.* **1994**, 41, 171-185.
10. Little, D. L.; Shaner, D. L. In *The Imidazolinone Herbicides*; Shaner, D. L., O'Connor, S. L., Ed; CRC Press: Boca Raton, FL, 1991, pp 53-70.
11. Shaner, D. L.; Reider, M. L. *Pest. Biochem. Physiol.* **1986**, 25, 248-257.
12. Anderson, P. C.; Hibberd, K. A. *Weed Sci.* **1985**, 33, 479-483.

13. Shaner, D. L.; Anderson, P. C.; Stidham, M. A. *Plant Physiol.* **76:** 545-546 (1984).
14. Shaner, D. L.; Anderson, P. C. In *Biotechnology in Plant Science*; Zaitlin, M.; Day, P. R.; Hollaender, A., Eds.; Academic Press: Orlando, Florida, 1985, pp 287-299.
15. Anderson, P. C.; Georgeson, M. *Genome* **1989**, 31, 994-999.
16. Newhouse, K., Singh, B., Shaner, D.; Stidham, M. *Theor. Appl. Genet.* **1991**, 83, 65-70.
17. Sathasivan, K.; Haughn, G.W.; Murai, N. *Plant Physiol.* **1991**, 97, 1044-1050.
18. Hattori, H.; Brown, D.; Mourad, G.; Labbe, H.; Ouellet, T.; Sunohara, G.; Rutledge, R.; King, J.; Miki, B. L. *Mol. Gen. Genet.* **1995**, 246, 419-425.
19. Muhitch, M. J.; Shaner, D. L.; Stidham, M. A. *Plant Physiol.* **1987**, 83, 451-456.
20. Shaner, D. L.; Singh, B.K.; Stidham, M. A. *J. Agric. Food Chem.* **1990**, 38, 1279-1282.
21. Weirsma, P.A.; Schmiemann, M. G.; Condie, J. A.; Crosby, W. L.; Moloney, M. M. *Mol. Gen. Genet.* **1989**, 219: 413-420.
22. Ouellet, T.; Rutledge, R. G.; Miki, B. L. *The Plant J.* **1992**, 2, 321-330.
23. Keeler, S. J.; Sanders, P.; Smith, J. K.; Mazur, B. J. *Plant Physiol.* **1993**, 102, 1009-1018.
24. Stidham, M. A.; Singh, B. K. In *The Imidazolinone Herbicides*; Shaner, D. L., O'Connor, S. L., Ed; CRC Press: Boca Raton, FL, 1991, pp 71-90.
25. Schmitt, G.K.; Singh, B. K. *Pest. Sci.* **1990**, 30: 418-419.
26. Singh, B.K.; Szamosi, I.T.; and Shaner, D.L. . In *Amino Acids and Their Derivatives in Higher Plants*, R.M. Wallsgrove, Ed; Cambridge University Press: Cambridge, U.K., 1995, pp. 59-75.
27. Shaner, D.L.; Singh, B.K. In *Herbicide Activity: Toxicology, Biochemistry and Molecular Biology*; Roe, R.M.; Burton, J.D.; Kuhr, R.J.; Eds.; IOS Press: Washington, D.C., 1997, pp 69-110.

Chapter 5

The Discovery of Imazamox, a New Broad-Spectrum Imidazolinone Herbicide

Thomas M. Brady, Barrington Cross, Robert F. Doehner, John Finn, and David L. Ladner

Cyanamid Agricultural Research Center, American Cyanamid Company, P.O. Box 400, Princeton, NJ 08543–0400

The new soybean herbicide, imazamox, **1** differs significantly from the other imidazolinone herbicides because of its soil residual behavior, allowing rotational planting of imidazolinone-sensitive crops. This behavior was the key to its discovery. In order to determine if soil persistence could be related to structure, an early set of imidazolinone analogs had been field tested. A synthesis program was subsequently undertaken to optimize activity and to find active compounds with reduced hydrolytic or metabolic stability. Some of these newer analogs had much less pre- than postemergence greenhouse activity, and we hypothesized which substituents were responsible for their soil deactivation. The results pointed to advanced studies of the highly active 5-methoxymethyl compound, **1**. In soil, **1** undergoes degradation to the 5-carboxy analog, **9**, a compound with low herbicidal activity.

With the discovery of the first imidazolinone herbicides by Dr. Marinus Los and commercial introduction of several products in the last ten years, one might wonder what would drive an effort to add yet another soybean-selective analog to the marketplace. In contrast to the initially introduced compounds in the series, second-generation compounds are guided by the luxury of experience and a firmer understanding of market needs. As a consequence, the discovery process is driven by somewhat different criteria from that of a totally new work area. Completely new areas of chemistry are usually discovered because of potency and commercialized if a lead can be improved through analog synthesis to a field-active material with selectivity in a useful crop. In the discovery of imazamox, field activity and soybean selectivity of imidazolinones had already been accomplished; the mission was to find an new analog which additionally had specific soil residual characteristics. The commercial reason for doing so were to expand the use of the imidazolinone chemistry into markets where existing compounds were excluded because of followcrop injury. This included the usual soybean / corn crop rotations in colder climates as well as areas where soybeans are followed by very sensitive crops such as canola or sugarbeet. Our belief was that this could be accomplished through optimization of the chemical structure, and through the use of predictive screening methods and metabolism studies for selecting field candidates.

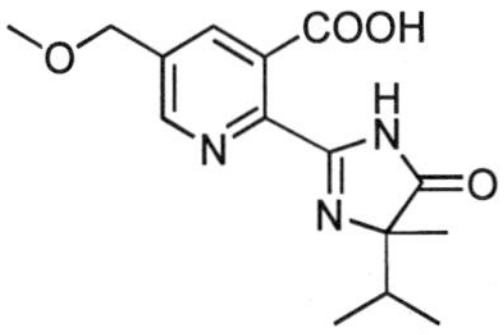

1
imazamox

The commercialized imidazolinone herbicides are shown in Fig. 1. The differences in structure among these compounds are the key factors for determining their end utility. In other words, the substitution pattern of the benzene or pyridine ring residues is directly responsible for selectivity or lack thereof, and, as will be shown for imazamox, for soil residual characteristics.

Figure 1. The Imidazolinone Herbicides

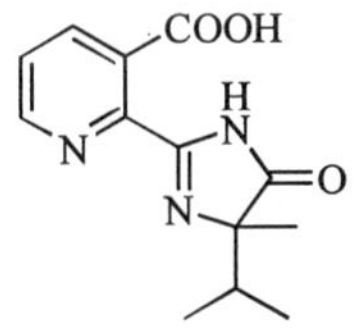

2
imazapyr

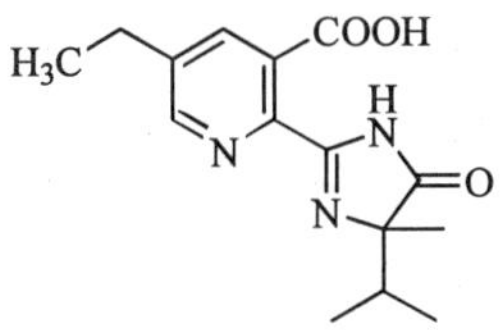

3
imazethapyr

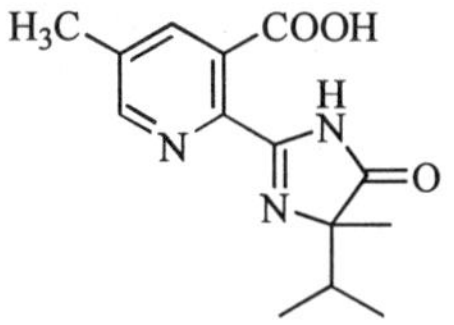

4
(Cadre® Herbicide)

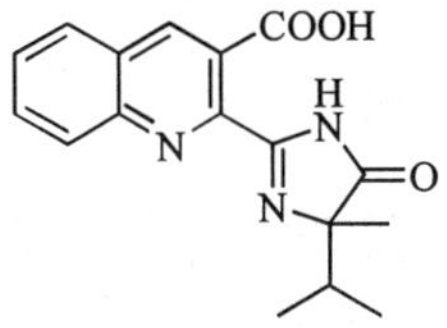

5
imazaquin

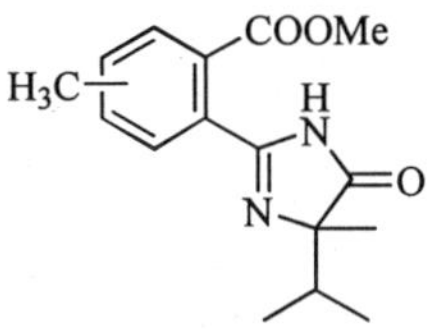

6
imazamethabenz methyl

Design Considerations

Fortunately, a great deal of structure-activity information had been gathered during the development of the first imidazolinone herbicides and there had been some important findings and soil metabolism studies of the commercial compounds. From this information, several points were key to the design of a shorter residual compound.

First was the knowledge that in imazethapyr, **3**, and in imazamethabenz methyl, **6**, deactivation in plants occurs by hydroxylation of the alkyl side chain (1). This hydroxylated product is then further metabolized to non-herbicidal components within the plant. We believed that oxidation of a side chain might provide a mechanism for soil deactivation as well.

® Registered trademark, American Cyanamid Company

Figure 2. Metabolic Pathways of Imidazolinones

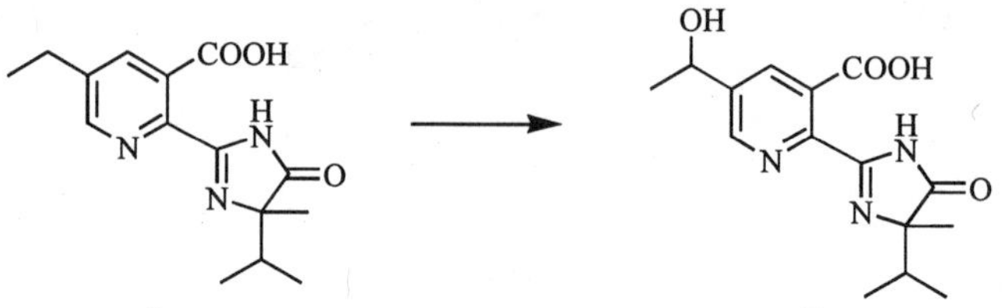

3 7

Secondly, structure-activity relationships (SAR's) had been determined and showed that functional groups which were already oxygenated, such as aldehydes **8** and carboxylic acids **9** were lower in activity than the corresponding un-oxidized analogs (2). This was most evident for preemergence herbicidal activity, lending support to the postulate that soil activity would be lower for such compounds.

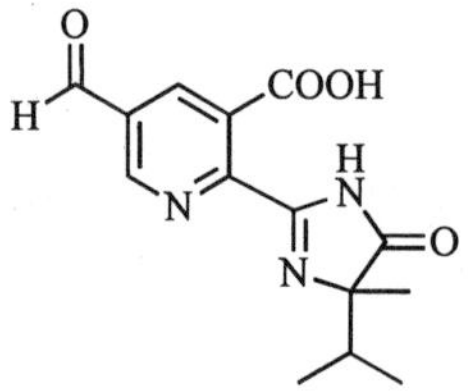

8

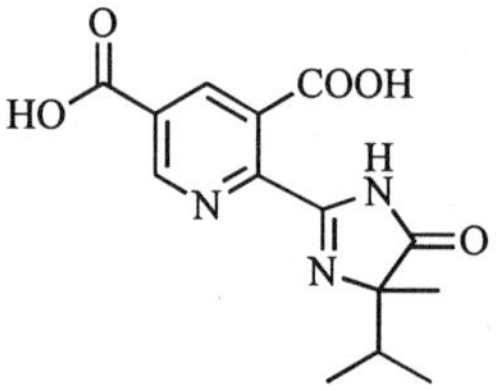

9

Finally, we felt it was important to concentrate the analog work on pyridine analogs which were 5- or 5,6-substituted. Although imidazolinone modifications elsewhere provide compounds which are converted to inactive compounds in the soil (3), we concluded from our knowledge of imidazolinone SAR's that such modifications also usually decreased the initial herbicidal activity below useful levels.

The strategy can be summarized as one which sought active herbicides of type I which could be converted to inactive metabolites of Type II in the soil. This parallels the profile of selective compounds, differing only in the location of the deactivation, i.e., soil instead of crops.

Figure 3. General Scheme for Deactivation of Imidazolinones

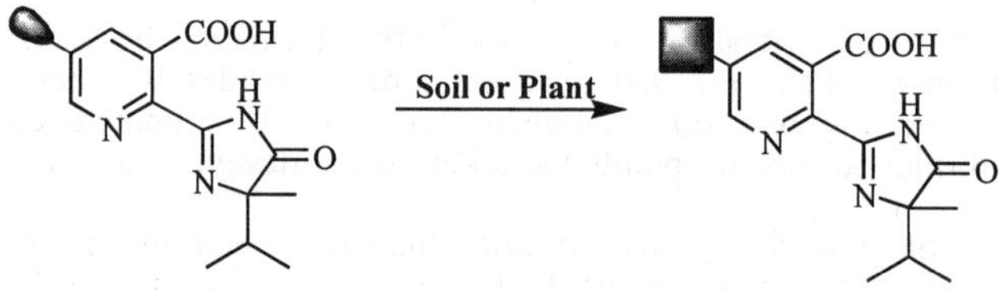

TYPE I **TYPE II**

Active Herbicide **Deactivated Metabolite**

Selection of Analogs and Screening

Some early experiments had been carried out in an attempt to determine if heterocyclic ring analogs of imazaquin were more susceptible to metabolism and safer to rotational crops. In this experiment, the six greenhouse-active compounds, prepared in support of patent filings, and shown in Figure 4, were applied postemergence as soybean-selective herbicides in Minnesota and Wisconsin. The following year, corn was planted in the same plot and the level of injury due to herbicide carry-over was measured. This experiment was based on the hypothesis that the heterocycles might be more likely to chemically break down than a benzene ring. Although we had no prior evidence to support this, it was known that sulfonyl ureas containing heterocyclic ring substituents are more prone to rapid soil metabolism, for example (4).

Figure 4. Heterocyclic-fused Imidazolinone Analogs

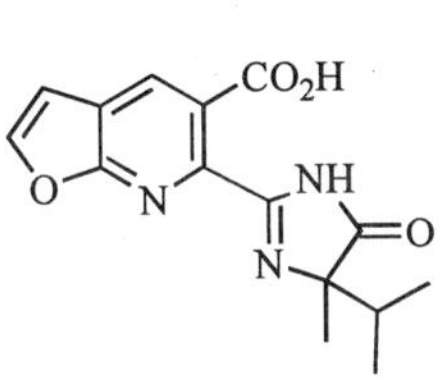

10

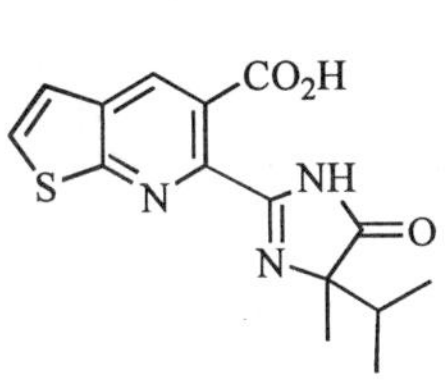

11

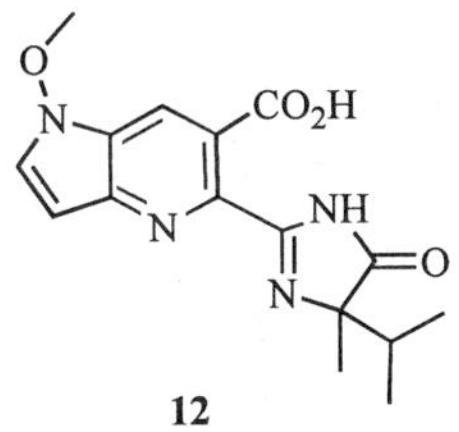

12

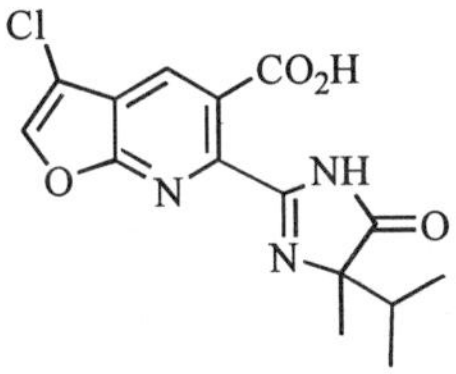

13

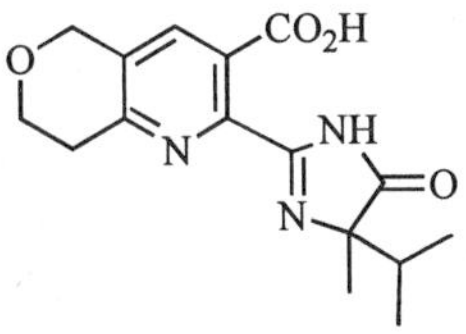

14

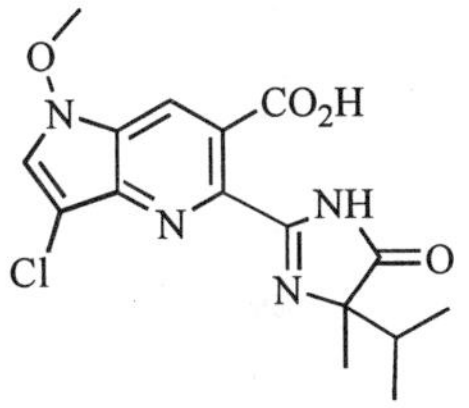

15

From this rather long-term experiment we concluded that some of these compounds were not particularly effective as herbicides in the field, and many of the more active ones caused unacceptable injury to the rotational corn crop. The only compound which looked somewhat promising was **14**, the pyranopyridine compound. The similarity of this structure to that of imazamox is now evident, and this indeed provided us an important clue at the time.

Field experiments, especially the kind just described, have inherent drawbacks with respect to the time required. It was important that laboratory experimental models be developed so that many more analogs could be examined for their potential as low soil residual field candidates. Indeed, many analogs had been prepared in support of patent filings, directed to some extent, by a QSAR model which suggested highly active 5-substituents (5). When short residual activity was also desired, consideration was given to two classes of compounds. The first class we believed would have a tendency to hydrolyze to less active compounds: these included masked carboxylic acids and protected aldehydes. The other class consisted of compounds whose substituents were postulated to provide some metabolic activation towards enzymatic oxidation: these included the methoxymethyl and methylthiomethyl compounds. The details of some of this work have already been reported (6). Table I includes the compounds which were evaluated in this investigation. The bioactivity of each shown for comparison in Table I. Both pre- and postemergence activity is shown. Some of these, such as the aldehyde, **8**, and carboxylic acid, **9**, were obviously inferior as herbicides. Noteworthy was the tendency of some, such as the 5-methylthiomethyl coompound, **25**, to be substantially more active post- than preemergence. We viewed this as confirmation that the profile was being influenced by structure, since most imidazolinones are more active preemergence.

Table I. Relative Pre- and Postemergence Herbicidal Activtiy and Soil Half-Life of Various 5-Substituted Pyridine Imidazolinones

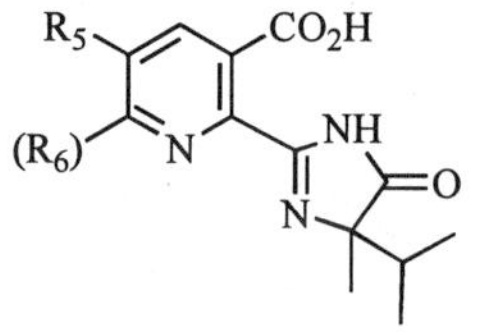

Compound	R_5	Relative Activity* Postemergence	Relative Activity* Preemergence	Half Life, Days
1	$MeOCH_2$-	10	8.5	22
5	5,6-benzo	4	4	42
8	HCO-	2	0	<1
9	HO_2C-	2	0	>20
14	5,6-pyrano[4,3]-	4	4.5	44
24	$HOCH_2$-	4	3	21
25	$MeSCH_2$-	2.5	0	1
26	EtO-	8.5	7	23
27	$MeOCH_2CH_2O$-	8.5	7	15
28	$MeOCH_2O$-	8	6	4
29	S-CH_2CH_2-S-CH-	2.5	2	2
30	O-CH_2CH_2-O-CH-	7	5.5	18
31	MeO-N=CH-	9.5	8.5	25
32	Me-N-N(Me)-CH-	4	4	<1
33	$(MeO)_2CH$-	8.5	7	25
34	MeCO-N-CH_2CH_2-O-CH-	2	0	>70
35	$(Me)_2NCO$-N-CH_2CH_2-O-CH-	2	0	20
36	MeO_2C-N-CH_2CH_2-O-CH-	2.5	2	>70
37	MeN-CH_2CH_2-O-CH-	4	4	<1
38	$(MeO)_3CH$-	5	4.5	<1
39	MeO_2C-	7	5	<1

* 10 = most active, 0 = least active

These compounds were then screened through a laboratory soil degradation assay. A half-life was calculated for the disappearance of parent compound and the results are summarized in Table I. Some half-lives were too short such as the ortho-ester and some were not sufficiently herbicidal such as the dithioketal. Those with promising profiles in both areas were selected for screening in the field. As noted previously, compound **14** had given some interesting field results and imazamox was a substantially more active compound with the same ether functionality.

Field testing of several compounds confirmed that imazamox was the most effective herbicide; the closest to it were the oxime and dimethyl acetate. The oxime was not selective in soybean, however, and the acetate had poorer residual control. The followcrop data confirmed that imazamox had a favorable half-life compared to imazaquin and imazethapyr, showing less reduction in shoot fresh weight of either corn or sugarbeets at 4x the use rate compared to 2x use rates for the standards. It was also evident that imazamox had outstanding postemergence activity on some species which further encouraged its commercial development despite the existence of imazaquin and imazethapyr.

The design and synthesis of herbicidal compounds with short residual properties was therefore successful. The subsequent testing and metabolism studies helped select imazamox as the compound with the optimum properties of residual weed control and rotational crop safety.

Synthesis of Imazamox

The synthesis of imazamox was accomplished by two methods, as shown in Figure 5. The 5-methyl pyridine 2,3-dicarboxylic anhydride **16** was chlorinated with N-chloro-succinimide to the monochloromethyl derivative **17** and it was then reacted with the aminomide and base to give the acid diamide. Displacement with methoxide gave the corresponding methoxymethyl compound which was then cyclized to imazamox with hydroxide. Alternatively, bromination of the diester **18** gave the monobromo diester **19** which reacted with methoxide to give **22**. Heating **22** with the diamide and *t*-butoxide in toluene in an adaptation of a route described for other imidazolinones (7) gave imazamox directly.

Figure 5. Synthesis of Imazamox

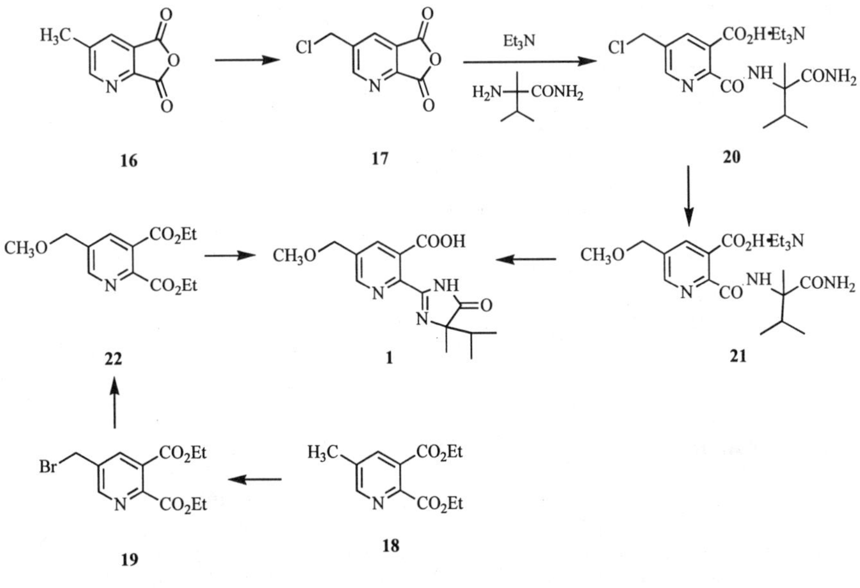

Physical Properties

The physical properties of this compound appear in Table II. Like other imidazolinones, it is a solid and a weak carboxylic acid, resulting in a technical material of low volatility. Solubilities were measured at pH 7.

Table II: Physical Properties of Imazamox

Molecular formula:	$C_{15}H_{19}N_3O_4$
Molecular weight:	305.3
Water Solubility:	4,160 ppm
Volatility:	$< 1.0 \times 10^{-7}$ torr
Melting point:	166.0 - 166.7°C
K_{ow}:	0.004 @ pH 7
pK_a:	2.3, 3.3 and 10.8
Common name:	imazamox

Also like other imidazolinones, imazamox has a favorable toxicological profile as shown in Table III. The values shown are for the highest dose tested.

Consistent with its structural features the compound is stable to hydrolysis but shows a fairly rapid degradation in soil under laboratory conditions. The half life is about 3-4 weeks. The dicarboxylic acid **9** has been identified as an initial soil metabolite. Soil degradation is assumed to be microbial. Detailed studies have been reported elsewhere (8).

Table III. Toxicological Profile of Imazamox

- Rat, Acute Oral LD_{50} (tech. and form.): >5000 mg/kg
- Rabbit Acute Dermal LD_{50} (tech.): >4000 mg/kg
- Rat, Acute Dermal LD_{50} (form.): >4000 mg/kg
- Rabbit, Eye (tech.): Moderately irritating
- Rabbit, Eye (form.): Non-irritating
- Rabbit, Skin (tech. and form.): Non-irritating
- Rat, Inhalation LC_{50} (tech.): >6/3 mg/L
- Rat, Inhalation LC_{50} (form.): >5.0 mg/L
- Guinea pig, Skin sensitization: Non-sensitizer
- Non-mutagenic in the following systems: Microbial Mutagenicity Assay (Ames) CHO/HGPRT
- Non-genotoxic in the following systems:
  - *in vitro* Chromosomal Aberrations in Chinese Hamster Ovary cells
  - *in vivo* Mouse Micronucleus Assay
- NOEL on Non-Target Species:
  - Mallard duck, Oral LD50: >2000 mg/kg
  - Bobwhite quail, Oral LD50: >2000 mg/kg

Mode of Action

The compound, like other imidazolinones, acts as an inhibitor of branched-chain amino acid synthesis, namely inhibition the AHAS enzyme. It is inherently active in the plant enzyme and does not require metabolic activation. While it is rapidly absorbed and translocated by the foliage, it is not as readily absorbed by the roots, which, in addition to soil degradation, accounts for some of its lower preemergence activity. The herbicidal symptoms are typical of AHAS inhibitors, namely cessation of growth, leading eventually to plant necrosis.

Crop Selectivity

The compound has selectivity in leguminous crops, primarily a result of metabolism in the tolerant species, but differences in absorption and uptake are significant between susceptible and tolerant species. This has been reported in detail elsewhere (8). Tolerance by IMI®Corn results from genetically-modified plant containing an AHAS enzyme not inhibited by this compound.

Summary

Imazamox represent a new herbicide with broad spectrum postemergence weed control at low use rates. Its structure gives a favorable half-life for rotational crops, while providing optimum soil persistence. Its discovery as a product candidate was the result of the synthesis and screening of a broad number of analogs, not only for their herbicidal activity but also for favorable soil persistence. While it still remains a difficult problem to predict a priority just how a compound will metabolize in the soil or the plant, it was gratifying that chemical principles of hydrolytic and oxidative potential could be taken into account for selection of analogs, and the experimental results in this case support some of the hypothesis used in the process.

Acknowledgments

The authors are indebted to Laura Quakenbush for the greenhouse testing and Shirley Rodaway for the soil degradation assay of these compounds; Dale Shaner, Berhane Tecle and the many other Cyanamid scientists who synthesized analogs, conducted field studies, and determined the ecotoxocological profiles. Finally, we wish to acknowledge Marinus Los whose initial creative inventions were directly responsible for the discovery of imazamox.

Literature Cited

1. Ladner, D.W., in: Stetter, J., (ed) *Chemistry of Plant Protection*, 10 Herbicides, Inhibitors of branched chain amino acid biosynthesis, pp 111-112.
2. Ladner, D.W., in: Stetter, J., (ed) *Chemistry of Plant Protection*, 10 Herbicides Inhibitors branched chain amino acid biosynthesis, pp 106-107.
3. Guaciaro, M.A., Los, M., Little, D.L., Marc, P.A., Quakenbush, L., (1992) in: Baker, D.R., Fenyes, J.G., Steffens, J.J. (eds) *Synthesis and Chemistry of Agrochemicals III.* ACS Symposium Series, 504. American Chemical Society, Washington, DC, pp 56-74.
4. Sionis, S.D., Drobny, H.G., Lefebre, P., Upstone, M.E. (1985) Brit Crop Prot Conf - Weeds - **1**:49
5. Cross, B., Ladner, D.W., (1992) in: Draber, W., Fujita, T. (eds) *Rational Approaches to Structure, Activity and Ecotoxicology of Agrochemicals* CRC Press, Boca Raton, FL Chap 14, pp 331-355.
6. Finn, J.M., Bosley, D., Rodaway, S.J., Quakenbush, L.M., (1995) in: Baker, D.R., Fenyes, J.G. and Busarub, G.S., (eds), *Synthesis and Chemistry of Agrochemicals IV*, ACS Symposium series, 584, American Chemical Society, Washington, DC pp 46-59
7. Ciba-Geigy (1987) EP 327 492A
8. Rodaway, S.J., N Central Weed Science Society Meeting, Grand Rapids, MI, December, 1994.

® Registered trademark, American Cyanamid Company

CONTROL OF WEEDS AND ACARIDS

Chapter 6

Synthesis and Herbicidal Activity of Aryloxyphenyl and Heterocyclic Substituted Phenyl *N*-Arylbenzotriazoles

A. D. Crews, M. E. Condon, and M. C. Manfredi

Cyanamid Agricultural Research Center, American Cyanamid Company, P.O. Box 400, Princeton, NJ 08543–0400

The title compounds are new low application rate pre- and post-emergence herbicides with safety on broadleaf crops. This chapter will describe the initial discovery, synthetic methodology and structure/activity relationships which were observed within the two structural classes.

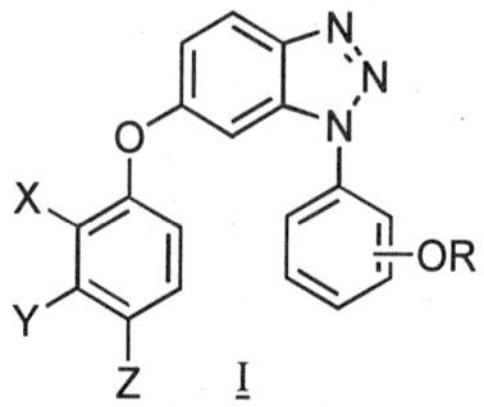

I

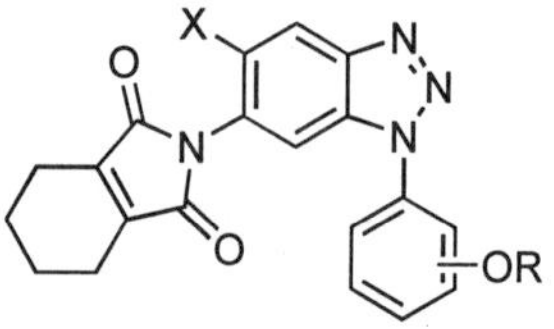

II

The level of herbicidal activity in structural class I is increased with phenoxy substitution containing combinations of perfluoroalkyl groups and halogens. Analogues of structural class II are most active when X is fluorine. Alkoxy functionality in both classes is preferred *ortho* to the N-Phenyl bond.

As part of a general program aimed at the structural elaboration of aryloxy substituted N-alkyl benzotriazoles (1), a small group of herbicidally active compounds in which a phenoxy spacer had been inserted between the benzotriazole nitrogen and the alkyl portion of the molecule were discovered (2).

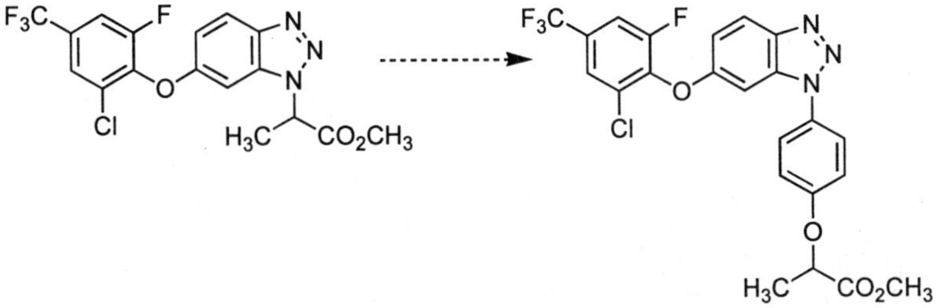

Our efforts aimed at defining SAR's within similar structural classes of PROTOX inhibiting herbicides have shown that aryloxy functionality can most often be interchanged with cyclic imide groups without detrimental effects to the biological profile. This was found to be the case with the title compounds and gave rise to cyclic imide substituted N-phenylbenzotriazoles (3).

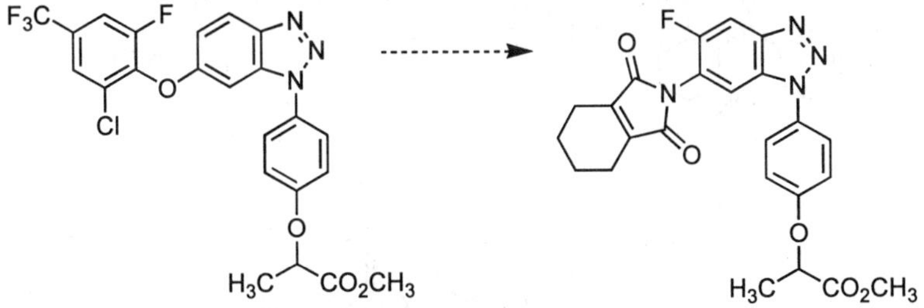

The herbicidal activity in both structural classes was followed up with a small analogue program which sought to study the effect of N-phenyl substitution as well as optimize regiochemistry of the appending group or groups with respect to the benzotriazole-phenyl bond.

Preparation of Aryloxy N-Phenyl Benzotriazoles

Arloxy benzotriazoles are prepared by two general methods described below.

From Nitrophenol. Nucleophilic displacement of an activated aryl halide with 3-nitrophenol (Scheme I) was carried out in either DMSO or DMF and potassium carbonate at 100°C providing the diphenyl ether **1**. Nitration of **1** with nitric acid and sodium nitrate in sulfuric acid at low temperature gave **2** which reacted with *o*-anisidine in refluxing acetonitrile to give the triaryl intermediate **3**. Iron reduction in acetic acid gave the *bis*-amino intermediate **4** which was subjected to standard diazotization conditions affording the benzotriazole **5**. Cleavage of the methyl ether with boron tribromide in dichloromethane followed by treatment with an alkylating agent and base in DMF gave the appropriately substituted N-phenyl benzotriazole **6**.

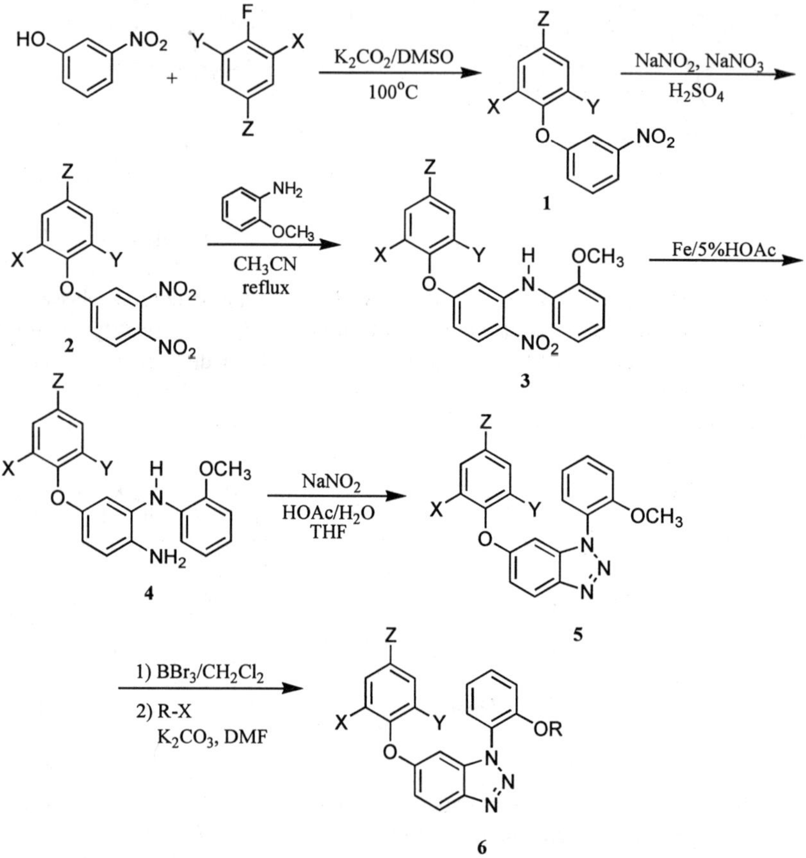

Scheme I

From 2,4-Difluoronitrobenzene. A second general method (Scheme II) could be employed when the aryloxy portion of the target is available as a nucleophilic phenol. Thus, 2,4-difluoronitro-benzene reacted with *o*-anisidine in DMSO at 95°C affording the diphenylamine **7**. A second nucleophilic displacement with a substituted phenol gave the intermediate **8** which could be reduced with either elemental iron in acetic acid or Raney® Nickel under H_2 in THF resulting in the intermediate **9**. Diazotization afforded the benzotriazole **10** which was treated with boron tribromide in dichloromethane followed by an alkylating agent and base in DMF to give the desired appropriately substituted aryloxy-N-phenylbenzotriazole **11**.

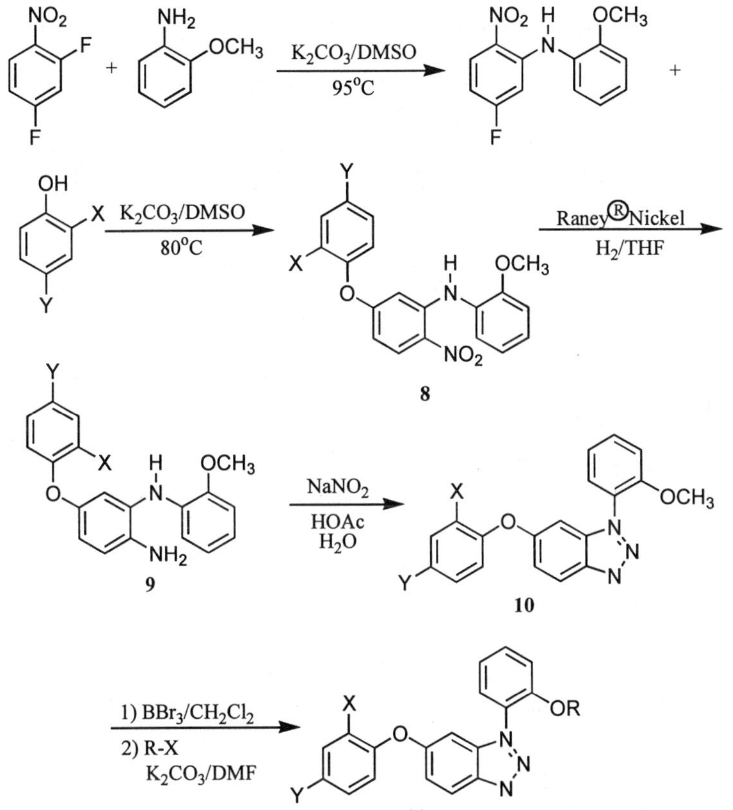

Scheme II

Preparation of Heterocyclic Substituted N-Phenyl Benzotriazoles

Heterocyclic substituted analogues were prepared via the general procedures outlined below.

From Substituted Anilines. The general method for preparation of heterocyclic substituted analogues (Scheme III) can begin with any number of commercially available anilines that are substituted in the 2- and 5-positions. Anilines were treated with acetic anhydride at low temperature to obtain phenyl acetamides **12**. A mixture of nitric and sulfuric acids in acetic acid afforded the nitrobenzene **13** which when treated with *o*-anisidine in refluxing acetonitrile provided the diphenyl amine **14**. Reduction of the nitro group with iron and acetic acid afforded the *bis*-amino intermediate **15** which was converted to the

N-phenyl benzotriazole **16** using standard diazotization conditions. Hydrolysis of the acetamide with acidic aqueous ethanol provided the aniline **17** which reacted with tetrahydrophthalic anhydride affording the imide substituted benzotriazole **18**. Cleavage of the methyl ether with boron tribromide in dichloromethane followed by standard alkylation conditions provided the desired heterocyclic substituted N-phenyl benzotriazole **19**.

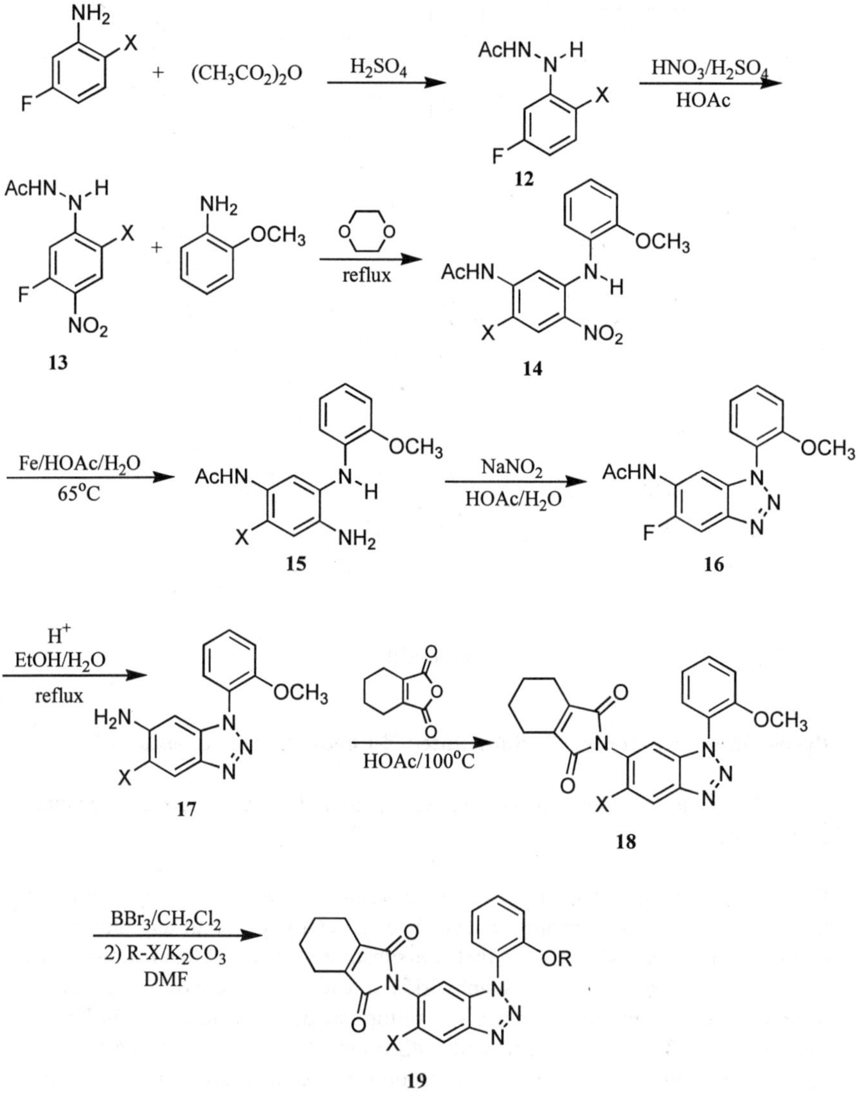

Scheme III

Herbicidal Activity

The herbicidal effects observed during our limited effort within both structural classes agreed with what is generally expected PROTOX symptomology. Both the aryloxy and cyclic imide analogues were most active when applied post-emergence at rates between 32 and 64g/ha. The aryloxy analogues seemed to be one rate more active than the imides. Pre-emergence treatments show the best promise for crop selectivity; however, application rates between 250 to 500g/ha were required to achieve acceptable control of target weed species.

Tables I and 2 are representative samples of herbicidal activity of some of the most active aryloxy and tetrahydrophthalimide analogues respectively. The ratings are based on a scale from 0-9. A rating of 0 indicates the complete absence of activity while 9 constitutes complete herbicidal control.

Table I. Post-emergence Herbicidal Activity at an Application rate of 32 g/ha

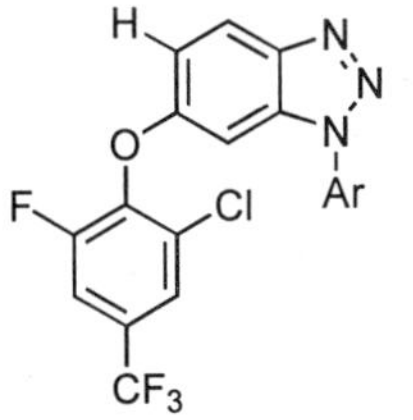

Ar	ABUTH	AMBEL	IPOSS	ECHCG	SETVI	GLXMAW
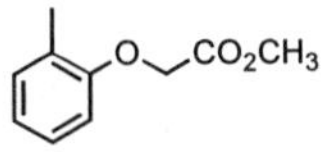	9	8	9	4	4	7
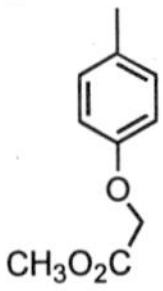	9	6	9	4	3	4
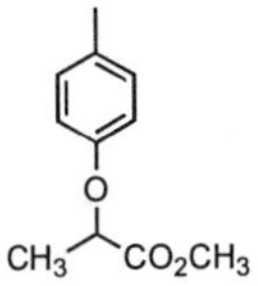	9	4	9	1	2	1

Structure-Activity Relationships

For the purpose of this discussion the effects of structure modification of two main regions of the aryloxy- as well as the heterocyclic substituted benzotriazoles series will be presented. Many assumptions were made based on previous knowledge of structure-activity relationships observed in the N-alkylbenzotriazole project (1).

Table II. Post-emergence Herbicidal Activity at an Application rate of 62 g/ha

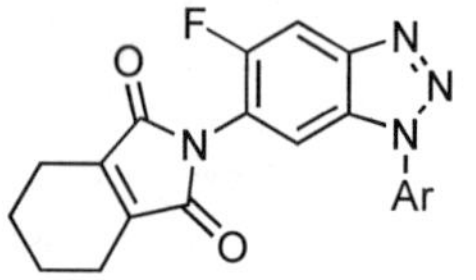

Ar	ABUTH	AMBEL	IPOSS	ECHCG	SETVI	GLXMAW
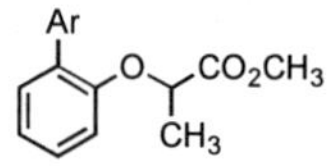	9	8	9	7	5	4
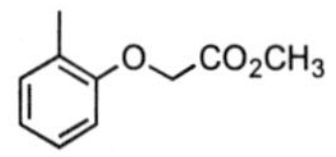	9	9	9	7	4	4
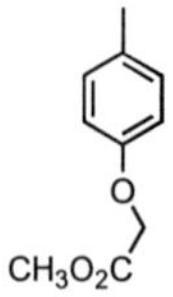	5	8	7	2	2	4

Aryloxy N-Phenylbenzotriazoles

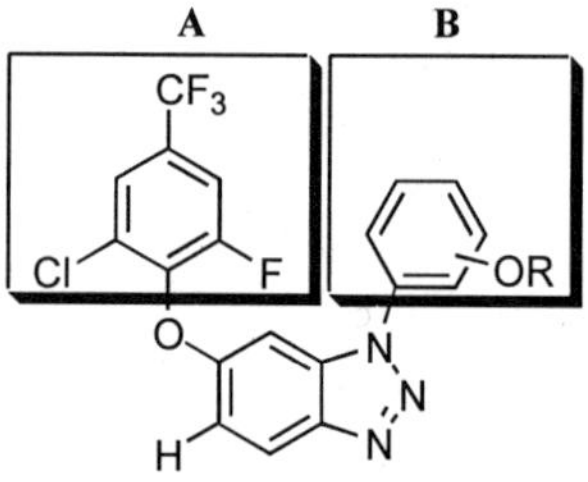

Region A. Of the phenyl substitution patterns investigated, the one which led to the most active series of compounds was the fluoro-chloro substituted benzotrifluoride depicted in the diagram above. Any attempt to deviate from this pattern led to compounds that were many rates less active both pre- and post-emergence.

Region B. Herbicidal activity was optimized when alkoxy substitution was *ortho* to the N-phenyl bond. Of the many different analogues prepared, the *o*-lacate delivered the highest level of herbicidal activity. Chain homologation and/or branching at the alpha carbon had a detrimental effect on both pre-and post-emergence herbicidal activity.

Heterocyclic Substituted N-Phenylbenzotriazoles

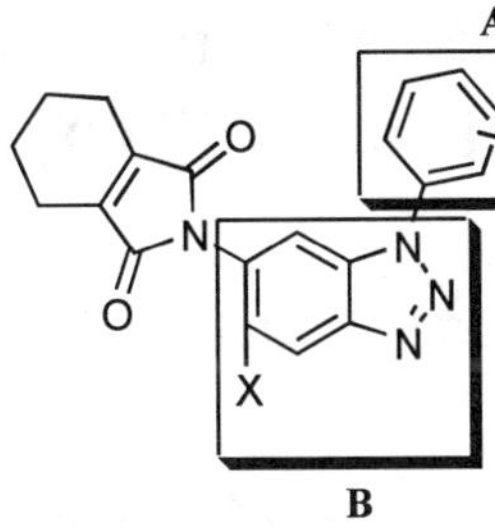

Region A. As was the case in the Aryloxy series, the *ortho*-lacate analogue was the most active. Branching and/or chain homologation of the alkyl ethers did not appreciably affect the level of observed herbicidal activity.

Region B. Fluorine substitution of the benzotriazole ring *ortho* to the heterocyclic substituent afforded the most active compounds within the series. However, the post emergence activity of the fluoro analogues was still one rate less active than similar aryloxy analogues.

Conclusions

The N-Phenylbenzotriazole protox inhibitors were discovered as a result of a limited effort to examine the effect of a penoxy spacers on the herbicidal activity of N-Alkylbenzotriazole herbicides (2,3). Three schemes for synthesis of the N-Phenylbenzotriazoles have been defined. The structure-activity relationships observed parallel that of known protox inhibiting herbicides. Post-emergence activity was observed at much lower rates than pre-, however, crop selectivity was most prevalent in pre-emergence applications. Given the fact that both structural classes of chemistry offered no distinct advantage over the commercial standards such as Acifluorofen and Sulfentrazone, the area was not further elaborated.

Literature Cited

1. Nielson, D. R., U.S. Patent 5,324,711, **1989**
2. Condon, M. E., U.S. Patent 5,484,762, **1996**
3. Condon, M. E., U.S. Patent 5,523,277, **1996**

Chapter 7

Synthesis and Herbicidal Activity of *bis*-Aryloxybenzenes: A New Structural Class of Protox Inhibitors Derived from N-Phenyl Benzotriazoles

A. D. Crews M. E. Condon, S. D. Gill, G. M. Karp, M. C. Manfredi, and J. H. Birk

Cyanamid Agricultural Research Center, American Cyanamid Company, P.O. Box 400, Princeton, NJ 08543–0400

The title compounds are new low application rate post-emergence herbicides with safety on broadleaf crops. The paper will describe the elaboration of existing chemistries which led to the discovery of the Aryloxybenzenes as well as the synthetic methodologies which were developed. A brief description of trends in structure-activity relationships which were observed will be given.

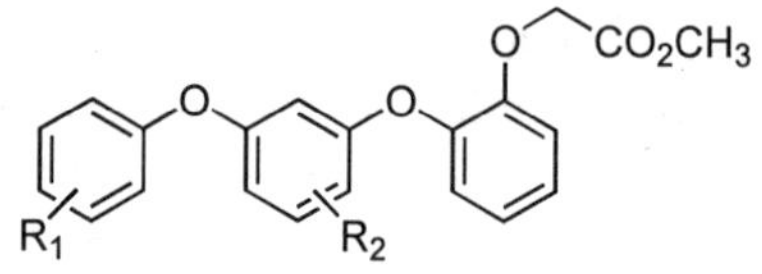

Figure 1 depicts an aniline-substituted diphenyl ether intermediate 1 which was prepared enroute to the N-phenylbenzotriazoles 2 (2) described in the previous chapter. When 2 was subjected to herbicidal evaluation, the results fostered sufficient interest to warrant the preparation of a small number of compounds such as the analogue 3. The level of herbicidal activity observed for 3 prompted an extensive synthesis program that ultimately afforded the *bis*-aryloxybenzenes 4 (1) as a potent new class of PROTOX inhibiting herbicides.

Preparation of *bis*-Aryloxybenzenes

Aryloxybenzenes are prepared by one of the four general methods described below.

From a Dinitro-Diphenyl Ether. (Scheme I) Nucleophilic displacement of one nitro group of diphenylether **5** with a methoxy-substituted phenol in refluxing acetonitrile provided the tri-aryl system **6**. Treatment of **6** with boron tribromide in dichloromethane afforded the phenol **7** which was alkylated with methyl

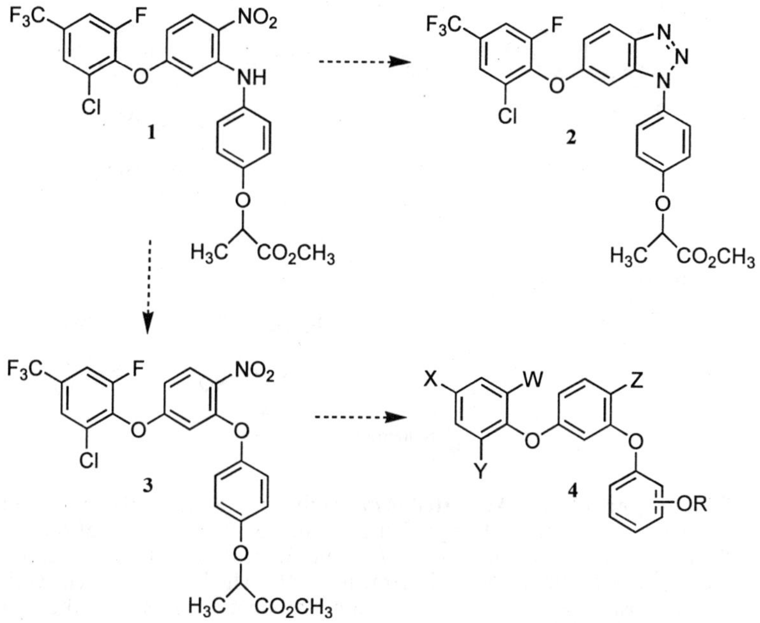
F3C
F
NO2
NH
Cl
1
O
H3C
CO2CH3
F3C
F
N
N
N
Cl
2
O
H3C
CO2CH3
F3C
F
NO2
Cl
3
O
H3C
CO2CH3
X
W
Z
Y
4
OR

FIGURE 1

bromoacetate and potassium carbonate in DMF to give the desired *bis*-aryloxybenzene **8**.

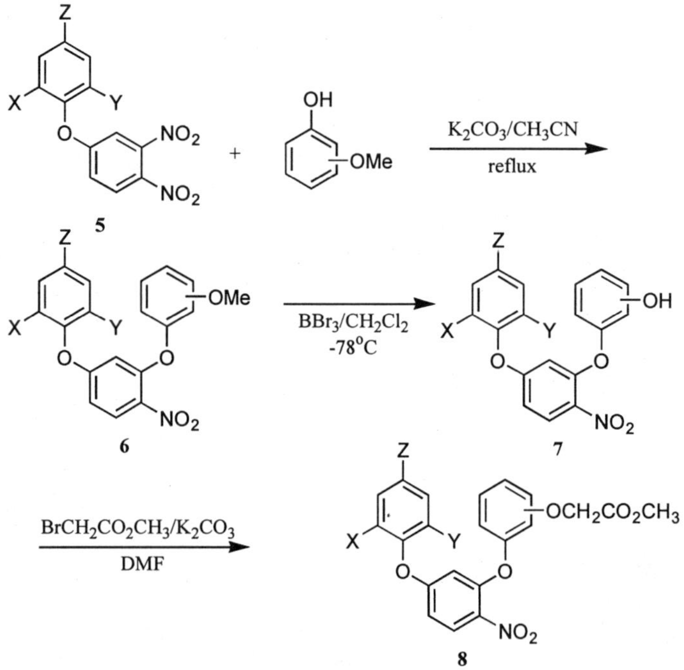

Scheme I

From 3-Fluorophenol and Activated Aryl Halides. (Scheme II) Nucleophilic displacement of an activated aryl halide with 3-fluorophenol afforded the diphenylether **9**. Nitration of **9** with 70% nitric acid in sulfuric acid at low temperature afforded the reactive intermediate **10** which was treated with an appropriately substituted phenol and potassium carbonate in DMF at 50^{0}C, giving the desired *bis*-aryloxybenzene **11**.

From 3-Fluoro-4-Nitrophenol (Scheme III) Protection of the phenol with dimethoxymethane, phosphorus oxychloride and dimethylformamide in toluene at 80^{0}C afforded the methoxymethyl ether **12**. The reaction of **12** with a substituted phenol and base in DMF at 100^{0}C gave diphenylether **13** which was treated with acid in refluxing methanol, giving phenol **14**. Treatment of phenol **14** with a reactive aryl halide and base in DMF at 100^{0}C provided the desired *bis*-aryloxybenzene **15**.

From 2,4-Difluoro Nitrobenzene (Scheme IV) The reaction of a substituted phenol with difluoro nitrobenzene in DMF at 100^{0}C gave the 4-fluoro-2-phenoxy nitrobenzene intermediate **16**. Treatment of **16** with a nucleophile such as hydroxypyrazole in DMF at elevated temperatures afforded the desired pyrazoloxy diphenyl ether **17**.

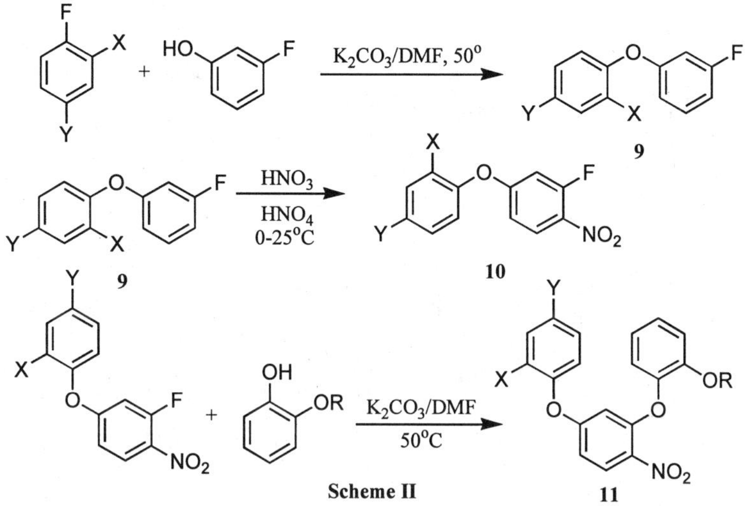

Scheme II

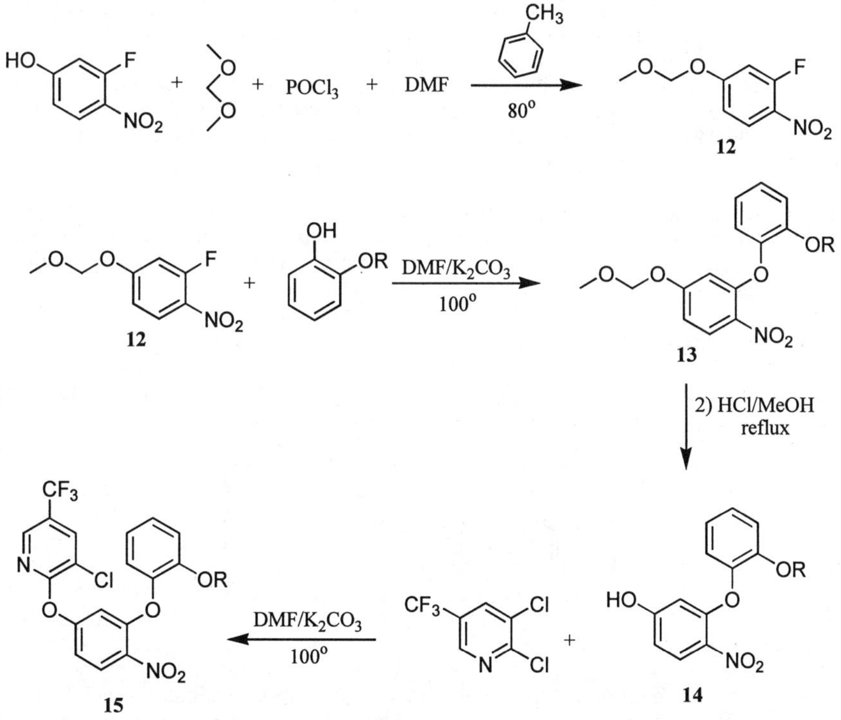

Scheme III

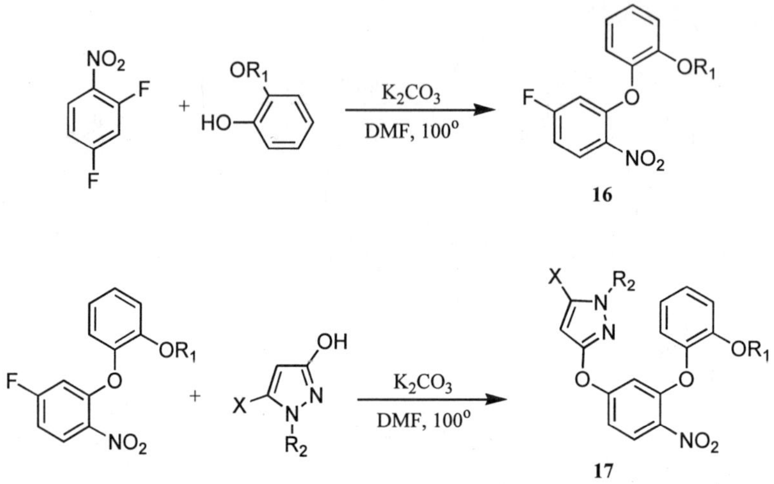

Scheme IV

Herbicidal Activity

The herbicidal effects observed were typical of known PROTOX inhibitors such as Acifluorfen and Sulfentrazone. In low dose post-emergence applications a rapid burn-down of foliage was generally observed within 12 hours. Post crop selectivity was observed in legumes to the extent that crops were able to overcome the initial burn and grow to maturity. Pre-emergence applications, although requiring much higher rates to achieve desired weed control, were much less phytotoxic with respect to grassy as well as broadleaf crops.

The following data are representative of the herbicidal activity observed when *bis*-Aryloxyobenzenes are applied post-emergence. The ratings are based on a scale from 0 to 9. A rating of 0 indicates a complete absence of activity while 9 constitutes complete control.

Structure-Activity Relationships

For the purpose of this discussion, the effects of structure modification to three main regions of the molecule will be discussed.

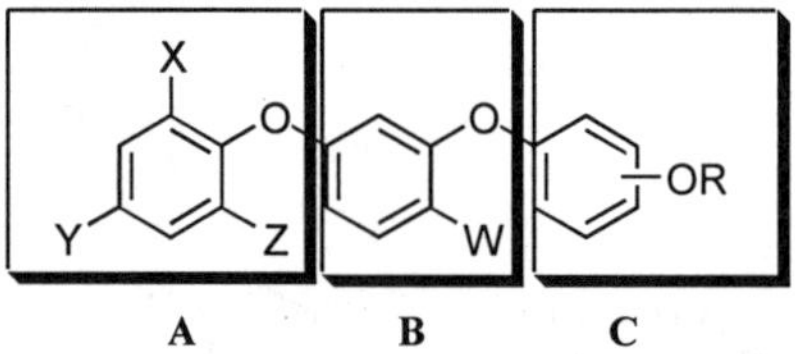

Region A. Of the aromatic and heteroaromatic systems investigated several trends in herbicidal activity became apparent. The most interesting A-rings were the substituted phenyls followed by heteroaromatic systems such as trifluoromethyl

Table I. Effect of Aromatic ring substitution on post-emergence Herbicidal Activity

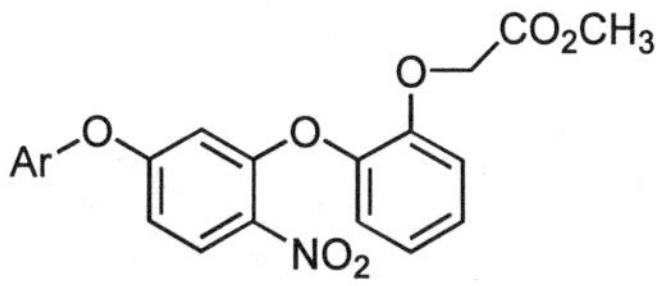

Aromatic Ring	Rate g/ha	ABUTH	AMBEL	CASOB	ECHCG	GLXMAW
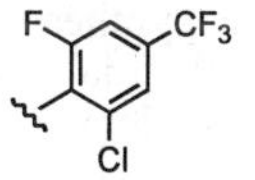	125	9	9	7	5	4
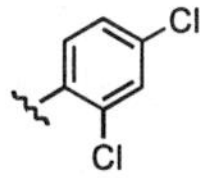	125	9	9	8	3	2
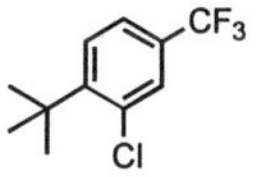	125	9	9	8	3	5
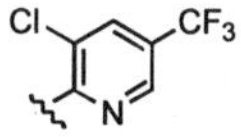	125	7	9	7	2	3
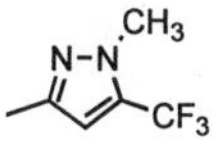	125	9	9	6	8	7

pyridines and pyrazoles. In general, the following relative order of activity was observed.

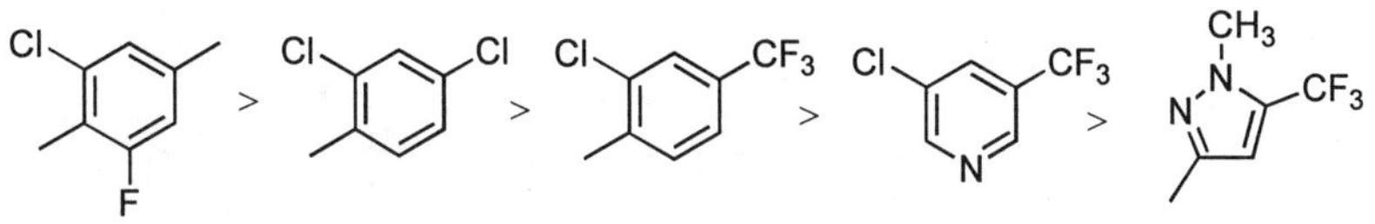

Region B. Electron withdrawing groups (W) such as nitro are required *ortho* to the B-C ether linkage. Regioisomeric compounds where the nitro is *ortho* to the A-B linkage are much less active. The A and C rings are always *meta* disposed about the B ring, while the replacement of nitro with an electron donating group such as amino had a negative effect on activity.

Region C. Optimum levels of activity are achieved when the ether functionality is located *ortho-* to the B-C linkage. Of the most active compounds, the following relative order of herbicidal activity as it relates to R-group substitution was observed.

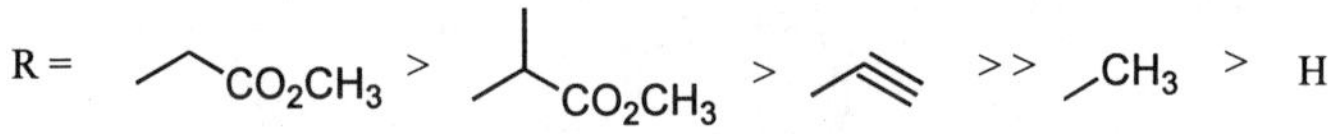

Conclusions

The *bis*-Aryloxybenzene (1) PROTOX inhibitors were discovered when an intermediate prepared in the process of elaborating the N-Phenylbenzotriazoles (2) was found to be herbicidally active. Four schemes for the synthesis of *bis*-Aryloxybenzenes have been defined. The highest level of herbicidal activity is observed in post-emergence applications; however, a greater degree of crop selectivity can be achieved with pre-emergence treatments.

Literature Cited

1. Condon, M. E., E.P. 647,612-A, **1995**
2. Condon, M. E., U.S. Patent 5,484,762, **1996**

Chapter 8

Synthesis and Herbicidal Activity of Fused Benzoheterocyclic Ring Systems

George Theodoridis, Jonathan S. Baum, Jun H. Chang, Scott D. Crawford, Frederick W. Hotzman, John W. Lyga, Lester L. Maravetz, Dominic P. Suarez, and Harlene Hatterman-Valenti

Agricultural Products Group, FMC Corporation, P.O. Box 8, Princeton, NJ, 08543

Fused benzoheterocyclic ring systems **1** represents a new class of highly active protoporphyrinogen oxidase (PPO) inhibitors. These compounds differ from traditional 2,4,5-trisubstitutedphenyl-1-heteroaryl PPO inhibiting herbicides in that positions 4 and 5 of the phenyl ring are linked together to form a fused heterocyclic ring. Synthesis, biological properties, and SAR are discussed.

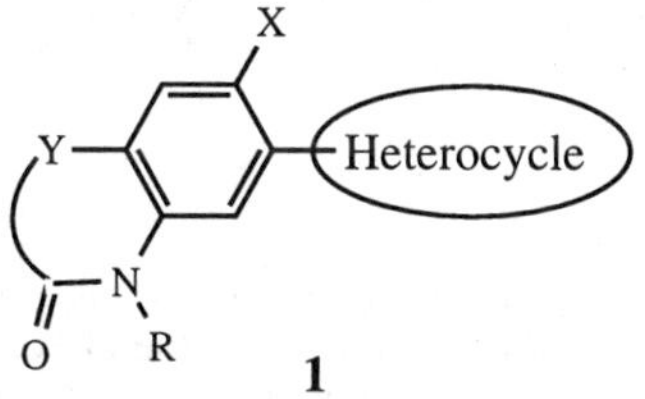

1

The past two decades have seen a great deal of work in the research and development of herbicides that act by inhibiting the enzyme protoporphyrinogen oxidase (Protox). These herbicides are represented by a wide range of diverse chemistries, of which the diphenyl ether class of herbicides such as compound **2** (*1*), and more recently the 2,4,5-trisubstitutedphenyl-1-heterocyclic area of herbicides such as compound **3** (*2*), have received a great deal of attention (*3*).

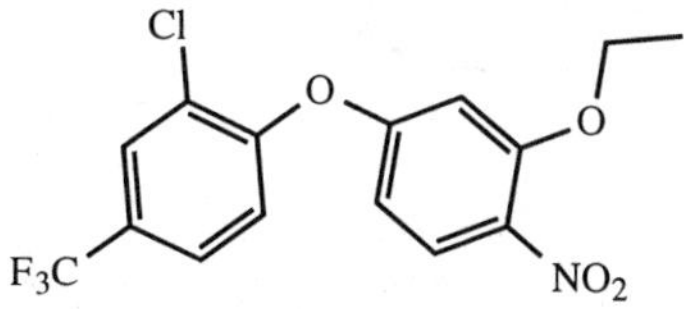

2

Oxyfluorfen

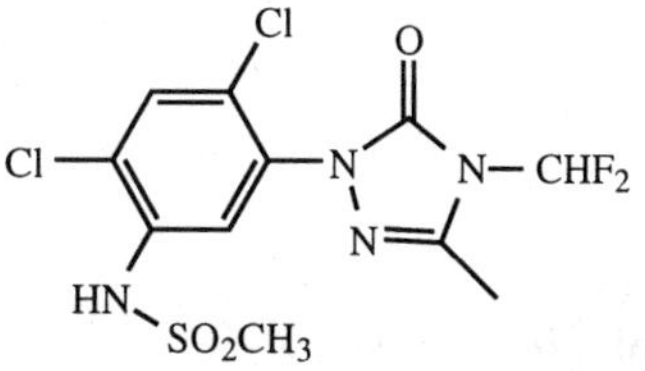

3

Sulfentrazone

Previous papers on the 2,4,5-trisubstitutedphenyl heterocyclic class **4** of Protox inhibitors have discussed the effects on herbicidal activity of various substituents at the 2,4, and 5 positions of the phenyl ring (*4-9*). The subject of this work is the synthesis and herbicidal activity of fused benzoheterocyclic ring system **5**, which has shown remarkable herbicidal activity at very low rates of application. These compounds, as will be discussed below, do not closely follow the SAR rules developed for the related 2,4,5-trisubstitutedphenyl heterocyclic **4**.

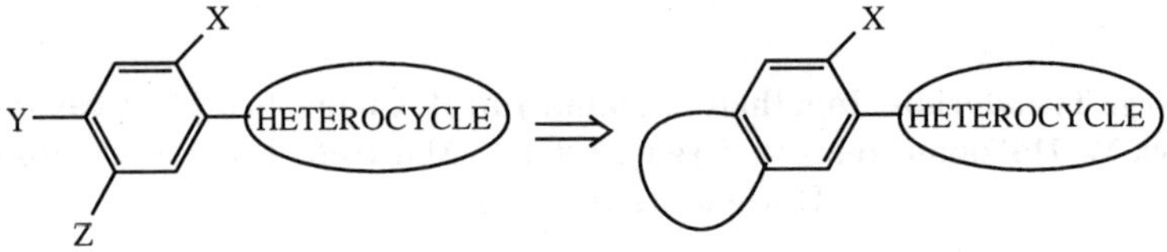

4 **5**

Figure 1. Fused benzoheterocyclic ring systems.

We will briefly discuss the biological activity of several fused benzoheterocyclic molecules **5**, where the heterocycle on the right side of the molecule is kept constant. We chose the tetrahydrophthalimide ring, compound **6**, as the constant right-side heterocycle.

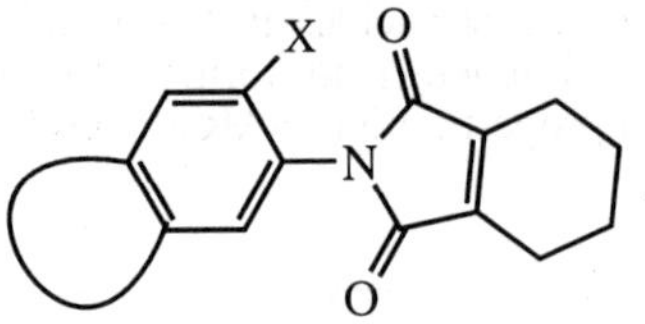

6

The main part of our discussion will focus on the synthesis and structure-activity of 7-heteroaryl-2(1H)quinolinones and the effect that various heterocycles have on their biological activity. To investigate the SAR at the various positions of the fused benzoheterocyclic ring we chose the quinolinone ring system **7**.

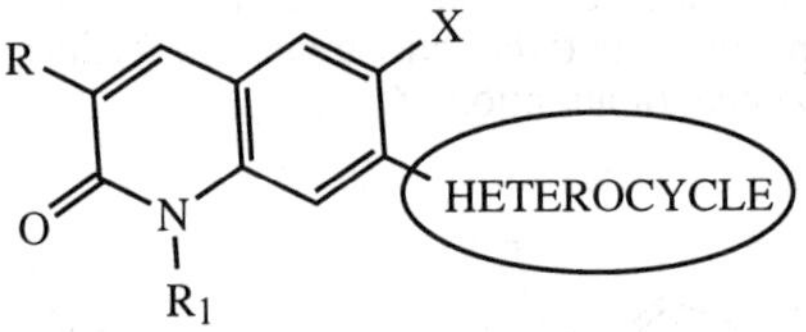

7

Biological Testing

The compounds described were tested preemergence and postemergence on various weeds and crops in the greenhouse The seeds of the plant test species were planted in furrows in steam-sterilized sandy loam soil contained in disposable fiber flats. A topping soil of equal portions of sand and sandy loam soil was placed uniformly on top of each flat to a depth of approximately 0.5 cm.

The flats were placed in a greenhouse and watered for 8-10 days, then the foliage of the emerged test plants was sprayed with a solution of the test compound in acetone-water containing up to 5 ml liter $^{-1}$ sorbitan monolaurate emulsifier/solubilizer. The concentration of the test compound in solution was varied to give a range of application rates. Phytotoxicity data were taken as percentage control, determined by a method similar to the 0-100 rating system described previously *(10)*, with 0% control of crops or weeds showing no effect relative to controls, and 100% control indicating complete crop or weed destruction. Biological data in Tables I-VIII are presented as the preemergence and postemergence application rates required to give 85% control as compared with untreated plants. In general, the 95% confidence interval for individual ED_{90} values in these tests is $ED_{90}/2$ to $ED_{90}x2$ (e.g., the CI for an ED_{90} of 30 g ha^{-1} is 15-60). The weeds species used in this study were morningglory, velvetleaf, johnsongrass, green foxtail, and cocklebur.

Synthesis of 7-Heteroaryl-2(1H)quinolinones

The synthesis of these molecules was quite challenging. In Approach 1 (Figure 2) we attempted to use known literature procedures to build the quinolinone ring first *(11,12)*, which could then be nitrated, reduced to the corresponding 7-amino-2(1H)quinolinone, which could in turn be used as the starting point for building the desired heterocycle, in this case a triazolinone **8**. This approach has several severe limitations, such as the limited number of R and R_1 groups that can be introduced at positions 1 and 3 of the quinolinone ring. An even more severe limitation turned out to be the lack of regioselectivity at the nitration step, which resulted in a mixture of all possible regioisomers.

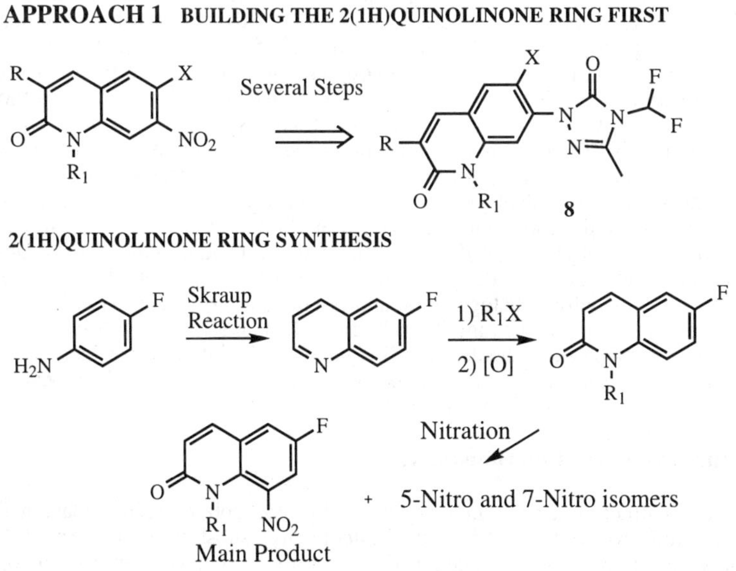

Figure 2. Synthesis of 1,3,6-trisubstituted-7-(heteroaryl)-2(1H)quinolinone.

The synthesis of 1,3,6-trisubstituted-7-(heteroaryl)-2(1H)quinolinone according to Approach 2, involved building the triazolinone ring first and then using the Knorr synthesis to build the quinolinone ring (*13*). It was hoped that the desired regioisomer would be obtained as a consequence of the greater steric hindrance of the undesired isomer (Figure 3). Unfortunately both isomers were obtained, with the undesired isomer **9** in greater abundance.

APPROACH 2 BUILDING THE 2(1H)QUINOLINONE RING LAST

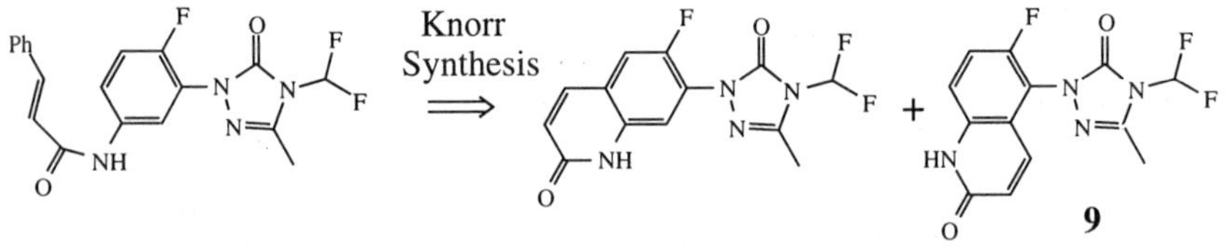

9

MIXTURE OF ISOMERS

Figure 3. Synthesis of 2(1H)quinolinone ring.

After the available literature procedures for the synthesis of 2(1H) quinolinones failed to provide a satisfactory synthetic route to the desired targets, we undertook the search for a novel approach for the design and synthesis of 2(1H)quinolinones. This new synthetic approach had to have a combination of desired characteristics, such as mild reaction conditions, and be general enough to introduce a wide range of groups at various positions, as well as be regiospecific. We arrived at such synthesis by applying the Meerwein reaction to generate a 2-halo-propionate carbon chain (Figure 4). Nitration with 70% nitric acid in concentrated sulfuric acid as solvent, followed by reductive cyclization with iron dust in acetic acid, gave the corresponding 3,4-dihydro-2(1H)quinolinone in excellent yields. At this point the 3,4-dihydro-2(1H)quinolinone can be converted to the corresponding 2(1H)quinolinone by the use of triethylamine in tetrahydrofuran. Alkylation at the nitrogen of the 2(1H)quinolinone is achieved by heating an alkyl halide with potassium carbonate in dimethylformamide (*14*). Alternatively, both dehydrohalogenation and N-alkylation of the 2(1H)quinolinone can be achieved in one step by heating 3,4-dihydro-2(1H)quinolinone with two equivalents of potassium carbonate and the desired alkyl halide in dimethylformamide.

The synthesis of the triazolinone, tetrazolinone, and other heterocyclic rings have been previously described in detail (*15-18*).

Structure-Activity Relationships

Our previous structure-activity investigations of protoporphyrinogen oxidase herbicides have centered around the 2,4,5-trisubstitutedphenyl substitution pattern (*6-9*). In summary, position 2 of the phenyl ring required a halogen group for optimum

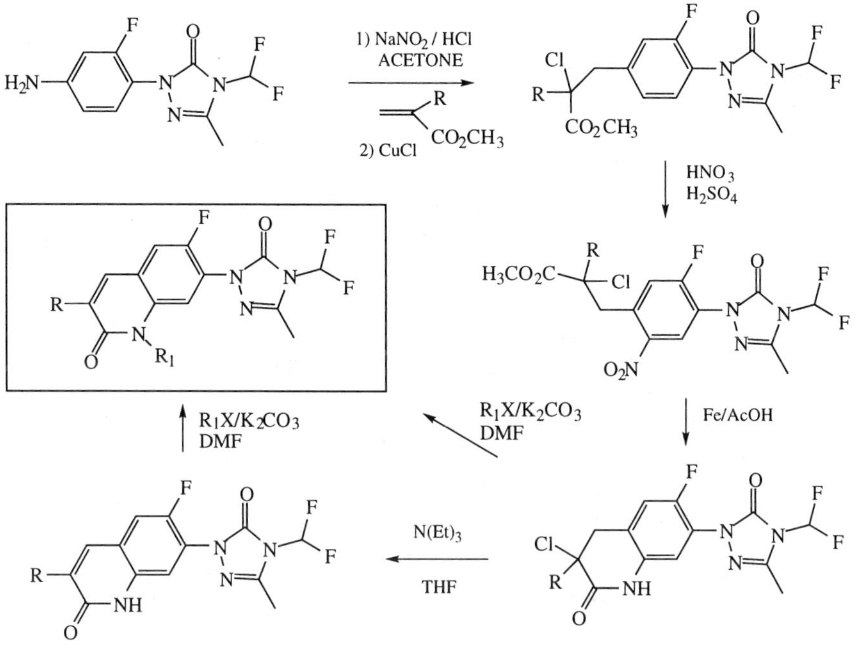

Figure 4. Novel synthesis of the 2(1H)quinolinone ring via a Meerwein approach.

biological activity, with fluorine providing the highest overall activity. Position 4 of the phenyl ring required a hydrophobic, electronegative group such as halogen for optimum activity, with both chlorine and bromine providing the best activity.

Electron donating groups such as alkoxy groups resulted in significant reduction of biological activity. Introducing a methoxy group at this position was predicted, in general, to result in lower biological activity. A good example of how the SAR of the 2,4,5-trisubstitutedphenyl system and that of the fused benzoheterocyclic ring system differ is shown in figure 5. Compound **10** required over 4000 grams/ha for morningglory control. When the 4 and 5 positions were tied together into a fused heterocyclic ring, the biological activity dramatically improved. Compound **11** controlled both velvetleaf and morningglory at 125 grams/ha (Table I).

Effect of the Fused Benzoheterocyclic Ring System on Biological Activity

Though a large number of fused benzoheterocyclic rings were investigated, we will be discussing only six different heterocycles. All these compounds have a tetrahydrophthalimide ring on the right side of the molecule (Figure 6). As shown in Table II, both five and six membered rings, compounds **11**, **12**, and **13**, resulted in good biological activity. Interestingly, reversing the position of the oxygen and the nitrogen of the 1,4-benzoxazin-3-one molecule **16** resulted in significant loss of biological activity. The benzoisoxazolinone ring **14** as well as the benzoxazole ring **15** also resulted in reduced biological activity.

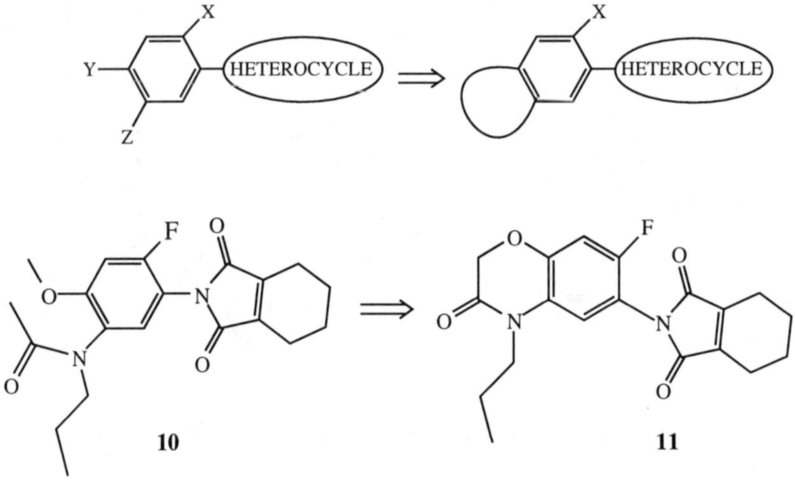

10 **11**

Figure 5. Comparison of open vs. fused ring analogs.

Table I. Comparison of the Biological Activity of Open vs. Fused Ring Analogs

Preemergence Biological Activity ED_{85} grams/ha		
Compound	Velvetleaf	Morningglory
10	2000	>4000
11	62.5	125

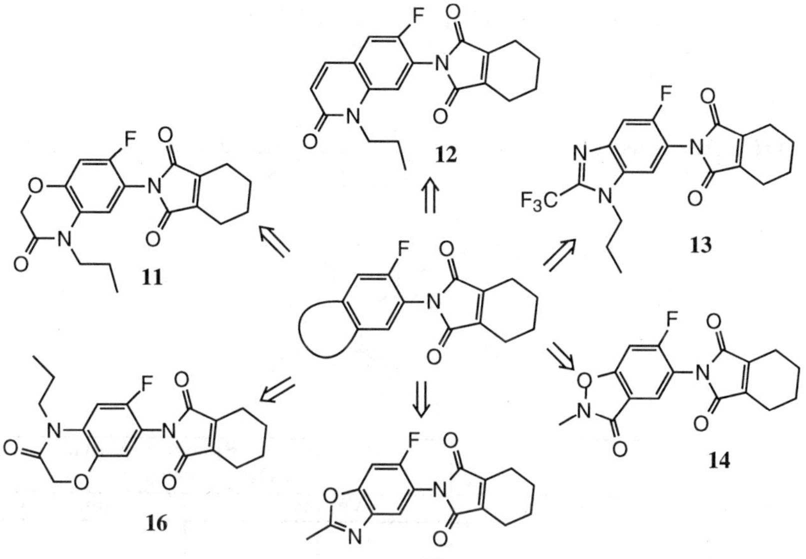

Figure 6. Effect of the fused benzoheterocyclic ring system on biological activity.

Table II. Effect of the Fused Benzoheterocyclic Ring System on Preemergence Biological Activity

	Preemergence Biological Activity ED_{85} grams/ha	
Compound	Velvetleaf	Morningglory
11	62.5	125
12	125	500
13	125	250
14	>2000	>2000
15	2000	>2000
16	>2000	>2000

Effect of Substituents at Position 6 of the Quinolinone Ring

Several groups were introduced at position 6 of the quinolinone ring, ortho to the pyrimidinedione heterocycle (Table III). The fluorine group, compound **17**, provided the best biological activity. It was interesting that the next most active

2(1H)quinolinone was compound **18,** which has a hydrogen at position 6 and not a chlorine. This was contrary to the structure-activity relationships previously developed for the aryl triazolinone ring system (*8,9*).

Table III. Effect of Substituents at Position 6 of the Quinolinone Ring on the Preemergence Biological Activity

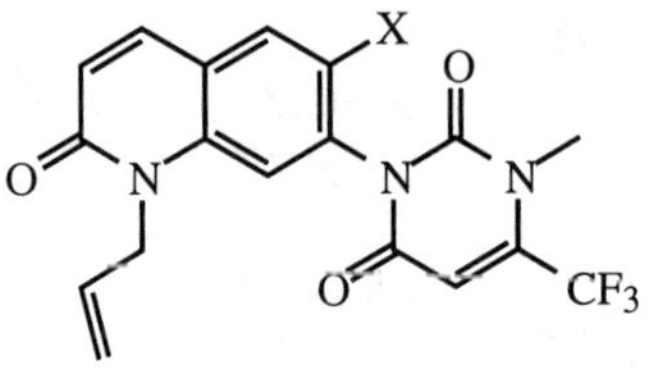

		Preemergence Biological Activity ED85 grams/ha	
Compound	X	Velvetleaf	Morningglory
17	F	1	3
18	H	3	10
19	Cl	10	100

We also investigated position 3 of the quinolinone ring. Compound **17** (R=H), and compound **20** (R=methyl), provided the best activity (Table IV).

Table IV. Effect of Substituents at Position 3 of the Quinolinone Ring on the Preemergence Biological Activity

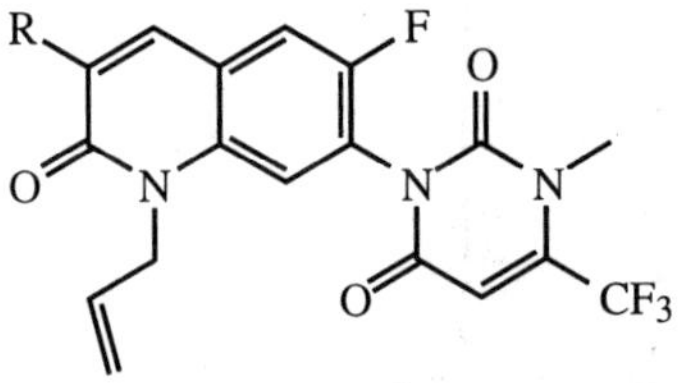

		Preemergence Biological Activity ED85 grams/ha	
Compound	R	Velvetleaf	Morningglory
17	H	1	3
20	CH_3	3	10
21	CF_3	100	300
22	CO_2CH_3	300	>1000

Several chemical groups were introduced at the nitrogen of the quinolinone ring (Table V). In general a wide range of substituents provided good biological activity. Steric factors do not seem to be important for biological activity, since both small groups such as methyl and ethyl, compounds **25** and **24**, as well as large groups such as benzyl, compound **29**, provided comparable levels of biological activity. Compound **17** (R= allyl) and compound **23** (R=propargyl) provided the highest level of biological activity.

Table V. Effect of Substituents at Position 1 of the Quinolinone Ring on Preemergence Biological Activity

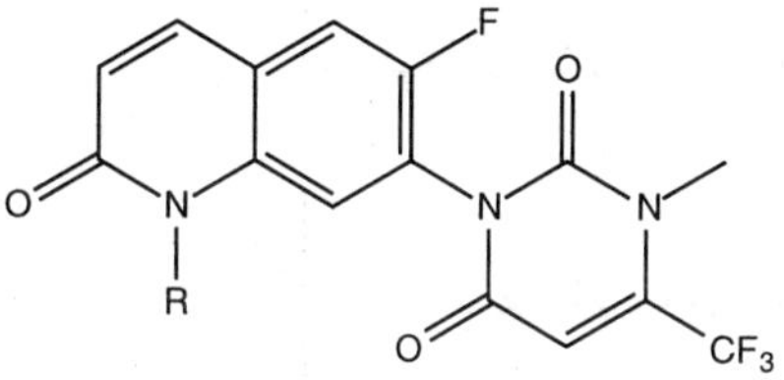

		Preemergence Biological Activity ED85 grams/ha	
Compound	R	Velvetleaf	Morningglory
23	$CH_2C{\equiv}CH$	1	3
18	$CH_2CH{=}CH_2$	1	3
24	CH_2CH_3	3	10
25	CH_3	10	10
26	$CH_2CH_2CH_3$	10	30
27	$CH_2CH(CH_3)_2$	10	30
28	H	30	30
29	$CH_2C_6H_5$	10	30
30	CH_2CN	10	30
31	$CH_2CO_2CH_3$	10	100

Hydrogenation of the double bond of the 2(1H)quinolinone ring resulted in compound **32**, which, though less active than its unsaturated analog **23**, remains very active (Table VI).

Table VI. Effect of Unsaturation of the Quinolinone Ring on Preemergence Biological Activity

		Preemergence Biological Activity ED_{85} grams/ha	
Compound		Velvetleaf	Morningglory
23	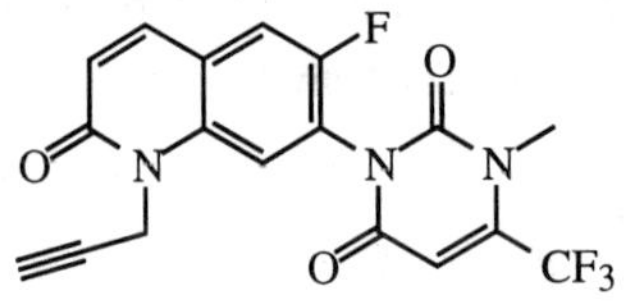	1	3
32	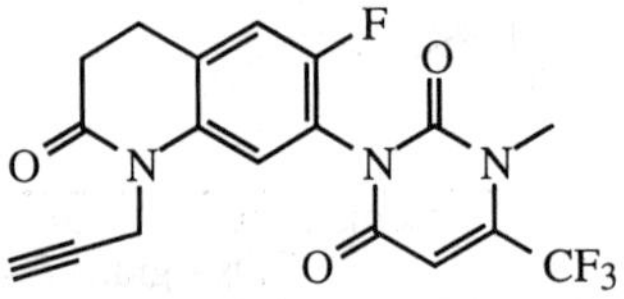	10	30

Several heterocyclic rings at position 7 of the 2(1H)quinolinone ring were investigated (Table VII). The pyrimidinedione **17** shows excellent preemergence biological activity, controlling both velvetleaf at 1 gram/ha, and morningglory at 3 gram/ha. The triazolinone ring **33** followed very closely in activity requiring 15 grams/ha to control both velvetleaf and morningglory. Both the tetrahydrophthalimide **34** and the tetrazolinone **35** rings were significantly less active.

In addition to their preemergence activity this class of PPO herbicides is also highly active when applied postemergence on a wide range of important weeds (Table VIII). When compound **36** was applied preemergence, at a rate of 10 grams/ha, velvetleaf, morningglory, cocklebur, and johnsongrass were controlled. Postemergence application of compound **36**, at 10 grams/ha, gave good control of velvetleaf, morningglory, and cocklebur, while grass weeds such as green foxtail and johnsongrass, at this rate, were not completely controlled.

Summary

The 1,3,6-trisubstituted-7-heteroaryl-2(1H)quinolinones described in this chapter are highly active pre- and postemergence herbicides, with a broad spectrum of weed control. 3-[1-(Propyn-2-yl)-6-fluoroquinolin-2-on-7-yl]-1-methyl-6-trifluoromethyl uracil **23** and 3-[1-(propyn-2-yl)-6-fluoroquinolin-2-on-7-yl]-1-amino-6-trifluoromethyluracil **36** were among the most biologically active representatives of this class of herbicides. The mechanism of action involves the inhibition of protoporphyrinogen oxidase, which results in the buildup of high levels of protoporphyrin IX, a photodynamic toxicant.

Table VII. Effect of the Heterocyclic Ring on Preemergence Biological Activity

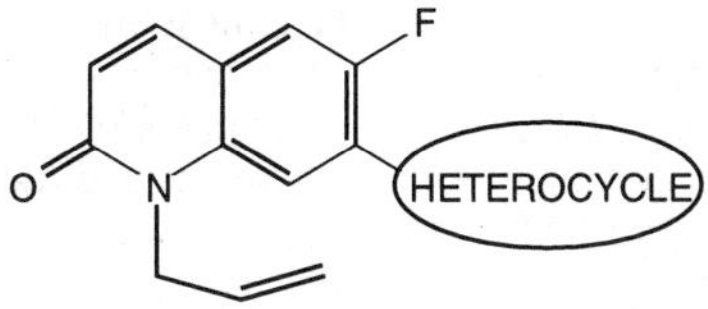

Preemergence Biological Activity ED$_{85}$ grams/ha			
Compound	Heterocycle	Velvetleaf	Morningglory
17	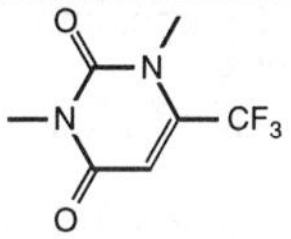	1	3
33	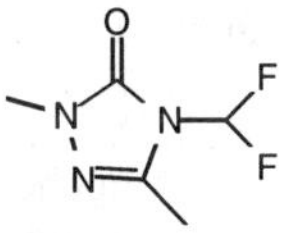	8	15
34	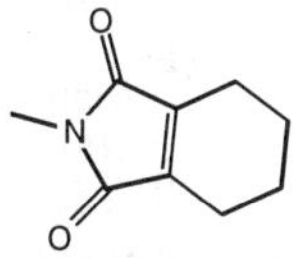	250	500
35	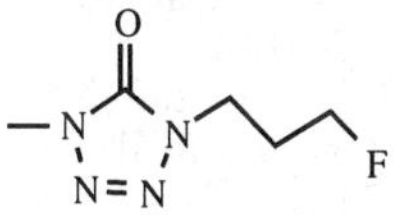	250	250

Table VIII. Biological Activity of 3-(1-Propargyloxy-quinolin-2-on-7-yl)-1-amino-6-trifluoromethyluracil

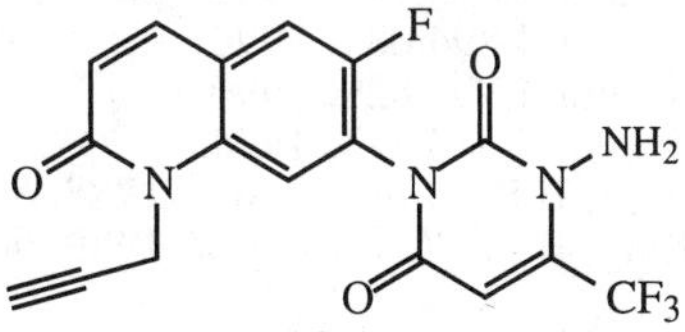

36

Greenhouse Biological Activity % Control at 10 grams/ha					
	Velvetleaf	Morningglory	Cocklebur	Green Foxtail	Johnsongrass
Preemergence	100	100	100	70	90
Postemergence	100	100	100	50	50

Acknowledgments

The authors would like to acknowledge the contributions of Blaik P. Halling, Debra A. Witkowski, and M. Joan Plummer for their work in elucidating the mechanism of action of these herbicides (*19, 20*), and to express their thanks to James T. Bahr and William A. Van Saun for their encouragement and advice. Finally, the authors acknowledge the support of FMC Corporation.

Literature Cited

1. Bayer, H.O.; Swithenbank, C.; Yih, R,Y. U.S. Patent 3,798,276, 1974.
2. Theodoridis, G. U.S. Patent 4,818,275, 1989.
3. Porphyric Pesticides: Chemistry, Toxicology, and Pharmaceutical Applications, edited by Constantin A. Rebeiz, ACS Symposium Series No. 559, American Chemical Society, Washington, D.C. 1994.
4. Ohta, H.; G.Jikihara, T.; Wakabayashi, K.; Fujita, T. *Pestic.Biochem.Physiol.* **1980,** *14*, pp 153-160.
5. Wakabayashi, K. *J. Pestic.Sci..* **1988**, *13*, pp 337-361.
6. Theodoridis, G.; Hotzman, F.W.; Scherer, L.W.; Smith, B.A.; Tymonko, J.M; Wyle, M.*J. Pestic.Sci..* **1990**, *30*, pp 259-274.
7. Theodoridis, G.; Hotzman, F.W.; Scherer, L.W.; Smith, B.A.; Tymonko, J.M; Wyle, M.J. In Synthesis and Chemistry of Agrochemicals III, edited by Baker, D.R.; Fenyes, J.G.; Steffens, J.J.; ACS Symposium Series No. 504; American Chemical Society; Washington, D.C. 1992, pp 122-133.
8. Theodoridis, G.; Baum, J.S.; Hotzman, F.W.; Manfredi, M.C.; Maravetz, L.L.; Lyga, J.W.; Tymonko, J.M; Wilson. K.R.; Poss, K.M.; Wyle, M.J. In Synthesis and Chemistry of Agrochemicals III, edited by Baker, D.R.; Fenyes, J.G.; Steffens, J.J.; ACS Symposium Series No. 504; American Chemical Society; Washington, D.C. 1992, pp 134-146.
9. Theodoridis, G. *J. Pestic.Sci..* **1997**, *50*, pp 283-290.
10. Frans, R.E.; Talbert, R.E.; *Research Methods in Weed Science*, 2nd edn, ed. B. Truelove, Auburn, AL, 1977, pp 15-23.
11. Sveinbjornsson, A.; Bradlow, H.L.; Aoe S.; Vanderwerf, C.A. *J.Org.Chem.* **1951**, *16*, pp 1450-1452.
12. Loev, B.; Broomall, Macko, E. U.S. Patent 3,555,030, 1971.
13. Manimaran, T.; Thiruvengadam, T.K.; Ramakrishnan, V.T. *Synthesis*, **1975**, 739.
14. Theodoridis, G.; Malamas, P. *J.Heterocyclic Chem.* **1991**, *28*, pp 849-852
15. Theodoridis, G. U.S. Patent 4,878,941, 1989.
16. Theodoridis, G. U.S. Patent 4,894,084, 1990.
17. Theodoridis, G. U.S. Patent 4,909,829, 1990.
18. Theodoridis, G. U.S. Patent 5,310,723, 1994.
19. Witkowski, D.A.; Halling, B.P. *Plant Physiol.* **1988**, *87,* 632.
20. Witkowski, D.A.; Halling, B.P. *Plant Physiol.* **1989**, *90*, 1239.

Chapter 9

3-Benzyl-1-methyl-6-trifluoromethyluracils: A New Class of Protox Inhibitors

Marvin J. Konz, Harvey R. Wendt, Thomas G. Cullen, Karen L. Tenhuisen, and Olga M. Fryszman

Agricultural Products Group, FMC Corporation, P.O. Box 8, Princeton, NJ 08543

3-Benzyl-1-methyl-6-trifluoromethyluracils, as a new class of herbicides, were developed from a weakly active Protox inhibitor, N-[(4-chlorophenyl)methyl]glutarimide. In preemergent applications, 3-[(2,3,5-trichlorophenyl)methyl]-6-trifluoromethyl-1-methylpyrimidine-2,4-dione controlled broadleafs and grasses at rates of 10-30 g/ha in greenhouse evaluations and at rates of 60-120 g/ha in field trials with corn as the most tolerant crop. The historical background of this discovery, structure-activity relationships and the synthesis of 3-benzyluracils is discussed.

N-[(4-Chlorophenyl)methyl]glutarimide (**1**) was screened in the excised cucumber (*1*) and algae assay (*2*) for intrinsic herbicidal activity. The biodata indicated **1** to be a weak Protox inhibitor. On greenhouse testing, the glutarimide was active in pre- and postemergent applications but only at a rate of 8 kg/ha.

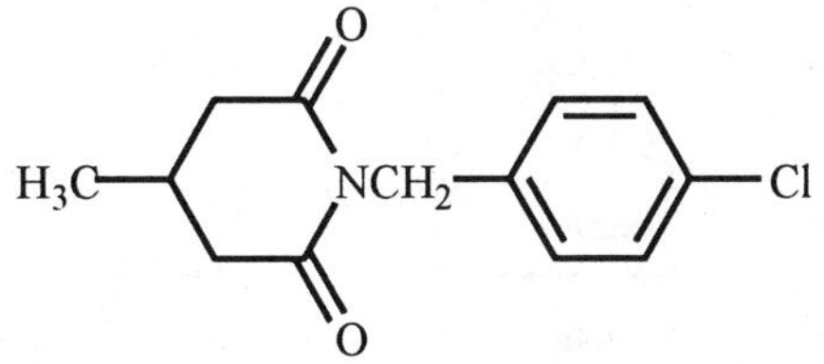

1

Weak PPO Inhibitor

- Excised Cucumber : pI_{50} 4.1
- Algae : pI_{50} 4.7

Although **1** was not very active, the structural comparison to two herbicide classes, 5-benzyloxy-1,3-dioxanes (**2**) (*3,4*) and 2-benzylisoxazolidine-3,5-diones (**3**) (*5*), was extremely interesting. The former is a mitotic inhibitor (Plummer, M. J., FMC Corporation, unpublished data) with grass toxicant properties whereas the latter is a carotenoid biosynthesis inhibitor with broadleaf and grass activity. It is quite unusual to observe a change in the mechanism of action which could be attributed to a change in the heterocyclic ring portion of the molecule. For this reason and the absence of reported herbicidal activity of N-benzylglutarimides, it was decided to prepare additional analogs for testing.

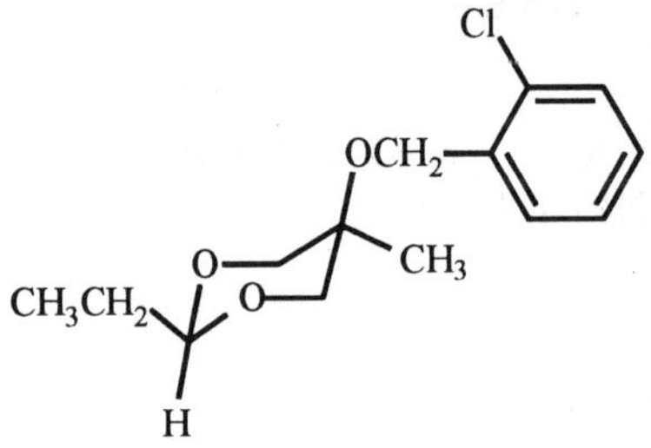

2

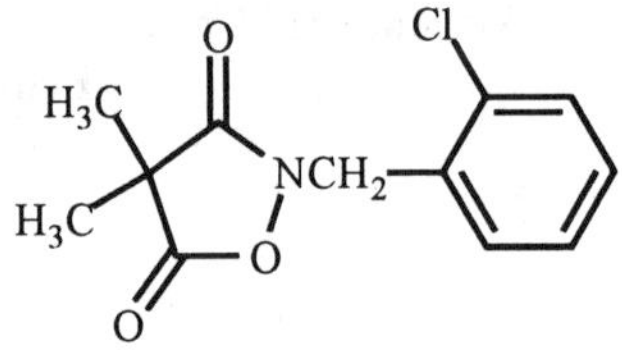

3

N-Benzylglutarimides

As the first approach, a chlorine probe strategy was employed to determine what positions of substitution in the aromatic ring would contribute to activity. For this purpose, ten compounds were prepared as shown in Scheme 1. The set of compounds were evaluated in both the excised cucumber and algae assays. The data is summarized in Table I.

The algae assay data was chosen for the Free-Wilson analysis as it met the criteria that the data should have a minimum spread of two orders of magnitude. The results of this analysis are shown in Table II. The conclusion was that substitution at the 3- and 5-positions had a significant contribution to activity whereas substitution at the 2- and 4-positions had a negligible effect. This is in contrast to other Protox inhibitors where substitution at the 2,4,5-positions of the aromatic ring is a criterion for biological activity.

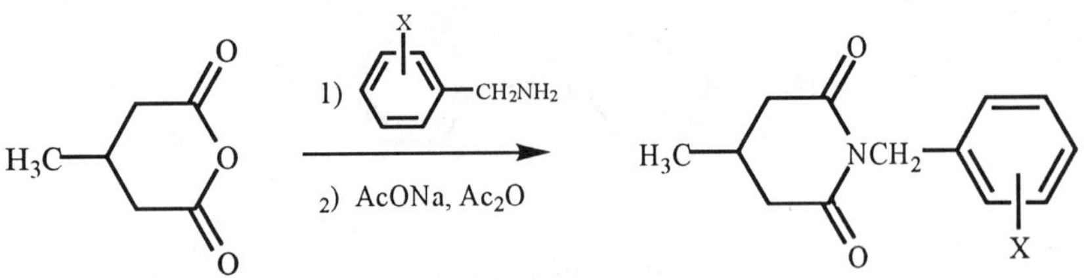

Scheme 1. Synthesis of N-Benzylglutarimides

Table I. Intrinsic Assay Data: N-Benzylglutarimides

Compound No.	X	Excised Cucumber pI_{50}	Algae pI_{50}
1	4-Cl	4.1	4.7
4	H	<4.0	NM
5	2-Cl	<4.0	3.9
6	3-Cl	4.2	4.8
7	2,3-DiCl	5.1	5.6
8	2,5-DiCl	<4.0	4.9
9	3,4-DiCl	4.8	5.6
10	3,5-DiCl	4.6	6.1
11	2,4,6-TriCl	4.0	<4.0
12	2,3,6-TriCl	<4.0	4.4
13	2,6-DiCl	4.1	<4.0

NM = Not Measurable

Table II. N-Benzylgutarimides: Free-Wilson Group Contributions

Total Compounds Defined: 10
Algae Assay, pI_{50}
Activity of Parent (theoretical): 4.41

	Substituent Position in Aromatic Ring				
Substituent	*2*	*3*	*4*	*5*	*6*
Other	0.07	0.45	0.12	0.36	-0.35
H	0.00(5)	0.00(6)	0.00(8)	0.00(9)	0.00(8)
Cl	-0.14(6)	0.90(6)	0.25(3)	0.72(2)	-0.72(3)

F = 7.556 [F-Table Val: (5% = 5.05) (1% = 10.97)]
(p = 0.02)
Coefficient of Multiple Correlation = 0.940

Of the glutarimides, compound **10** was selected for greenhouse testing. At the highest application rate (3,000 g/ha), it was not active in either pre- or post-emergent treatments.

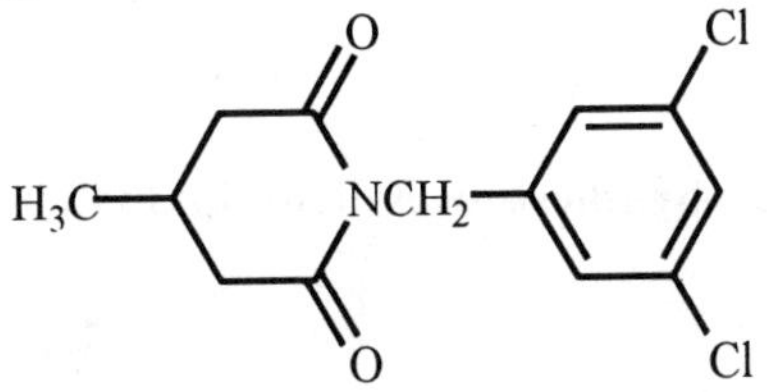

10

Although glutarimides were not active, the aromatic substitution pattern for intrinsic activity appeared to be unique. N-Benzyltetrahydrophthalimides (**A**) were selected to further examine this pattern for two reasons. First, the compounds could be prepared in one step from the tetrahydrophthalic anhydride and the appropriate benzylamine. Secondly, several chlorinated N-phenyltetrahydrophthalimides (**B**) were available for comparison, providing information as to the value of inserting a methylene bridge between the heterocyclic and aromatic rings. (*6,7*)

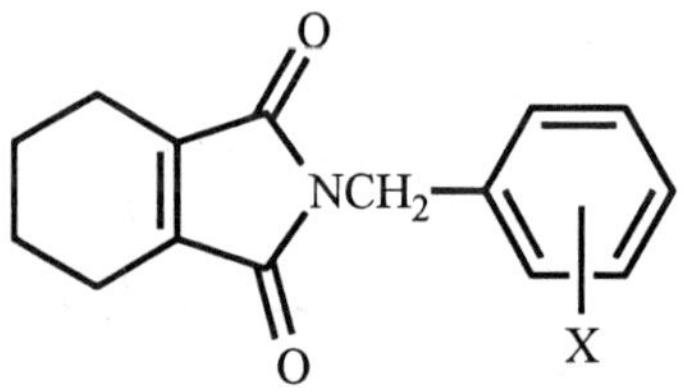

A

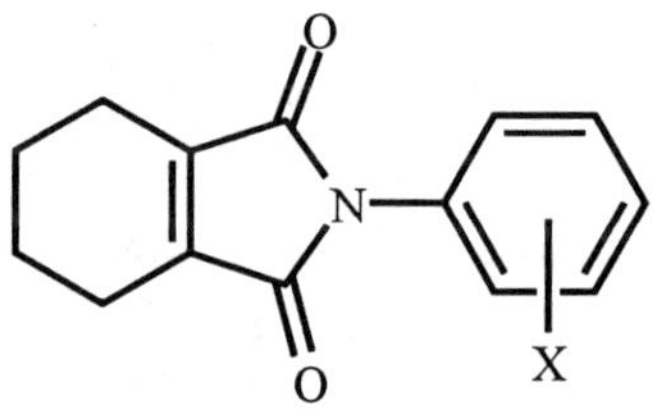

B

N-Benzyltetrahydrophthalimides

As in the glutarimides, a chlorine probe strategy was followed. Eleven N-benzylphthalimides were prepared by the reaction of 3,4,5,6-tetrahydrophthalic anhydride with the appropriate benzylamine in toluene and removal of the water formed by azeotropic distillation (Scheme 2). The intrinsic assay data for this set of compounds is contained in Table III.

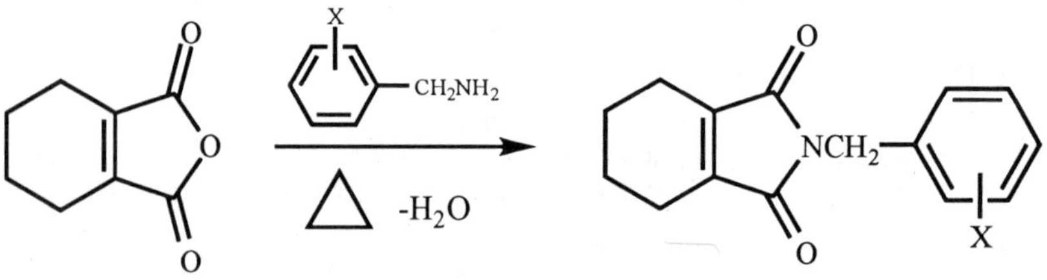

Scheme 2. Synthesis of N-Benzylphthalimides

Table III. Intrinsic Assay Data: N-Benzylphthalimides

Compound No	*X*	*Excised Cucumber* pI_{50}	*Algae* pI_{50}
14	4-Cl	5.6	6.0
15	3,4-DiCl	5.8	6.2
16	3,5-DiCl	5.8	7.0
17	H	4.8	4.6
18	2-Cl	4.0	5.1
19	3-Cl	5.7	6.2
20	2,3-DiCl	6.2	6.5
21	2,4-DiCl	NM	5.5
22	2,5-DiCl	<4.0	5.7
23	2,6-DiCl	NM	4.2
24	2,4,6-TriCl	<4.0	<4.0

NM = Not Measurable

The algae assay was also chosen for the Free-Wilson analysis of the N-benzylphthalimides. As in the glutarimides, substitution at the 3- and 5-positions of the aromatic ring again provided the most effective inhibition of Protox (Table IV). By comparing algae data, these compounds were more active inhibitors than the respective N-phenyltetrahydrophthalimides analogs (Table V), supporting a conclusion that the substituent pattern and methylene bridge are favorable structural parameters.

N-[(2,3-Dichlorophenyl)methyl]-3,4,5,6-tetrahydrophthalimide (**20**) was selected for herbicide evaluation. In postemergent applications, compound **20**, at 300 g/ha, provided 75-80% control of the broadleaf and grass test species. Crop tolerance (corn, wheat, soybean), however, was marginal. The preemergent activity of **20** was the most interesting. Grass control (johnsongrass, green foxtail) was 75% at 300 g/ha with corn tolerance at the highest application rate (3,000 g/ha). For broadleaf control, a rate of 3,000 g/ha was required. This shift in weed control spectrum and the potential for corn tolerance suggested that the N-benzylheterocycles as Protox inhibitors may provide an opportunity to obtain a corn tolerant herbicide.

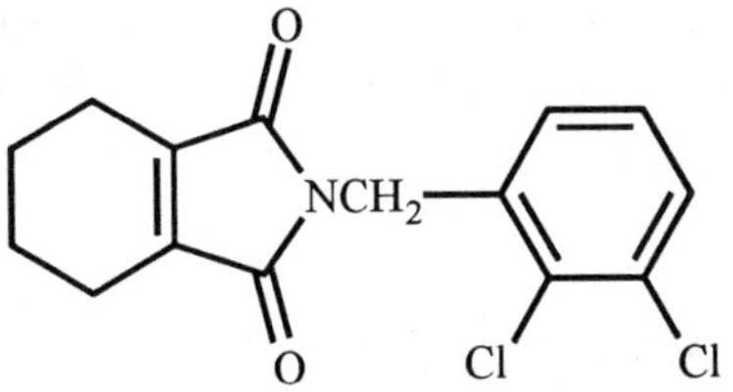

20

Table IV. N-Benzyltetrahydrophthalimides: Free-Wilson Group Contributions

Total Compounds Defined: 11
Algae Assay, pI_{50}
Activity of Parent (theoretical): 5.06

	Substituent Position in Aromatic Ring				
Substituent	*2*	*3*	*4*	*5*	*6*
Other	0.06	0.57	0.15	0.33	-0.66
H	0.00(5)	0.00(7)	0.00(7)	0.00(9)	0.00(9)
Cl	0.11(6)	1.15(4)	0.30(4)	0.66(2)	-1.32(2)

F = 8.728 [F - Table Val: (5% = 5.05) (1% = 10.97)]
(p = 0.02)
Coefficient of Multiple Correlation = 0.947

Table V. N-Benzyl- and N-Phenyltetrahydrophthalimides: Algae Assay

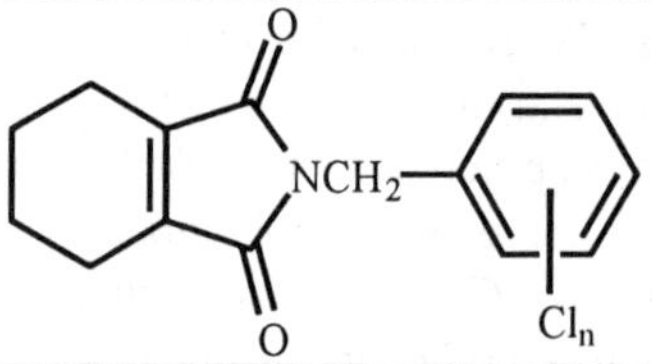

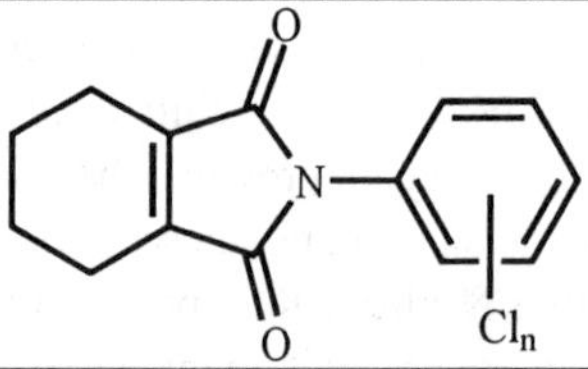

Cl_n	Algae, pI_{50}	Cl_n	Algae, pI_{50}
2-Cl	5.1	2-Cl	<4.0
3-Cl	5.7	3-Cl	4.3
2,3-DiCl	6.5	2,3-DiCl	NM
3,4-DiCl	6.2	3,4-DiCl	4.7
3,5-DiCl	7.0	3,5-DiCl	4.0

NM = Not Measurable

3-Phenyl-6-trifluoromethyluracils (**C**) with the appropriate aromatic substitution, are active at low application rates (<30 g/ha) and provide excellent broadleaf control. However, as in other Protox inhibitors, grass control is an issue. These observations and the preemergent data for compound **20** suggested the synthesis of 3-benzyl-6-trifluoromethyluracils (**D**) as candidate herbicides would be more advantageous than pursuing an optimization strategy for the benzyltetrahydrophthalimides.

3-Benzyl-6-trifluoromethyluracils

3-Benzyl-6-trifluoromethyluracils were prepared from substituted benzoic acids (Scheme 3) through either the benzylnitriles (Method A) or benzylphthalimides (Method B). The latter method is preferred as this sequence eliminated the low

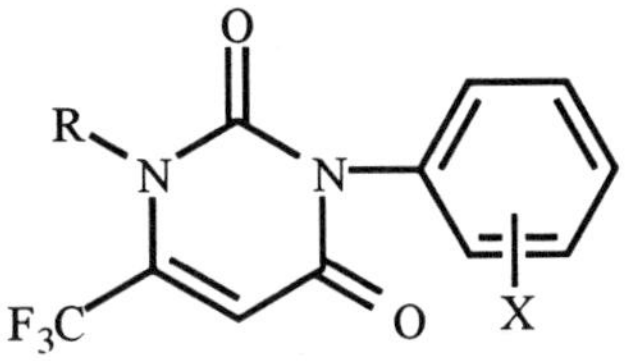

C

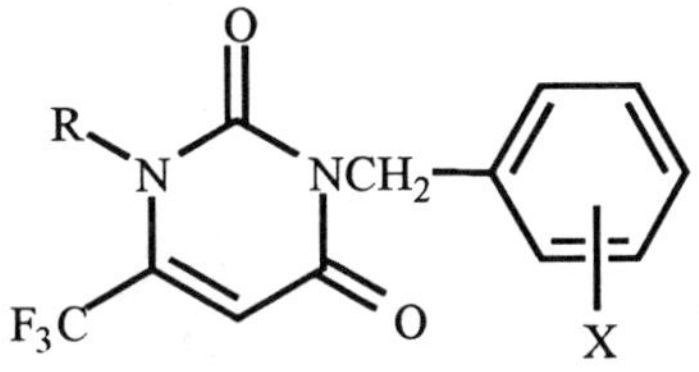

D

yields encountered in the synthesis of multisubstituted phenylacetonitriles and the potential hazards associated with the use of azides. By either method, however, the benzyl isocyanates were contaminated with by-products that could not be removed without extensive decomposition of the isocyanate. Therefore, the crude isocyanates were reacted with the 3-aminocrotonate to yield 3-benzyl-6-trifluoromethyluracils (**E**) in greater than 95% purity. The overall yields of **E** from either the phenacyl chlorides or benzylamines were from 55 to 65%. Compound **E** was then alkylated to provide the herbicide targets **D**.

Prior to the synthesis of a chloro probe set, three 3-benzyl-6-trifluoromethyluracils (**25**, **26**, **27**) were prepared for herbicide testing. The preemergent data for these compounds and for their respective 3-phenyluracils (**28**, **29**, **30**) is contained in Table VI. Compounds **25** and **26** were more active on grass than broadleaves but were overall less effective than the 3-phenyluracils. Compound **27** was dramatically different. It was active on grasses and broadleaves at 300 g/ha whereas the 3-phenyluracil (**30**) was inactive at the highest application rate. Corn tolerance in these tests was marginal at weed control rates for both classes of compounds. This data, however, supported previous observations that the structure-activity relationships of benzylheterocycles would be significantly different from that of phenylheterocycles.

With herbicidal activity having been established, the chlorine probe set was completed. The intrinsic assay data for this set of compounds is contained in Table VII. The excised cucumber results were selected for the Free-Wilson analysis. The results of the analysis are shown in Table VIII. As compared to the two previous benzylheterocycles, the 2-position of the aromatic ring has become significant and the relative contribution to activity was of the order: 3-position > 2-position > 5-position. This result was significantly different from that of 3-phenyl-6-trifluoromethyluracils, where, historically, the optimum substitution pattern is the 2,4,5-positions of the aromatic ring.

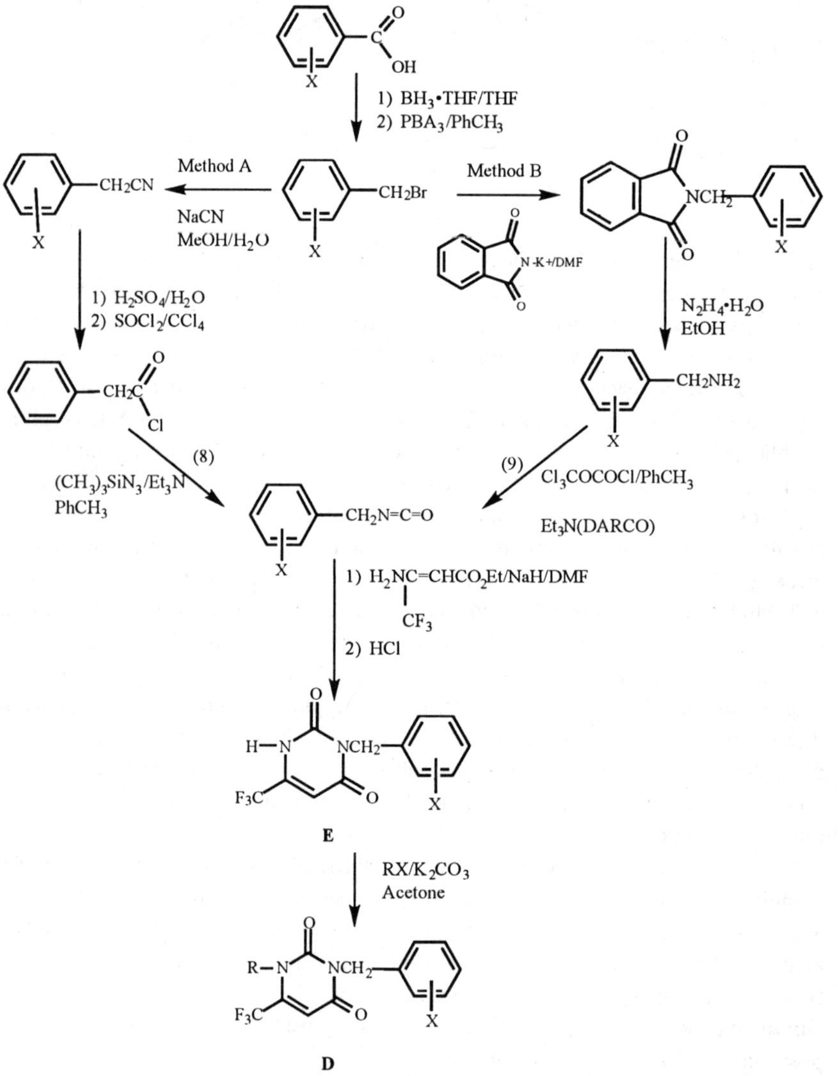

Scheme 3. Synthesis of 3-Benzyl-6-trifluoromethyluracils

Table VI. Preemergent Weed Control of 3-Benzyl- and 3-Phenyl-6-trifluoromethyluracils

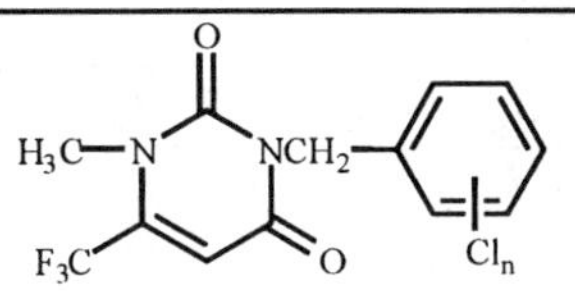

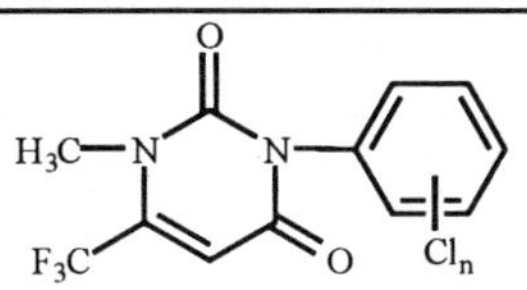

Compound No	*Cl_n*	*Rate (g/ha) for ≥ 85% Control* Grass	*Broad-leaf*	*Compound No*	*Cl_n*	*Rate (g/ha) for ≥ 85% Control* Grass	*Broad-leaf*
25	4-Cl	300	1000	**28**	4-Cl	100	30
26	3,4-DiCl	300	1000	**29**	3,4-DiCl	100	100
27	3,5-DiCl	300	300	**30**	3,5-DiCl	0% Control at 1000g	

Test Species: Velvetleaf, morningglory, chickweed, green foxtail, johnsongrass

In addition to the chlorine probe set, several compounds were prepared to further define structure-activity relationships (Table IX). From the herbicidal responses (intrinsic assays, greenhouse evaluations), the general structure indicated by **F** probably represents the major structural components for activity.

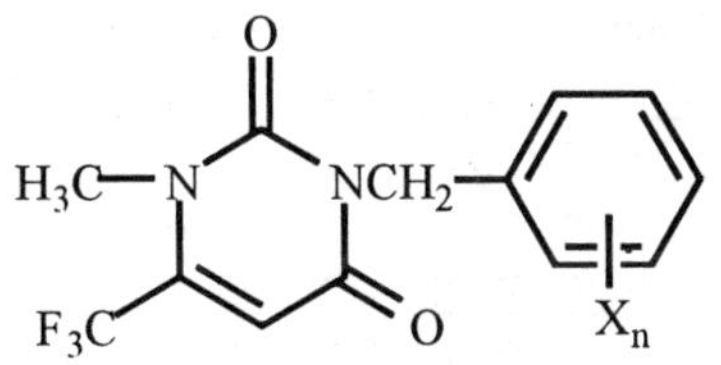

F

The 3-benzyl-6-trifluoromethyluracils in Table VII were evaluated in the greenhouse for preemergent activity. Of these, compound **40** was the most effective (Table X). Broadleaf and grass control was obtained at rates of 10-30 g/ha with tolerance toward corn at these weed control rates. In field trials, a similar weed spectrum was observed at application rates of 60-120 g/ha. However, corn injury did occur in the form of leaf necrosis and this injury was sufficient to preclude further development.

Conclusions

3-Benzyl-1-methyl-6-trifluoromethylpyrimidine-2,4-diones are a new class of Protox inhibitors which were developed from a weakly active herbicide lead. A

Table VII. Intrinsic Assay Data: 3-Benzyl-6-trifluoromethyl-1-methylpyrimidine-2,4-diones

Compound No	*X*	*Excised Cucumber* pI_{50}	*Algae* pI_{50}
25	4-Cl	6.5	6.4
26	3,4-DiCl	6.3	7.5
27	3,5-DiCl	6.9	8.2
31	H	5.5	5.6
32	2-Cl	5.7	5.7
33	3-Cl	6.4	6.4
34	2,4-DiCl	5.9	6.0
35	2,5-DiCl	6.6	6.8
36	2,6-DiCl	4.7	NM
37	2,3-DiCl	7.8	7.8
38	2,3,6-TriCl	4.5	4.5
39	2,3,4-TriCl	7.9	8.3
40	2,3,5-TriCl	8.5	8.8

NM = Not Measurable

Table VIII. 3-Benzyl-6-trifluoromethyl-1-methylpyrimidine-2,4-diones: Free-Wilson Group Contributions

Total Compounds Defined: 13
Excised Cucumber Assay, pI_{50}
Activity of Parent (theoretical): 5.45

	Substituent Position in Aromatic Ring				
Substituent	*2*	*3*	*4*	*5*	*6*
Other	0.36	0.53	0.11	0.34	-1.08
H	0.00(5)	0.00(6)	0.00(9)	0.00(10)	0.00(11)
Cl	0.73(8)	1.07(7)	0.22(4)	0.68(3)	-2.16(2)

F = 5.356 [F-Table Val: (5% = 3.97) (1% = 7.46)]
(p = 0.02)
Coefficient of Multiple Correlation = 0.890

Table XI. 3-Benzyluracils: Structure-Activity Studies

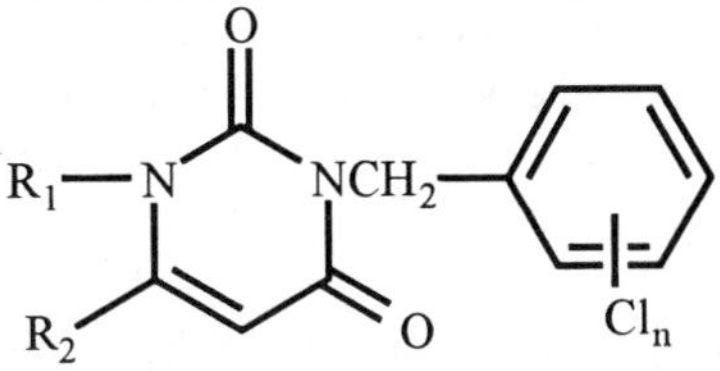

Structural Modifications		*Herbicidal Response*	
R_1:	H; NH_2; CH_3; $FCH_2CH_2CH_2$ $HC{\equiv}CCH_2$; CH_2CO_2Et	H - Inactive	CH_3; NH_2 - Best
R_2:	CH_3; CF_3	CH_3 - Inactive	CF_3 - Active
Benzyl		Phenethyl or α-CH_3	Inactive
Pyrimidine Ring		Reduction of double bond	Reduced Activity

Table X. Preemergent Herbicidal Activity: Greenhouse Evaluation

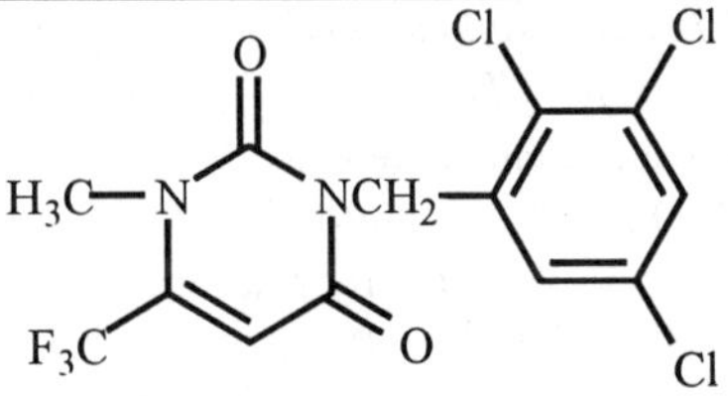

40

Weeds Controlled (≥ 85%) at Rates of 10-30g/ha	
Broadleafs	*Grasses*
Velvetleaf, morningglory	Crabgrass, foxtails
Pigweed, bindweed	Johnsongrass
Nightshade, kochia	Shattercane
Chickweed, smartweed	

key strategy was a chlorine probe set which established the position of substitution in the aromatic ring for optimal herbicidal activity. This substitution pattern was significantly different from Protox inhibiting phenylheterocycles (i.e., 2,3,5-positions vs. 2,4,5-positions). In preemergent applications and under green house conditions, these compounds controlled broadleaf and grass species at low application rates with tolerance toward corn. However, corn injury that was observed in field trials limited the potential for further development.

Acknowledgments

We are indebted to Mrs. Marjorie Plummer for the intrinsic assay data, to Drs. Harlene Hatterman-Valenti, Frederick Hotzman, Steven Hart and John Tymonko for greenhouse and field data, and to Dr. Jim Bahr for discussions and suggestions pertinent to the program.

Literature Cited

1. Simmons, K.A., Dixson, J.A., Halling, B.P., Plummer, E.L., Plummer, M.J.; *J. Agric. and Food Chem,* **1992** *40,* pp. 297-305.
2. Bar-Nun, S., Wallach, D., Ohad, I.; *Biochim Biophys Acta,* **1972**, 267, pp. 138-148.
3. Konz, M.J. (FMC Corporation), US Patent 4,207,088, **1980**.
4. Konz, M.J. (FMC Corporation), US Patent 4,302,238, **1981**.
5. Chang, J.H., Konz, M.J., Aly, E.A., Sticker, R.E., Wilson, K.R., Krog, N.E., Dickinson, R.R., *"3-Isoxazolidinones and Related Compounds: A New Class of Herbicides" in Synthesis and Chemistry of Agrochemicals, ACS Symposium Series No. 355,* American Chemical Society, Washington, DC, **1987**, pp. 10-23.
6. Dayan, F.E., Duke, S.O., *Pesticide Outlook,* **1996**, *5,* pp22-27.
7. Nagano, E., Hashimoto, S., Yoshida, R., Matsumoto, H., Kamosha, K. (Sumitomo Chemical Co., Ltd), Eur. Pat. Appl. 83055, **1983**; *Chem. Abstr.,* **1983** *99:175591.*
8. Washburne, S.S., Peterson Jr., J.R.; *Syn. Comm.,* **1972**, 2, pp. 227-230.
9. Ozaki, S., Ike, M., Mori, H., (Mitsu Toatsu Chemicals Inc.), Japan, *Kokai 7818, 515,* **1978**; *Chem. Abstr.,* **1978**, *89:42426.*

Chapter 10

Synthesis and Herbicidal Activity of 2-Alkylcarbonylphenyl Sulfamoylureas

M. E. Condon[1], T. E. Brady[1], D. Feist[1], P. M. Harrington[1], T. Malefyt[1], P. A. Marc[1], L. Quakenbush[1], D. Shaner[1], J. M. Lavanish[1,2], and B. Tecle[1]

[1]Cyanamid Agricultural Research Center, American Cyamamid Company, PO Box 400, Princeton, NJ 08543–0400

[2]Barberton Technical Center, PPG Industries, P.O. Box 31, Barberton, OH 44203

2-Alkylcarbonylphenyl sulfamoylureas represent a new class of highly active pre- and postemergence herbicides. This paper describes the initial discovery by PPG Industries, and the synthetic development of sulfamoylureas at the Cyanamid Agricultural Research Center.

The discovery by one company of novel chemistry which exerts its herbicidal effect via a new mode of action at low rates is a significant breakthrough which is noticed by most competitors. Certainly the discovery of the branched-chain amino acid biosynthesis inhibitors, the imidazolinones and the sulfonylureas, ranks as one of the most significant events in the history of herbicide discovery research. Undoubtedly it came as no surprise to DuPont that subsequent to the discovery of sulfonylureas virtually every agricultural chemical company engaged in synthesis programs patterned after this new herbicide class. This chapter will focus on the synthesis and herbicidal activity of one of these areas, the sub-class of sulfamoylureas referred to in the title. The work described was carried out by two different groups - it was initiated by researchers at PPG Industries in Barberton, Ohio, and subsequently carried on by workers at the Cyanamid Agricultural Research Center in Princeton, New Jersey.

Generic structure **1** typifies some of the first sulfonylureas which were commercialized, such as chlorsulfuron and sulfometuron methyl. The herbicidal sulfamoylurea class described in this chapter is included in generic structure **2**, and is differentialed from that of the sulfonylureas by the presence of the aniline nitrogen in the bridge.

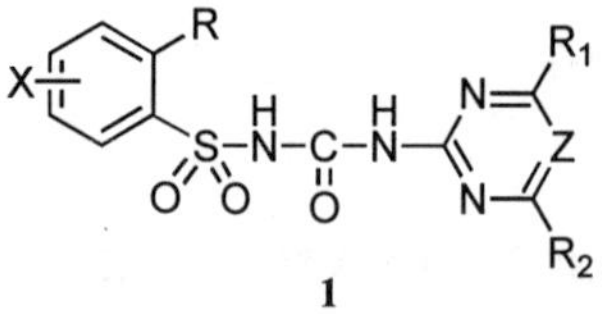

1
Sulfonylurea

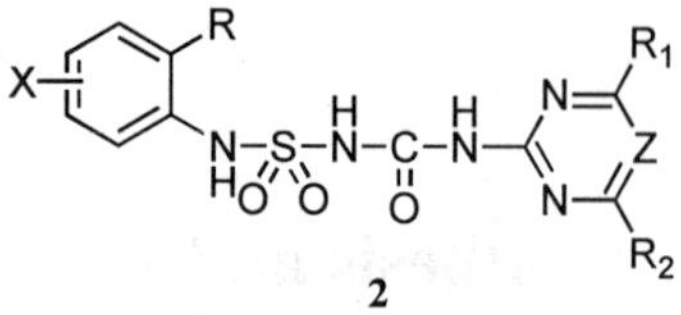

2
Sulfamoylurea

Several agrochemical companies have carried out synthesis in the sulfamoylurea area. Much of the initial effort was guided by what was known about structural preferences in the sulfonylurea area. Initially, close structural analogs of the sulfonylureas were prepared. This is implied by the generic structure **2** for sulfamoylureas. Thus, the types of substituents R which were incorporated into the *ortho* position were the ones incorporated into the sulfonylureas. The effect on herbicidal activity of using this strategy to explore the sulfamoylurea class is illustrated by the example in Table 1.

The potent postemergence activity of chlorsulfuron **3** on four representative broadleaf weeds is evident at a rate of 16 g/ha in the greenhouse. The dramatic effect of introduction of the extra nitrogen in the bridge is clearly shown by sulfamoylurea analog **4**, which at 1000 g/ha, or 60X the rate shown for chlorsulfuron shows very little activity. A number of comparisons such as these were made with similar findings. Undoubtedly, it was results such as these which led many workers to view sulfamoylureas as poor imitations of sulfonylureas.

Table I. Sulfamoylureas vs. Sulfonylureas

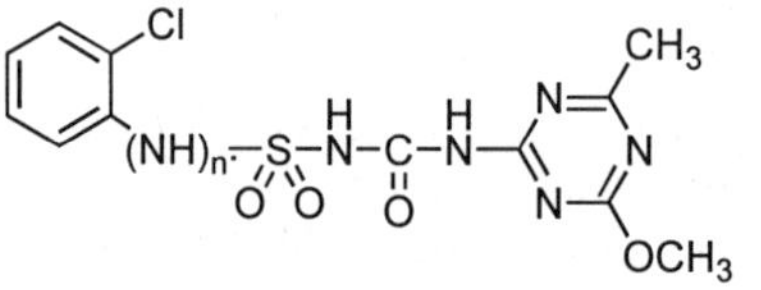

Postemergence Broadleaf Weed Control*

Species	**3** (n = 0) @ 16 g/ha	**4** (n = 1) @ 1000 g/ha
Velvetleaf	9	0
Field	9	0
Bindweed	9	0
Morningglory	9	4
Wild Mustard		

* Rating Scale: 0 = 0% Control; 9 = 100% Control

PPG was one of the companies which explored sulfamoylureas when the potent activity of the sulfonylureas became known. One of the PPG chemists made the significant observation that placement of an alkyl ketone in the *ortho* position led to analogs with much better activity than that observed with those having typical sulfonylurea substituents. A follow-up synthesis program allowed PPG to capitalize on this observation and to establish a proprietary position based on 2-alkylcarbonyl sulfamoylureas **5** (*1-6*). PPG's synthesis program led to PPG-2473, **6**, an analog which controlled broadleaf and some grass weeds both pre- and postemergence in wheat.

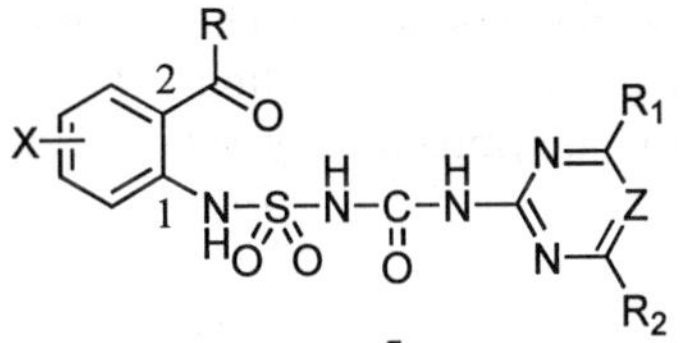

5

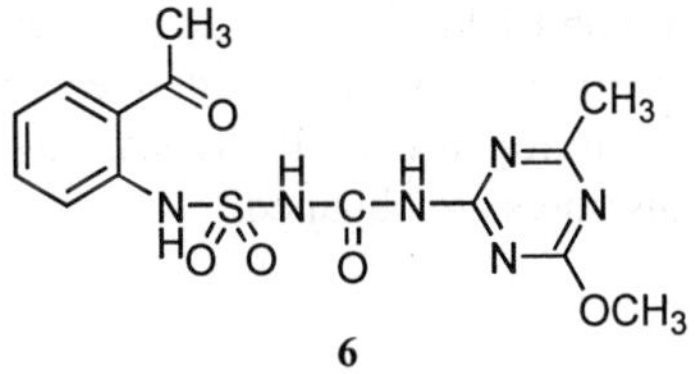

6
PPG-2473

At the time when American Cyanamid acquired this area of chemistry from PPG, this was the lead in the area. But after extensive field testing in cereals, it became clear that the level of crop injury was not acceptable. This prompted additional synthesis effort at Cyanamid, primarily with the goal of looking for increased cereal selectivity and for crop utilities which had not been the focus of PPG.

The structural modifications made as part of these two synthesis programs are summarized in Figure 1. The intent of this chapter is not to give a detailed accounting of all structure-activity studies carried out by both groups; rather selected comments will be made about all of them. The primary focus will be on variation of the alkyl ketone moiety, and derivitization and replacement of the ketone unit. The generic structure in Figure 1 implies some structural features which are presented without any supporting data. Early on in PPG's synthesis program it became clear that, as was the case with sulfonylureas, *ortho* disposition of phenyl substituents gave highest activity, and that the most effective heteroaryl rings were pyrimidines and triazines.

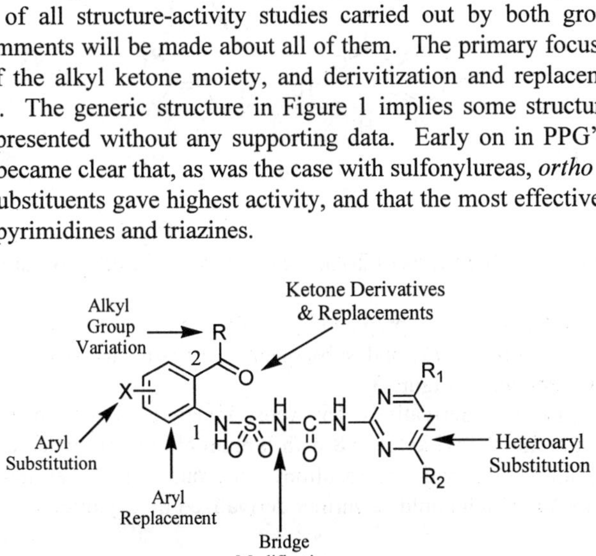

Figure 1. Summary of Structural Modifications

Synthesis

The synthesis of 2-alkylcarbonylphenyl sulfamoylureas **9** was accomplished using a one pot reaction of aminopyrimidines or -triazines with chlorosulfonylisocyanate followed by treatment with 2-aminophenyl ketones **8** in the presence of base. The 2-aminophenyl ketones **8** were conveniently prepared by the method of Sugasawa (*7*), which involves treatment of anilines **7** with nitriles in the presence of boron trichloride and aluminum chloride. The aniline and boron trichloride first generate an anilinodichloroborane which coordinates to the nitrile and ultimately delivers the acyl group to the position *ortho* to the aniline nitrogen (*8*). If the starting aniline is unsymmetrically substituted, two regioisomers are obtained.

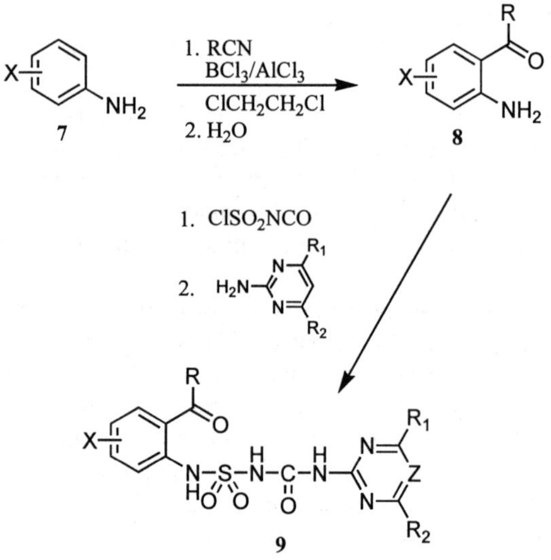

Figure 2. Synthesis of 2-Alkylcarbonyphenyl Sulfamoylureas

Ketal derivatives **12** were prepared by ketalization of nitroketones **10** followed by reduction to anilines **11**, and subsequent use in the usual one-pot coupling procedure as depicted in Figure 3.

Alcohols **14** and derivatives **15** were prepared as shown in Figure 4. Reduction of aminophenyl ketones **8** with lithium aluminum hydride followed by coupling of anilines **13** with chlorosulfonyl isocyanate and an amino heterocycle gave alcohols **14**, which could be further derivatized by acylation to yield analogs **15**.

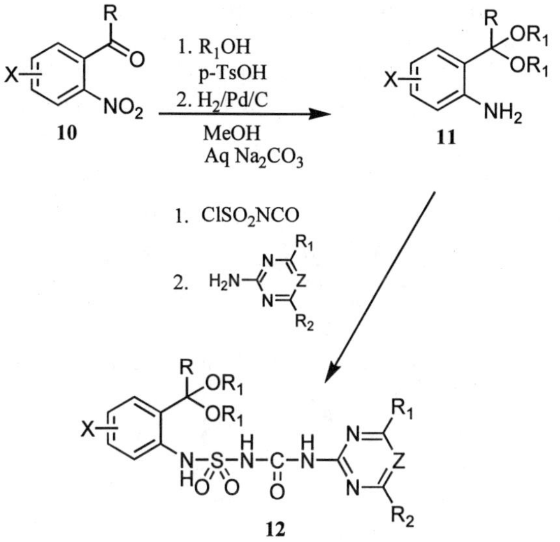

Figure 3. Synthesis of Ketal Derivatives

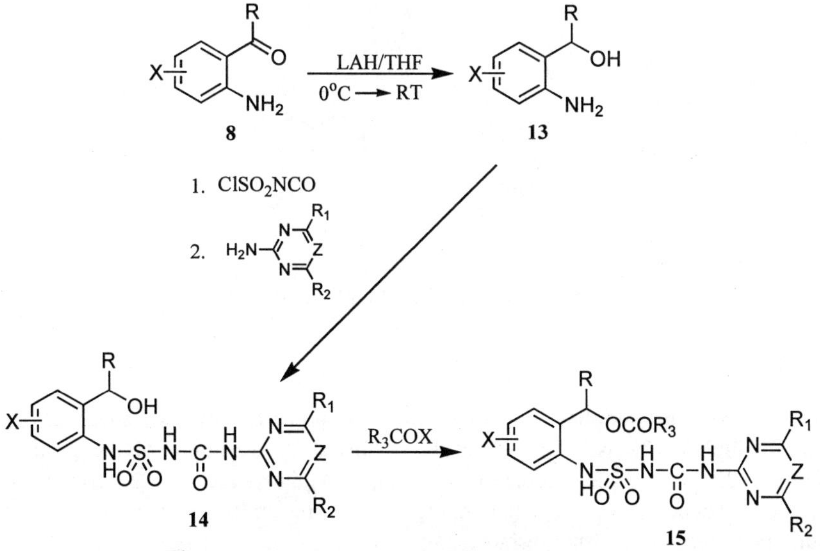

Figure 4. Synthesis of Alcohols and Derivatives

Sulfoxide **20** and sulfone **21** analogs of the ketones were prepared from o-chloronitrobenzene **16** as depicted in Figure 5. Reaction with thiols in the presence of base afforded o-alkylthio derivatives **17**, which were oxidized selectively to sulfoxides and sulfones with one and two equivalents of meta-chloroperbenzoic acid, respectively. Reduction to the anilines **18** and **19** followed by the usual one-pot coupling procedure afforded the analogs indicated.

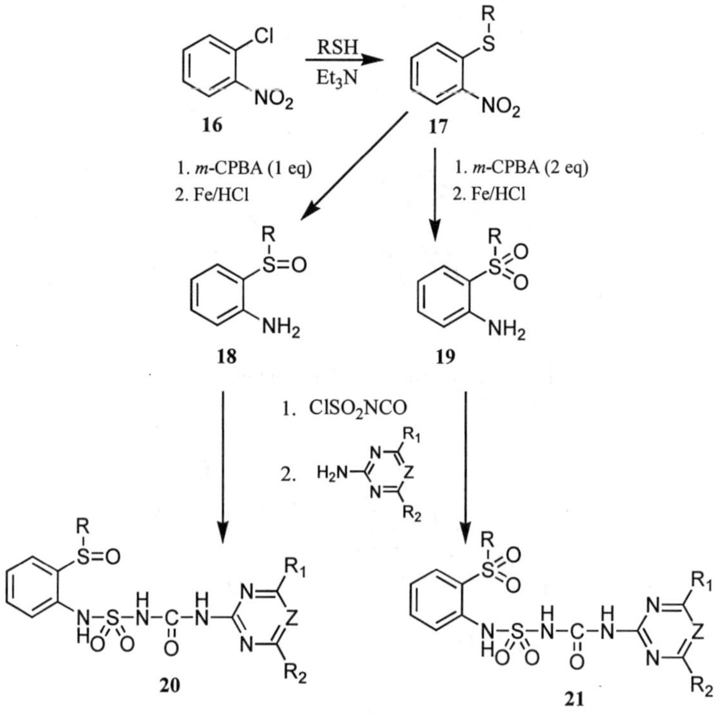

Figure 5. Synthesis of Sulfoxides and Sulfones

A series of thiophene analogs **25** was prepared by acylation of 3-bromothiophene **22** with acid chlorides followed by treatment of the product ketones **23** with azide, reduction with triphenylphosphine and hydrolysis to aminothiophenes **24** as depicted in Figure 6. The amines were converted to sulfamoylureas in the usual fashion.

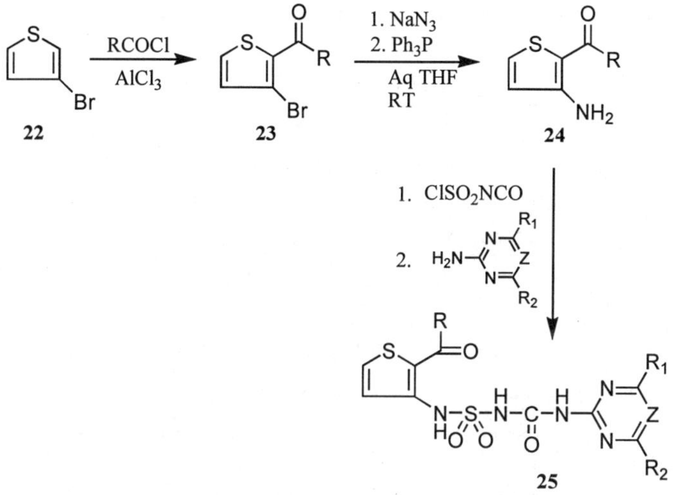

Figure 6. Synthesis of Thiophene Analogs

Biological Evaluation

The sulfamoylureas were evaluated in the greenhouse in a series of advanced evaluations. Analogs were active both pre- and postemergence, with a tendency for more activity postemergence, and were more active on broadleaf weeds than on grass weeds. The greenhouse data presented in the following tables use the following rating scale: a control rate of 0 represents 0% control, and a rating of 9 represents 100% control. These materials were also evaluated as inhibitors of acetohydroxyacid synthase (AHAS); some are quite potent, with I_{50}'s in the same range as the sulfonylureas.

Structure-Activity Relationships

Alkyl Ketones. One structural feature which was investigated rather thoroughly was that of the alkyl ketone moiety. Table II shows the postemergence activity @ 32 g/ha of a representative series of alkyl ketones in the dimethoxypyrimidine series. Average broadleaf weed control rates for the six weeds shown, along with crop ratings at the same rate show that within a relatively close structural series, broadleaf weed control varies from modest (n-Pr) to good (c-Pr). Moreover, while injury to upland rice, wheat, and barley at this rate is acceptable, injury to corn is not, with the possible exception of the Et analog.

Table II. Activity of Alkyl Ketones - Pyrimidine Series

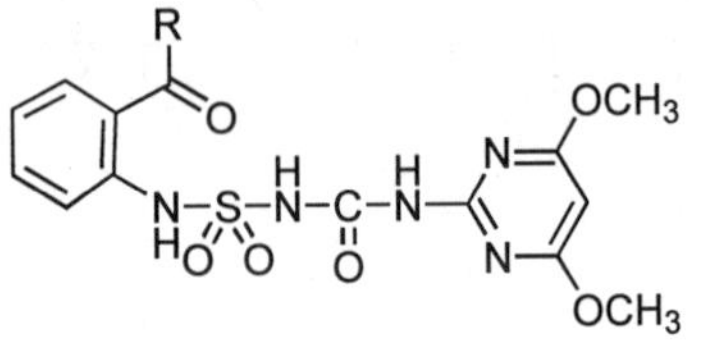

Postemergence Broadleaf Weed Control @ 32 g/ha

Species	R = -Me	-Et	*n*-Pr	*i*-Pr	-c-Pr
BL Weed Ave.*	6.2	7.6	4.7	5.3	8.0
Rice, Tebonnet	0.5	0.0	0.0	0.0	0.0
Wheat, Spring	0.0	0.0	0.0	1.0	0.0
Barley, Spring	0.0	0.0	0.0	1.0	0.0
Corn, Field	3.4	0.8	1.5	4.0	3.3

* Velvetleaf, Ragweed, Sicklepod, Morningglory, Mustard, Cocklebur

Table III shows the postemergence activity @ 32 g/ha of the same series of alkyl ketones in the dimethoxytriazine series. Although the levels of broadleaf weed control are comparable, the crop selectivity is not nearly as good for the triazine series as the pyrimidine series.

Table III. Activity of Alkyl Ketones - Triazine Series

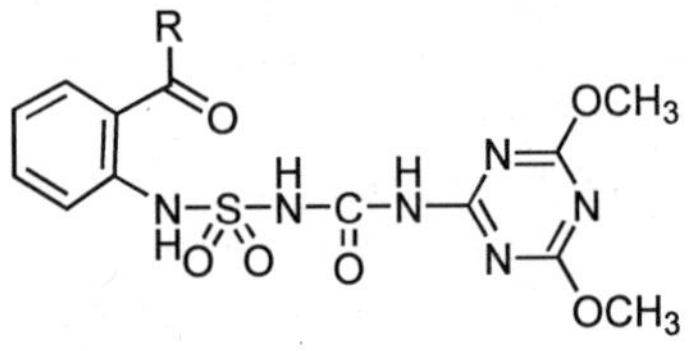

Postemergence Broadleaf Weed Control @ 32 g/ha

Species	R = -Me	-Et	*n*-Pr	*i*-Pr	-c-Pr
BL Weed Ave.*	5.4	8.4	6.5	4.0	8.3
Rice, Tebonnet	4.5	2.0	5.0	3.0	4.0
Wheat, Spring	1.0	1.0	0.0	1.0	2.0
Barley, Spring	0.5	1.0	1.0	2.0	2.0
Corn, Field	0.5	4.0	7.0	5.0	6.0

* Velvetleaf, Ragweed, Morningglory, Mustard

Cycloalkyl Ketones. Table IV shows a set of homologous cycloalkyl ketones in the dimethoxypyrimidine series. As the size of the ring is increased, herbicidal activity against four representative broadleaves decreases markedly, as does *in vivo* potency against AHAS. Curiously, the cyclopentyl analog has an I_{50} of 6 nM against the isolated AHAS enzyme but poor greenhouse activity at higher rates than the one shown here. Analogs in which the R group is aryl were prepared and were uniformly inactive.

Table IV. Activity of Cycloalkyl Ketones

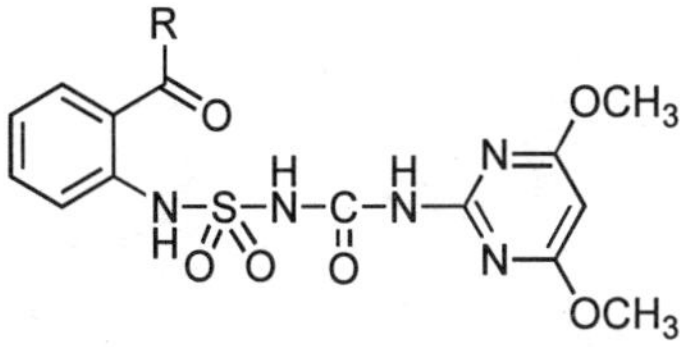

Postemergence Broadleaf Weed Control @ 63 g/ha

Species	R =		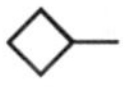	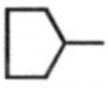	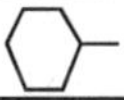
Velvetleaf		9	0	0	0
Ragweed		8	4	0	0
Morningglory		8	7	0	0
Mustard		9	9	0	0
AHAS I_{50} (nM)		0.4	2.0	6.0	2100

Ketals. Table V shows that ketals were comparable in activity to ketones, but offered no particular advantage.

Table V. Activity of Ketals

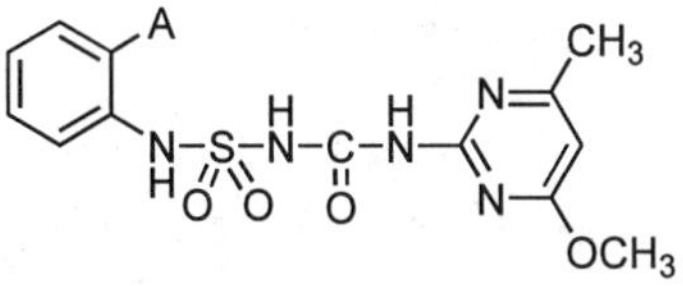

Postemergence Broadleaf Weed Control @ 500 g/ha

Species	R =	$-\overset{O}{\overset{\|}{C}}-CH_3$	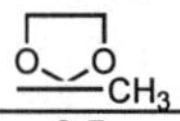
Velvetleaf		9.0	8.7
Ragweed		8.6	7.0
Morningglory		4.5	3.5
Mustard		8.8	9.0

Alcohols and Derivatives. In general, alcohols and acylated derivatives were less active than the corresponding ketones and ketals as shown in Table VI. In cases where alcohols approach the parent ketone in activity, there is no advantage from the potency, selectivity, or synthetic accessibility standpoint (*9*).

Table VI. Activity of Alcohols and Derivatives

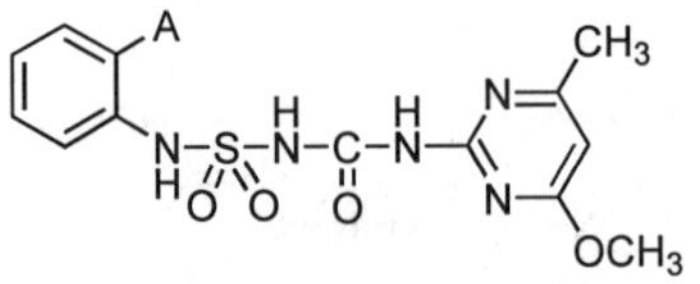

Postemergence Broadleaf Weed Control @ 500 g/ha

Species	A =	$-C(CH_3)(OCH_3)_2$	$-CH(CH_3)-OH$	$-CH(CH_3)\cdot OAc$
Velvetleaf		9.0	1.7	0.7
Ragweed		9.0	7.7	4.7
Morningglory		6.0	1.8	0.5
Mustard		9.0	7.3	9.0

Sulfoxides. Sulfoxides and sulfones were investigated as potential surrogates for the o-acyl moiety. Table VII gives one pair of analogs which illustrates the trend toward lower activity observed when this substitution was made. During the course of this investigation, similar work was disclosed by Bayer (*10*).

Table VII. Activity of Sulfoxides and Sulfones

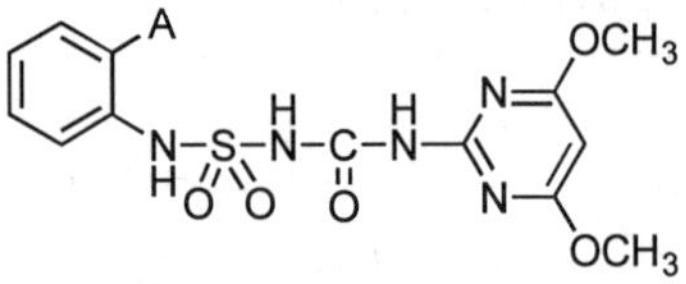

Postemergence Broadleaf Weed Control @ 63 g/ha

Species	A =	$-C(=O)-CH_3$	$-S(=O)-CH_3$	$-S(O)_2-CH_3$
Velvetleaf		9.0	7.0	9.0
Ragweed		8.0	2.0	3.5
Morningglory		3.3	0.0	1.5
Mustard		9.0	8.0	9.0

Heteroaryl Substitution. Table VIII shows a set of substituted pyrimidines arranged in order of increasing activity. In general, trends observed with triazines and pyrimidine substitution in the sulfamoylurea series parallel those observed in the sulfonylurea area, e.g., the best in this set are the methyl and methoxy substituted. Curiously, substitution with difluoromethoxy groups led to a substantial decrease in activity.

Table VIII. The Effect of Heteroaryl Substitution on Activity

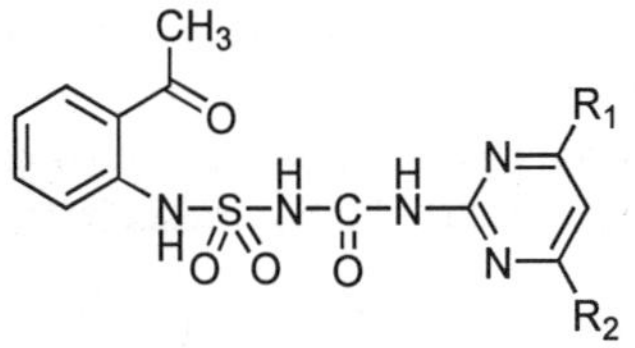

Postemergence Broadleaf Weed Control @ 500 g/ha					
R_1	R_2	VL	RW	MG	WM
H	H	0.0	0.0	0.0	0.0
Cl	Cl	0.0	0.0	0.0	0.0
Cl	Me	0.0	0.0	0.0	0.0
Cl	OMe	1.0	8.0	0.0	1.0
Me	Me	5.5	6.5	0.0	6.5
Me	OMe	9.0	8.6	7.1	8.8
OMe	OMe	9.0	9.0	8.6	9.0

Aromatic Substitution. In general, substitution on the aromatic ring decreased herbicidal activity as illustrated in Table IX. In cases where substituted analogs approach the parent in activity (*11*), there is no advantage from the potency, selectivity or the synthetic accessibility standpoint.

Table IX. The Effect of Aromatic Substitution on Activity

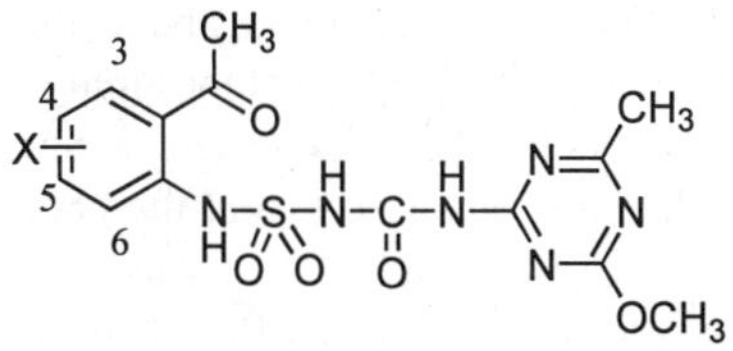

Postemergence Broadleaf Weed Control @ 500 g/ha				
X	VL	RW	MG	WM
H	9.0	8.0	7.0	9.0
4-OMe	0.0	6.5	0.5	9.0
4-SO_2Me	0.0	0.0	0.0	0.0
4-Cl	6.3	8.3	1.4	8.7
4-F	3.0	5.7	3.3	8.7
4-Me	0.0	0.0	0.0	0.0
5-Cl	3.0	9.0	6.0	8.0
6-COMe	5.0	7.3	6.5	8.7

Thiophene Analogs. Shown in Table X are two analogs in which the o-substituted phenyl ring is replaced by that of a heterocycle, represented by thiophene. This strategy was used successfully by DuPont in the development of the sulfonylurea thifensulfuron methyl (*12*). As the data indicates, when employed in the sulfamoylurea area, less active analogs resulted. During the course of this work thiophene-based sulfamoylureas were disclosed by Bayer.[13]

Table X. Activity of Thiophene Analogs

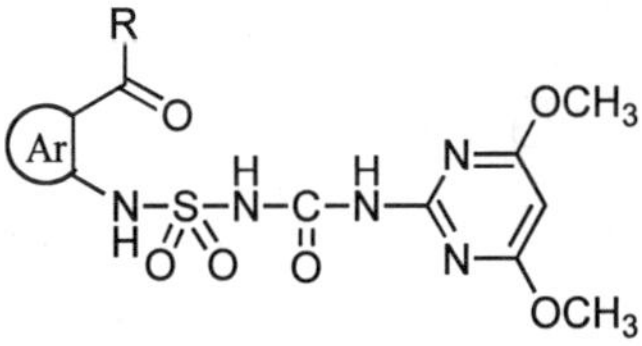

Postemergence Broadleaf Weed Control @ 63 g/ha

	Ar = Phenyl		Thienyl	
Species R =	Me-		Me-	
Velvetleaf	9	9	0	0
Ragweed	8	8	0	5
Morningglory	7	8	0	5
Wild Mustard	9	9	0	0

Utility in Transplanted Rice. As mentioned above the effect on activity of variation of the alkyl group was investigated thoroughly. On the basis of this work a small number of analogs had been identified as potential cereal herbicides - each had its own strengths and weaknesses, and the choice of the best was not easy. It wasn't until utility in transplanted rice was investigated that a clearer choice was possible - this was a utility which PPG had not evaluated. Table XI shows greenhouse postemergence weed control data in transplanted rice for a set of four alkyl ketones in the dimethoxypyrimidine series. The highest rate at which the rice is safe is contrasted with the rates required to control two important paddy rice weeds: barnyardgrass and Cyperus serotinus. Data such as these, in combination with upland data pointed to the cyclopropyl ketone as having the best selectivity and weed control. This ketone (AC 322140) has been extensively tested in the field, and is currently under development for use in both rice and cereals (*14-18*).

Table XI. Activity of Cycloalkyl Ketones in Transplanted Rice

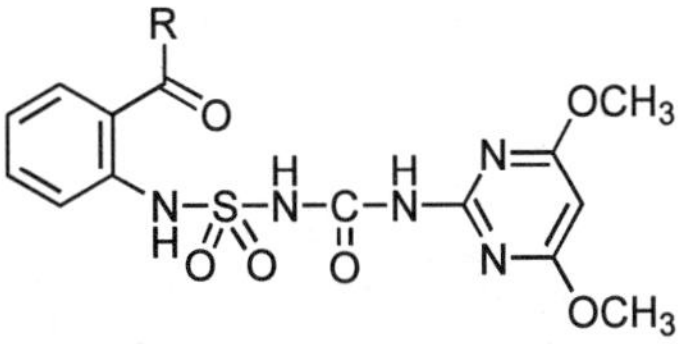

Postemergence Broadleaf Weed Control (g/ha)

Species R =	Me-	Et-		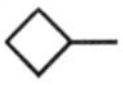
Rice Safe Rate	63	500	500	125
BG Control Rate	63	63	32	250
Cyp Control Rate	32	32	16	125

Table XII lists some key weeds controlled at 45-60 g/ha in transplanted rice, 2-15 days post transplant.

Table XII. Weed Control in Rice with AC 322140

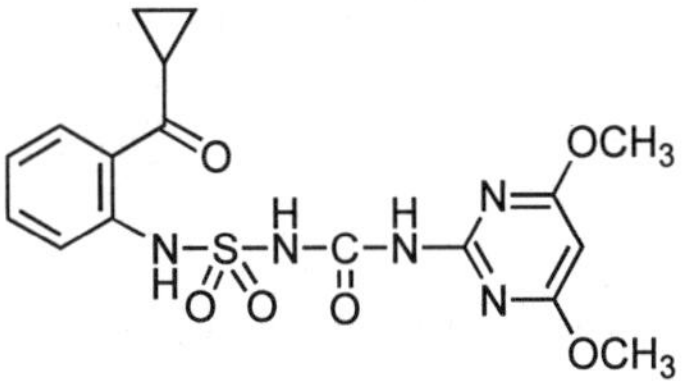

AC 322140

Cyperus serotinus	*Cyperus difformis*
Eleocharis congesta	*Elatine triandra*
Elkeocharis kuroguwai	*Lindernia annua*
Sagittaria pygmaea	*Lindernia procumbens*
Sagittaria trifolia	*Monochoria vaginalis*
Scirpus juncoides	*Rotala indica*

Utility in Cereals

AC 322140 has been field tested extensively in European cereals. At rates of 25-50 g/ha, AC 322140 controlled a number of important broadleaf weeds, with excellent tolerance in winter wheat, winter barley, and durum wheat at rates up to 150 g/ha. Spring barley showed good tolerance to preemergence applications, and slight to moderate injury from postemergence applications. AC 322140 showed good activity when applied in the spring. Table XIII lists a number of broadleaf weeds controlled by AC 322140 pre- and postemergence at 25-50 g/ha.

Table XIII. Weed Control in Cereals with AC 322140

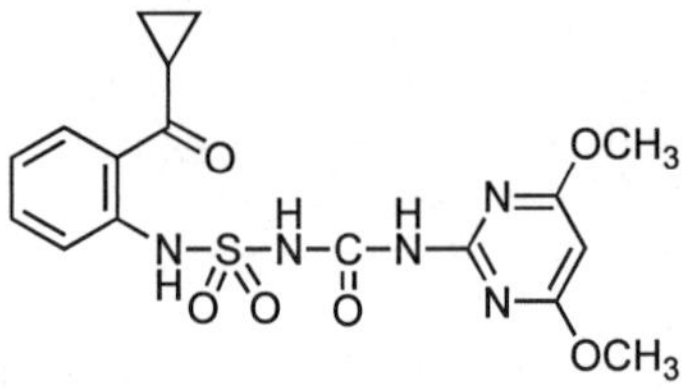

AC 322140

Anagalis arvensis	*Galium aparine*
Brassica napus	*Matricaria spp.*
Capsella bursa-pastoris	*Sonchus spp.*
Fumaria officiinalis	*Veronica spp.*

Rate of Hydrolysis of [^{14}C] AC 322,140

In connection with the development of AC 322140 its physical properties were investigated. Table XIV shows the rate of hydrolysis as a function of pH. The rate of hydrolysis of AC 322,140 is pH dependent - it is hydrolyzed more rapidly at lower pH than at higher pH. Since AC 322,140 is a weak acid, ionization of the acidic proton in the urea bridge at higher pH makes the herbicide more resistant to hydrolysis.

Table XIV. Rate of Hydrolysis of [^{14}C] AC 322,140

	Days After Incubation						
pH	0	1	3	5	7	14	21
3	100	69	17	0	0	0	0
5	100	72	20	12	10	0	0
6	100	82	60	52	45	18	5
7	100	93	89	90	89	82	78
8	100	97	96	94	96	93	90

Effect of N-Methylation

DuPont's tribenuron methyl is an N-methyl derivative of metsulfuron methyl. It is hydrolyzed 10-25 times faster than metsulfuron methyl, primarily due to the increased lability of the adjacent amide bond under acidic conditions (*19*). This property dramatically alters its soil residual properties. When this strategy was applied to AC 322,140 a relatively uninteresting analog was produced. It was subsequently discovered that the primary mode of hydrolysis of AC 322140 involved cleavage of the sulfamoyl unit and not the urea.

Table XV. Effect of N-Methylation on Activity

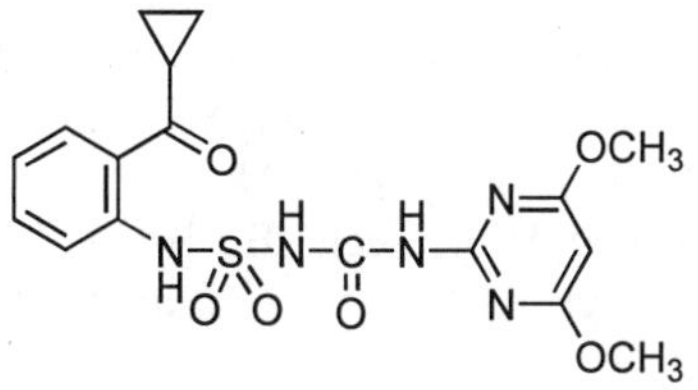

Species	R = H @ 63 g/ha	R = Me @ 500 g/ha
Velvetleaf	9	0
Ragweed	8	0
Morningglory	8	0
Wild Mustard	9	9

Rating Scale: 0 = 0% Control; 9 = 100% Control

Summary

In summary, the observation made by a PPG chemist, i.e., that 2-alkylcarbonyl analogs in the sulfamoylurea series were typically more potent than those which resulted from incorporation of the usual sulfonylurea *ortho* substituents, was exploited by PPG, patented, and further developed by American Cyanamid. Among other things this effort led to AC 322140, or cyclosulfamuron, which is currently in development at American Cyanamid for broad spectrum weed control in rice and cereals.

Literature Cited

1. Van Gemert, B., U. S. Patent 4,559,081, 1985.
2. Van Gemert, B., U. S. Patent 4,602,939, 1986.
3. Van Gemert, B., U. S. Patent 4,622,065, 1986.
4. Van Gemert, B., U. S. Patent 4,666,508, 1987.
5. Van Gemert, B., U. S. Patent 4,696,659, 1987.
6. Van Gemert, B., U. S. Patent 4,471,762, 1988.
7. Sugasawa, T.; Toyoda, T.; Adachi, M.; Sasakura, K. *J. Org. Chem.*, **1978**, *100*, 4842-4852.
8. Douglas, A. W.; Abramson, N. L.; Houpis, I. N.; Karady, S.; Molina, A.; Xavier, L. C.; Yasuda, N. *Tetrahedron Letters*, **1994**, *37*, 6897-6810.
9. Condon, M. E.; Harrington, P. M., U. S. Patent 5,464,808, 1995.
10. Riebel, H.-J.; Santel, H.-J.; Dollinger, M., Ger. Patent Appl. DE 4,302,702, 1994.
11. Condon, M. E.; Harrington, P. M., U. S. Patent 5,492,884, 1996.
12. Sionis, S. D.; Drobny, H. G.; Lefebvre, P.; Upstone, M. F., Proc. *Br. Crop Protection Conf., Weeds*, 1985, 49.

13. Riebel, H.-J.; Muller, K.-H.; Findeisen, K.; Santel, H.-J.; Lunssen, K.; Schmidt, R. R., PCT Patent Application WO 93/24482, 1993.
14. Brady, T. E.; Condon, M. E.; Marc, P. A., U. S. Patent 5,009,699, 1991.
15. Murai, S.; Hasui, H.; Kawai, K.; Kimpara, M.; Suzuki, M. *Proc. 14th Asian-Pacific Weed Sci. Soc. Conf.,* **1993**, 154-158.
16. Quakenbush, L.; Rodaway, S. J.; Tecle, B.; Brady, T. E.; Lapade, B.; Marc, P.; Condon, M. E.; Malefyt, T. *Proc. 14th Asian-Pacific Weed Sci. Soc. Conf.*, **1993**, 36-40.
17. Condon, M. E.; Brady, T. E.; Feist, D.; Malefyt, T.; Marc, P.; Quakenbush, L. S.; Rodaway, S. J.; Shaner, D. L.; Tecle, B. *Proc. Br. Crop Protection Conf., Weeds,* **1993,** 141-146.
18. Rodaway, S. J.; Tecle, B.; Shaner, D. *Proc. Br. Crop Protection Conf., Weeds,* **1993**, 3, 239-246.
19. Ferguson, D. T.; Schehl, S. E.; Hageman, L. H.; Lepone, G. E.; Carraro, G. A., *Proc. Br. Crop Protection Conf., Weeds,* **1985**, 43.

Chapter 11

1*H*-1,4-Benzodiazepine-25-diones and Related Systems: Synthesis and Herbicidal Activity

Gary M. Karp, Mark C. Manfredi, Michael A. Guaciaro, Philip M. Harrington, Pierre Marc, Charles L. Ortlip, Laura S. Quakenbush, and Iwona Birk

Cyanamid Agricultural Research Center, American Cyanamid Company, PO Box 400, Princeton, NJ 08543–0400

The synthesis and structure-activity relationships of a series of 1*H*-1,4-benzodiazepine-2,5-diones is described. Many of the analogs exhibit moderate levels of both pre- and postemergence herbicidal activity. These compounds have been found to act by inhibition of photosynthesis by blocking PSII electron transport. In addition, a number of closely related ring systems were prepared by modification of both the benzene and diazepine ring and were also evaluated for herbicidal activity. In nearly all cases there was a significant drop in activity.

As part of a broad-based random screening process carried out some years ago at the Agricultural Products Research Division, a small number of benzodiazepinediones were found to have herbicidal properties. The benzodiazepinediones, exemplified by **1**, were originally prepared as part of an anxiolytic screening program conducted by American Cyanamid Company's Medical Research Division (previously Lederle Laboratories, now part of Wyeth-Ayerst Research) (*1,2*). Based on the herbicidal activity of these compounds, a limited synthesis effort was initiated. This program was abandoned shortly thereafter, however, in preference to other areas of research. Recently, the benzodiazepinediones were re-evaluated as potential corn selective herbicides based on the lead compound **1**. At an early stage of the more recent synthesis effort, **2** was prepared and was found to have improved activity, both pre- and postemergence. The observation that the C-ring derived from the proline ester of **1** was not a prerequisite for herbicidal activity was advantageous, as it would simplify the synthesis of potential analogs. In order to establish an effective structure-activity relationship in this class a large number of analogs were prepared (*3,4*). The initial focus of the synthesis effort consisted of substituent modification in the diazepine ring (N-1, C-3 and N-4) as well as regiochemical variation at all four benzenoid positions (C-6 through C-9). In addition to benzodiazepine-2,5-dione analogs, this effort was ultimately expanded to include several closely related ring systems. Structural modifications included aryl ring replacement (pyridine and thiophene) and modification of the seven-membered diazepine ring, including the addition of heteroatoms and variation in the ring size.

Benzodiazepinedione **2** was found to strongly inhibit photosynthetic electron transport in both *in vitro* and *in vivo* systems. Furthermore, the site of action is believed to be the D1 protein of PSII as triazine-resistant plants were found to be resistant to compound **2** as well (*5*).

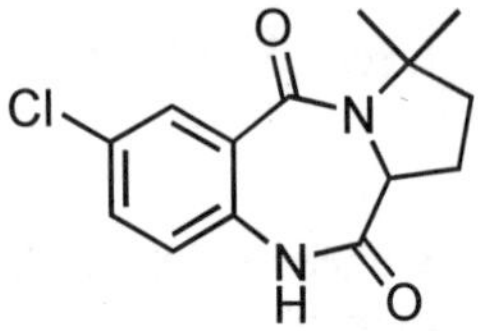

1

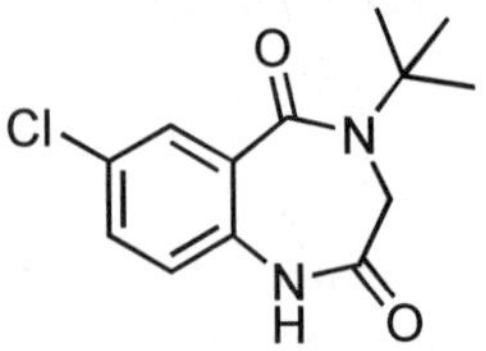

2

Synthesis.

Preparation of 1,4-Benzodiazepine-2,5-dione analogs. Many of the benzodiazepinedione analogs were prepared as shown in Figure 1. Commercially available *o*-nitrobenzoic acids were converted to the respective acid chlorides **3** and treated with N-alkylglycinate esters **4** to afford *o*-nitrohippuric esters **5**. Reduction of the nitro group gave crude *o*-aminohippuric esters which were readily cyclized under acidic conditions to afford the desired benzodiazepinediones **6** in good yield.

For many other analogs, the desired substituents were introduced after benzodiazepinedione ring formation due to the lack of available starting materials. For example, selective C-7 iodination and bromination took place with unsubstituted benzodiazepinedione **7** to afford **8** and **9**, respectively (Figure 2) (*6*). When C-7 is already substituted with a chlorine atom *(e.g.*, **2**) bromination is directed exclusively to C-9, although more forcing conditions are required. More complex benzenoid substituents could be introduced by palladium-mediated coupling reactions of the bromo- and iodobenzodiazepinediones. Vinyl, alkynyl, aryl and heteroaryl substituents were introduced in this manner (*6*). Benzodiazepinediones bearing a hydrogen at N-1 could be readily N-alkylated, as reported previously (*4*).

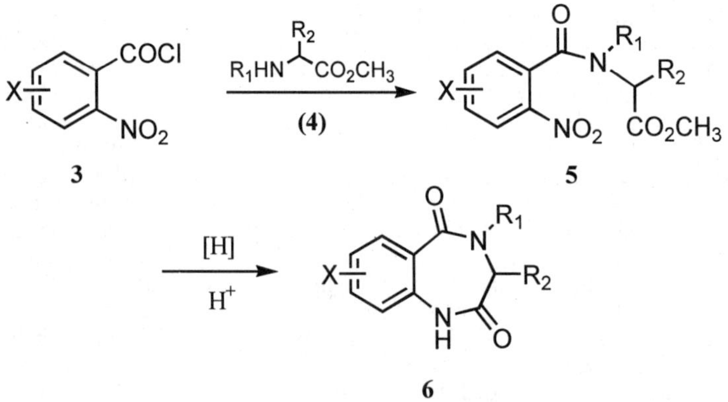

Figure 1. Preparation of 1,4-benzodiazepine-2,5-dione ring system from substituted *o*-nitrobenzoyl chlorides.

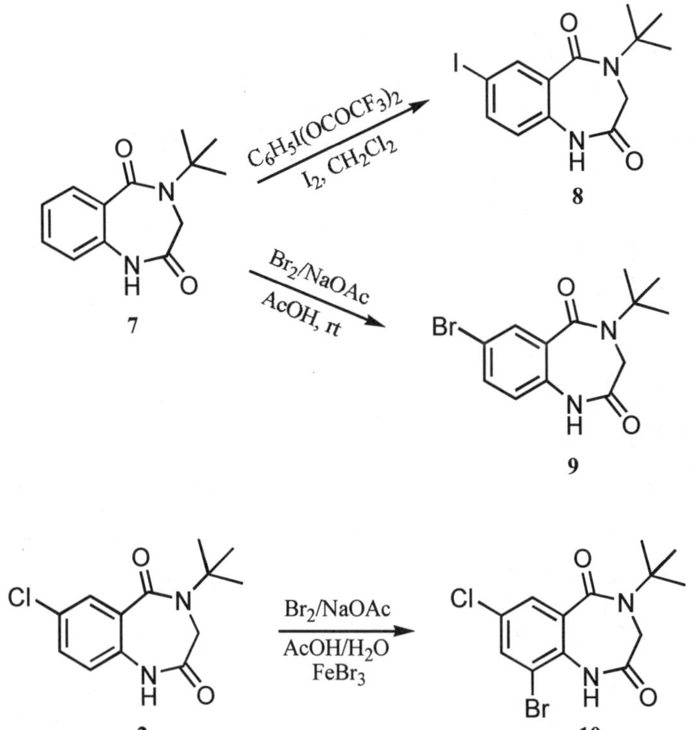

Figure 2. Selective halogenation reactions of benzodiazepinediones.

Diazepine ring modification. We chose to examine the effect of structural changes to the seven-membered diazepine ring on the resultant herbicidal activity. Three ring systems were examined in which changes were made to the benzodiazepinedione C-3. One system examined was the 1,3,4-benzotriazepine-2,5-dione ring system resulting from the replacement of the C-3 carbon with a nitrogen atom. Other ring systems studied were the 1,5-benzodiazocine-2,6-diones and 1,2,3,4-tetrahydroquinazoline-2,4-diones, resulting from the insertion of an additional methylene group between C-3 and N-4, and the removal of the C-3 methylene group, respectively.

1,3,4-Benzotriazepine-2,5-dione ring system. The interesting level of herbicidal activity of benzodiazepinedione **2** prompted the preparation of the 3-aza analogs **14-16** (Figure 3). Treatment of *o*-nitrobenzoyl chloride **3** with the acetone hydrazone of *t*-butylhydrazine gave the N-protected *o*-(nitrobenzoyl)hydrazine **11**. Acid-catalyzed N-deprotection gave **12**. Nitro reduction afforded the *o*-(aminobenzoyl)hydrazine **13** which cyclized to the parent **14** upon treatment with trichloromethyl chloroformate (diphosgene) (*7*). N-alkylation of **14** gave the 3-alkyl analogs **15** and **16** exclusively.

1,5-Benzodiazocine-2,6-dione ring system. Ring expansion of the benzodiazepinedione at C-3 by incorporation of an additional methylene unit gives rise to the 1,5-benzodiazocine-2,6-dione ring system. The *o*-nitrobenzoyl chloride **3**

was treated with methyl N-*t*-butylpropionate to give the nitro ester **17** (Figure 4). Nitro reduction afforded the amino ester **18**. Attemps to force closure of **18** to the benzodiazocine ring system under acidic or basic conditions failed. The desired product could be prepared, however, by conducting ester hydrolysis of **18** to give the amino acid **19** followed by treatment of **19** with DCC to give **20**.

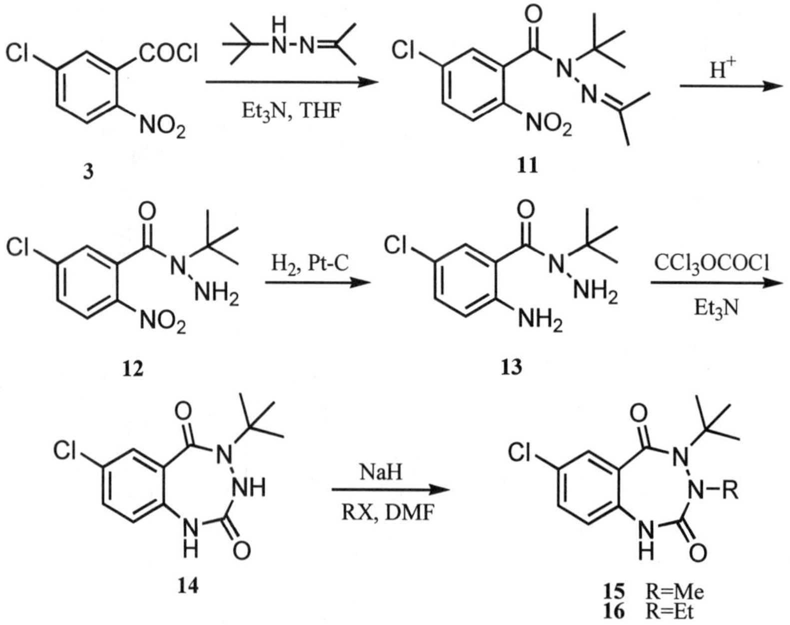

Figure 3. Preparation and alkylation of 1,3,4-benzotriazepine-2,5-diones.

1,2,3,4-Tetrahydroquinazolin-2,4-dione ring system. Removal of the C-3 carbon of the benzodiazepinedione leads to the six-membered tetrahydroquinazolin-2,4-dione ring system (Figure 5). Treatment of *o*-nitrobenzoyl chloride **3** with *t*-butylamine afforded the nitrobenzamide **21**. The cyclization precursor, **22**, was obtained by nitro reduction. Reaction of **22** with diphosgene followed by treatment of the resultant intermediate with base gave **23**.

Aryl ring replacement. A brief investigation was carried out whereby the phenyl ring of the benzodiazepinedione system was replaced with heteroaromatic moieties. Two such heterocyclic systems were examined: The 1H-pyrido[2,3-e][1,4]diazepine-2,5-dione and the 1H-thiopheno[3,2-e][1,4]-diazepine-2,5-dione ring systems.

1H-pyrido[2,3-e][1,4]diazepine-2,5-dione ring system. One regioisomeric pyridodiazepinedione was prepared resulting in replacement of the benzodiazepinedione C-9 with a nitrogen atom (Figure 6). Entry into this series was initiated with the copper promoted amination of 2-chloronicotinic acid **24**, affording N-benzylaminonicotinic acid **25** (*8*). The carboxylic acid functionality of **25** was converted to the acid chloride by treatment with thionyl chloride and then treated with methyl N-*t*-butylglycinate to afford the nicotinamide **26**. Catalytic transer hydrogenolysis caused N-debenzylation and concomitant cyclization to the pyridodiazepinedione **27**.

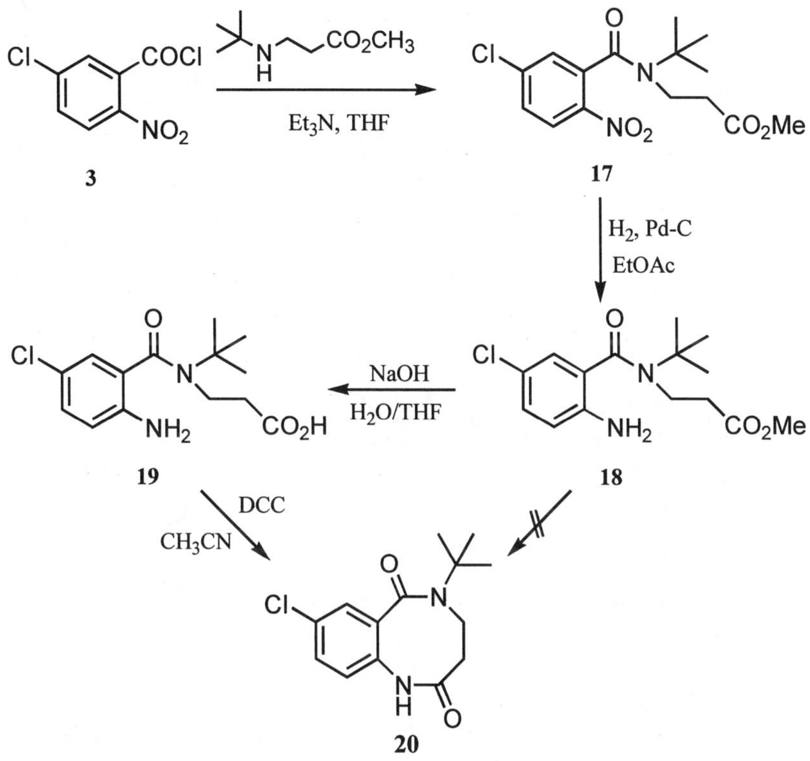

Figure 4. Preparation of 1,5-benzodiazocine-2,6-dione ring system.

1H-Thiopheno[3,2-e][1,4]diazepine-2,5-dione ring system. A single regiochemical thiophenodiazepinedione series was examined (Figure 7). Methyl 3-amino-2-thiophenecarboxylate **28** was N-protected as the N-*o*-nitrobenzyl analog **29**. Ester hydrolysis afforded **30**. Reaction of the amino acid **30** with thionyl chloride followed by treatment with methyl *tert*-butylglycinate gave **31**. Reductive debenzylation then gave the desired product, **32**.

Biological Activity

The benzodiazepinediones and related analogs were evaluated in the greenhouse in a standard protocol consisting of a variety of broadleaf and grass weeds and crops, both pre- and postemergence. The analogs were applied at rates of 1000 - 32 g/ha to the various plant species as shown in Table I.

Structure-Activity Relationships. The benzodiazepinediones as a class are more efficaceous against broadleaf weeds than grass weeds. The structure-activity relationship comparisons shown below, therefore, are based on post- and preemergence broadleaf data. The herbicidal activity is represented as percent broadleaf control averaged over all the broadleaf weeds at 250 g/ha. A "0" designates no activity at 250 g/ha but some activity was observed at higher rates. A compound that was inactive at the highest rate tested (generally 1000 g/ha) is denoted with an "*i*". The most important structural changes studied include substitution at the

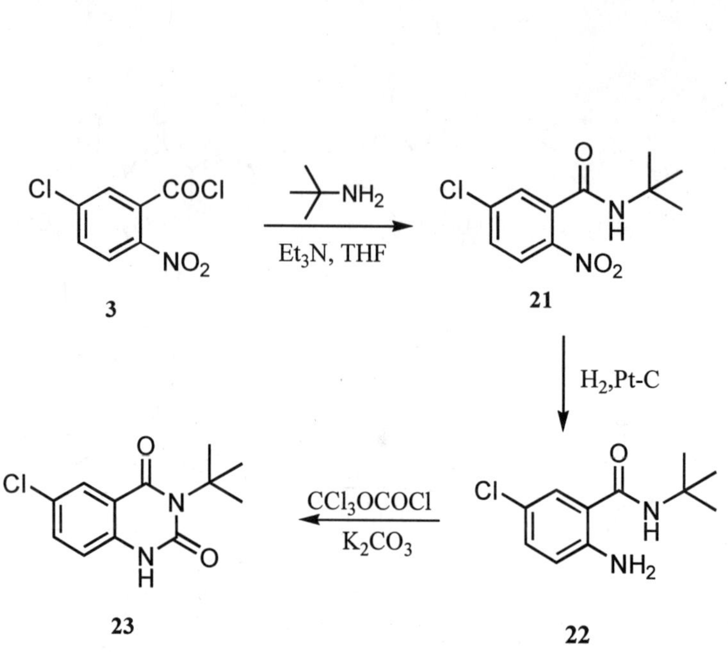

Figure 5. Preparation of 1,2,3,4-tetrahydroquinazolin-2,4-dione ring system.

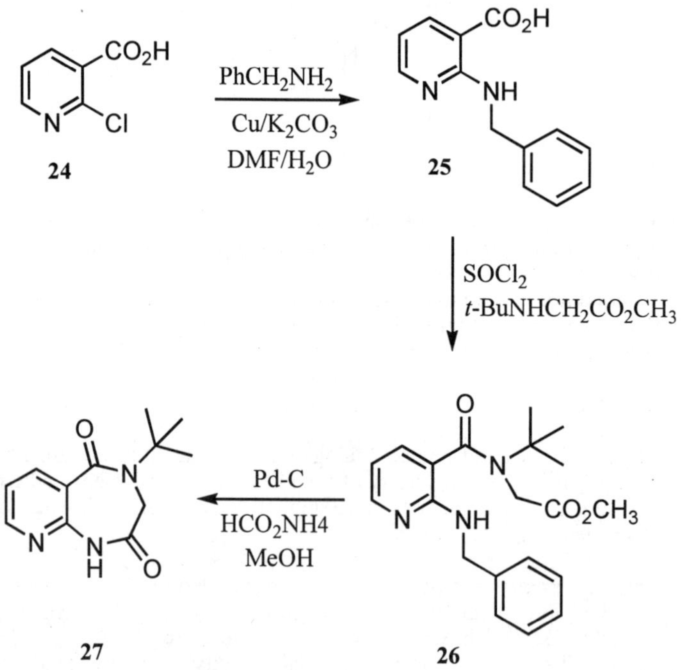

Figure 6. Preparation of 1H-pyrido[2,3-e][1,4]diazepine-2,5-dione ring system.

benzodiazepinedione N-1, C-3 and N-4, modification of the substituent and its regiochemistry in the benzene ring (C-6 through C-9), modification of the 1,4-diazepine ring and replacement of the benzene ring with other aromatic moieties.

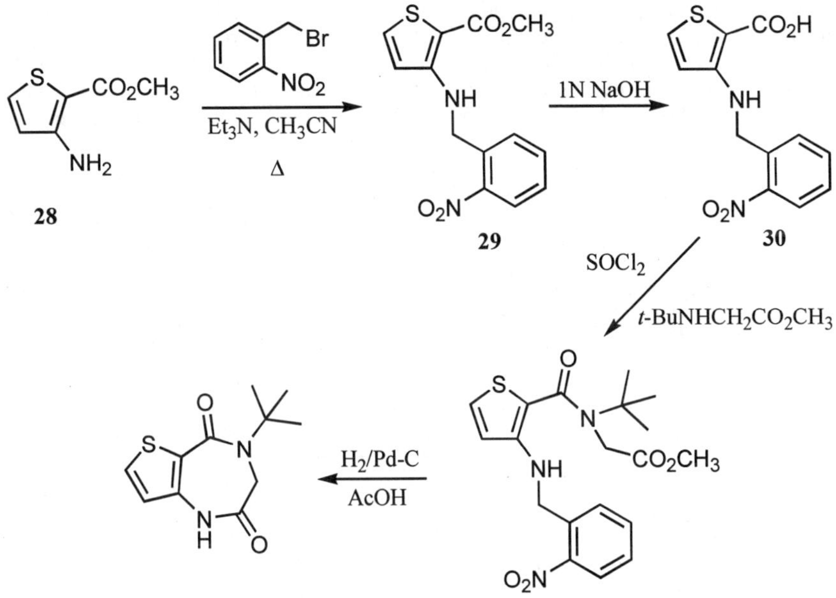

Figure 7. Preparation of 1H-thiopheno[3,2-e][1,4]diazepine-2,5-dione ring system.

Table I. Species Utilized In Standard Greenhouse Evaluation

Broadleaf Weeds	Grass Weeds	Crops
Bindweed	Barnyardgrass	Cotton
Lambsquarters	Green Foxtail	Corn
Morningglory	Wild Oats	Rice, Tebonnet
Wild Mustard		Soybeans
Black Nightshade		Sugarbeets
Ragweed		Sunflower
Velvetleaf		Spring Wheat

Substitution at N-3 and C-4. Herbicidal activity is highly dependent on the bulk of the acyclic substituent at N-4 as is shown in Table II. The most potent compound was the N-*t*-butyl analog which controlled both 80% of broadleaves postemergence and 95% of broadleaves preemergence at 250 g/ha. The N-neopentyl analog was slightly less active, controlling 70% and 75% of broadleaf weeds, respectively. With the exception of the *t*-amyl compound, decreasing the steric bulk of the substituent at N-4 was accompanied by decreasing potency. When the N-4 neopentyl compound was substituted at C-3 with a methyl group all activity was lost. This was unexpected since herbicidal activity is maintained in the pyrrolobenzodiazepine-diones, *e.g.*, **1**, which contain branching at this carbon. Presumably, C-3 branching in the acyclic series must affect the benzodiazepinedione conformation versus the more sterically constrained pyrrolobenzodiazepinedione. Even so, it is difficult to rationalize the drastic effect on activity. In a homologous series of C_3 through C_6 cycloalkyl substituents at N-4, maximum activity is observed with the cyclobutyl analog. Substitution at N-4 with variously substituted aromatic moieties afforded inactive compounds.

Table II. Benzodiazepinedione C-3 and N-4 Substitution vs. Activity

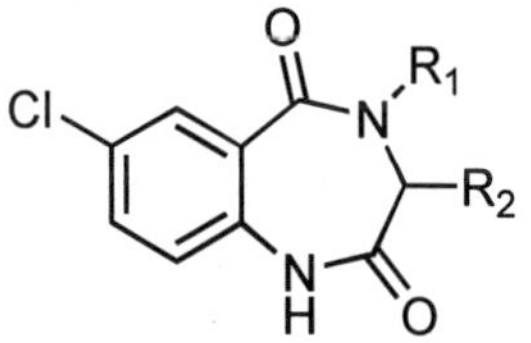

% Broadleaf control at 250 g/ha

R_1	R_2	Postemergence	Preemergence
t-butyl	H	80	95
neopentyl	H	70	75
t-amyl	H	20	45
i-propyl	H	20	20
propyl	H	30	35
allyl	H	0	20
propargyl	H	0	0
neopentyl	CH_3	*i*	*i*
cyclopropyl	H	20	0
cyclobutyl	H	60	55
cyclopentyl	H	20	10
cyclohexyl	H	30	10
3,5-$F_2C_6H_3$	H	*i*	*i*
3-$(CF_3)C_6H_4$	H	0	0
4-$(OCH_3)C_6H_4$	H	*i*	*i*

Substitution at N-1. Increasing the bulk of the substituent at N-1 leads to decreasing activity as shown in Table III.

Aryl Substitution. Extensive substituent modification in the aryl ring (C-6 through C-9) was carried out. Initially, a set of methyl substituted analogs were prepared and evaluated to determine which isomer elicits maximum activity. The order of herbicidal activity was found to be 7-methyl > 9-methyl >> 6-methyl (Table IV). We subsequently determined substitution at C-8 to have a deleterious effect on herbicidal activity (*4*). Based on these observations nearly all subsequent aryl substituent modifications were carried out at C-7 and C-9. Replacement of the 7-methyl substituent with an acetylene moiety resulted in a drop in herbicidal activity. Upon saturation of the acetylene moiety to the vinyl and ethyl analogs activity was completely lost.

The herbicidal activity varies greatly with the halogen substituent at C-7. The 7-bromo analog was slightly less active than the 7-chloro analog. The 7-fluoro and iodo analogs were less active still. The unsubstituted benzodiazepinedione **7**, shown for comparison, falls in between the fluoro and the bromo analogs in activity.

Substitution at C-7 with a number of additional electron withdrawing and donating groups (nitro, cyano, amino, hydroxy and methoxy) afforded compounds with poor activity as did substitution with aryl and heteroaryl moieties at both C-7 and C-9 (*6*).

A series of compounds were prepared to examine the effect of C-9 substitution on compounds containing 7-chloro or 7-methyl substitution. As the examples in Table IV demonstrate, introduction of a methyl group at C-9 causes only a slight decrease in activity. Introduction of a halogen atom, on the other hand, causes a significant decrease in herbicidal activity.

Table III. Benzodiazepinedione N-1 Substitution vs. Activity

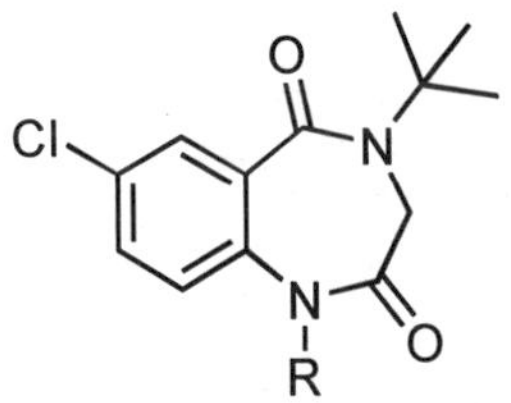

% Broadleaf control at 250 g/ha

R	Postemergence	Preemergence
H	80	95
CH_3	35	55
p-methoxybenzyl	*i*	*i*

Modification of the 1,4-diazepine ring. A brief effort was undertaken to explore the effect of changes to the 1,4-diazepinedione ring on herbicidal activity. The changes to the basic nucleus included the addition of heteroatoms and variation in ring size. (Table V). In the homologous series of six- to eight-membered benzo-fused systems, the parent benzo-1,4-diazepine-2,4-dione was somewhat more active postemergence than the 1,2,3,4-tetrahydroquinazolin-2,5-dione while the former was much more active preemergence. The benzo-1,5-diazocine-2,6-dione was inactive both pre- and postemergence. The 1,3,4-benzotriazepine-2,5-diones were inactive at the highest rates tested.

Aryl replacement. The phenyl ring of the parent benzodiazepinedione **7** was replaced with a pyridine and thiophene moiety (**27** and **32**, respectively) to assess the effect of introducing a heterocyclic isostere. In both instances all herbicidal activity was lost.

***In vitro* PSII assay.** In addition to whole plant activity a majority of the benzodiazepinediones and related compounds were assayed for *in vitro* photosystem II activity in the Hill Reaction. Potassium ferricyanide reduction of spinach thylakoid membranes were used to determined I_{50} values.

In general, the compounds possessing the highest level of whole plant activity were also the most potent *in vitro* (Table VI). Compounds substituted at C-7 with chloro, bromo and methyl (all containing 4-*t*-butyl) or at N-4 with neopentyl were highly active PSII inhibitors, with I_{50} values ranging from 0.2-1.7 μM. This compares to the *in vitro* activity of the commercial standard atrazine. Occasionally, the *in vitro* activity did not correlate with whole plant activity. The 7-fluoro analog exhibited moderate broadleaf activity in the greenhouse but had poor *in vitro* activity (500 μM). Possibly, the herbicidal activity is, at least in part, attributable to some other mode of action. In contrast, the 7-iodo analog represents an example of a compound with poor greenhouse activity but good *in vitro* activity (4.6 μM). It is reasonable that the weak whole plant activity may be due to poor uptake or subsequent metabolic inactivation.

Table IV. Aryl Substitution vs. Activity

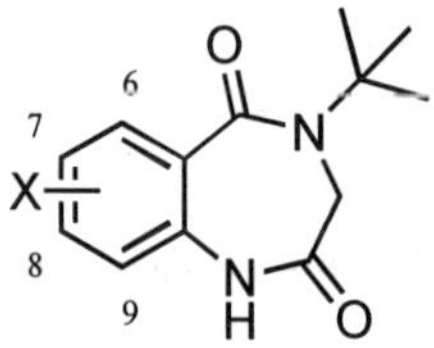

% Broadleaf control at 250 g/ha

X	Postemergence	Preemergence
7-Me	65	90
9-Me	30	60
6-Me	*i*	*i*
7-acetylene	30	40
7-vinyl	*i*	*i*
7-ethyl	*i*	*i*
7-F	15	45
7-Cl	80	95
7-Br	75	65
7-I	5	0
H	45	55
7-Cl	80	95
7-Cl, 9-Me	65	70
7,9-Cl_2	*i*	*i*
7-Cl, 9-Br	*i*	*i*
7-Me	65	90
7,9-Me_2	60	50
7-Me, 9-Cl	10	15

Conclusions

The structure-activity profile of the 1,4-benzodiazepine-2,5-diones demonstrates the rather limited substitution pattern that is tolerated for effective weed control. Substitution at N-1 and C-3 (in the bicyclic series) and benzenoid substitution at C-6 and C-8 has a detremental effect on herbicidal activity. The most active compounds were derived from substitution at N-4 with bulky moieties (*t*-butyl and neopentyl) and at C-7, primarily with chloro, bromo and methyl. Substitution at C-9 was limited to methyl. Attempts to alter the diazepine ring led to decreased activity. The few analogs prepared in an effort to introduce an isosteric replacement for the phenyl ring also led to decreased activity.

Table V. Effect of Diazepine Ring Modification on Activity

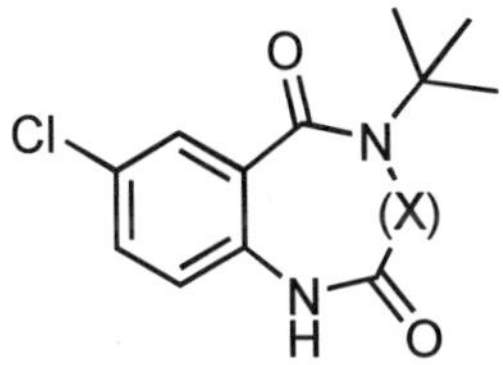

% Broadleaf control at 250 g/ha

X	Postemergence	Preemergence
-	55	0
CH_2	80	95
$(CH_2)_2$	*i*	*i*
NH	*i*	*i*
NMe	*i*	*i*
NEt	*i*	*i*

Table VI. *In Vitro* PSII Activity for Selected Analogs

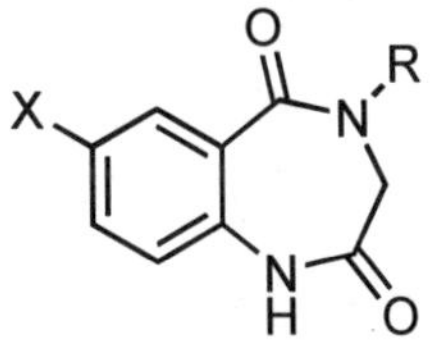

X	R	I_{50} (μM)
Cl	*t*-butyl	1.7
Cl	neopentyl	0.2
Br	*t*-butyl	0.5
CH_3	*t*-butyl	1.7
F	*t*-butyl	500
I	*t*-butyl	4.6

Literature Cited

1. Wright, W. B.; Brabender, H.; Greenblatt, E.; Day, I.; Hardy, R. *J. Med. Chem.*, **1978,** *21,* 1087.
2. Wright, W. U.S. 3,984,562.
3. Guaciaro, M. A.; Harrington, P. M.; Karp, G. M., U.S. 5,438,035 (1995).
4. Karp, G. M.; Manfredi, M. C.; Guaciaro, M. A.; Ortlip, C. L.; Marc, P.; Szamosi, I. T., *J. Agric. Food. Chem.*, **1997,** *45,* 493.
5. Singh, B. K.; Szamosi, I. T.; Dahlke, B. J.; Karp, G. M.; Shaner, D. L. *Pestic. Biochem. Physiol.*, **1996,** *56,* 62.
6. Karp, G. M. *J. Org. Chem.*, **1995,** *60,* 5814.
7. Karp, G. M. *J. Heterocycl. Chem.*, **1996,** *33,* 1131.
8. Coppola, G. M.; Fraser, J. D.; Hardtmann, G. E.; Shapiro, M. J. *J. Heterocycl. Chem.*, **1985,** *22,* 193.

Chapter 12

Synthesis and Structure-Activity/Selectivity Studies of Novel Heteroaryloxy, Aryloxy, and Aryl Pyridazines as Bleaching Herbicides

Michael S. South, Terri L. Jakuboski, Michael J. Miller, Mohammed Marzabadi, Susan Corey, Jack Molyneaux, Sarah Allgood South, Jane Curtis, Don Dukesherer, Steve Massey, Fenn-Ann Kunng, John Chupp, Robert Bryant, Kurt Moedritzer, Scott Woodward, Dare Mayonado, and Martin Mahoney

Monsanto Life Sciences, 800 North Lindbergh Boulevard, St. Louis, MO 63167

While exploring the novel cyclization reactions of 4-chloroazodienes with electron rich olefins we developed a new and general synthesis of substituted 3-phenylpyridazines. A number of these analogs were found to exhibit bleaching herbicidal activity (phytoene desaturase inhibition). Further methodology development coupled with analog synthesis led to the preparation of 3-heteroaryloxy and 3-aryloxypyridazines with increased unit activity and selectivity as well as good environmental properties. These compounds were found to be more active than current commercial standards on a number of important weed species, with selectivity in corn in the US and small grains in Europe. Greenhouse activity of the most active analogs ranged from 17-140 g/hectare on important narrow-leaf species.

The carotenoid biosynthetic pathway (Figure 1) has been a target for the discovery of new herbicidal agents for over 20 years. Much of this pathway is present only in the plant kingdom, which makes it attractive from a mammalian safety standpoint. Many new herbicidal agents have been investigated that inhibit certain enzymes in this pathway. In particular, the inhibition of phytoene desaturase serves as a target for many herbicide discovery programs throughout the ag industry and has recently been the topic of several reviews (*1-3*). We have developed a series of pyridazine herbicides that inhibit phytoene desaturase (*4*) that are among the most active and selective analogs discovered to date with this mode of action (*5-12*). Five of these analogs (Figure 2) have undergone extensive field testing and have exhibited selectivity in corn in the US and small grains in Europe. We refer to these compounds as AZOD herbicides since the original synthesis of the 3-phenylpyridazine lead is derived from a novel 4 + 2 azodiene cyclization reaction.

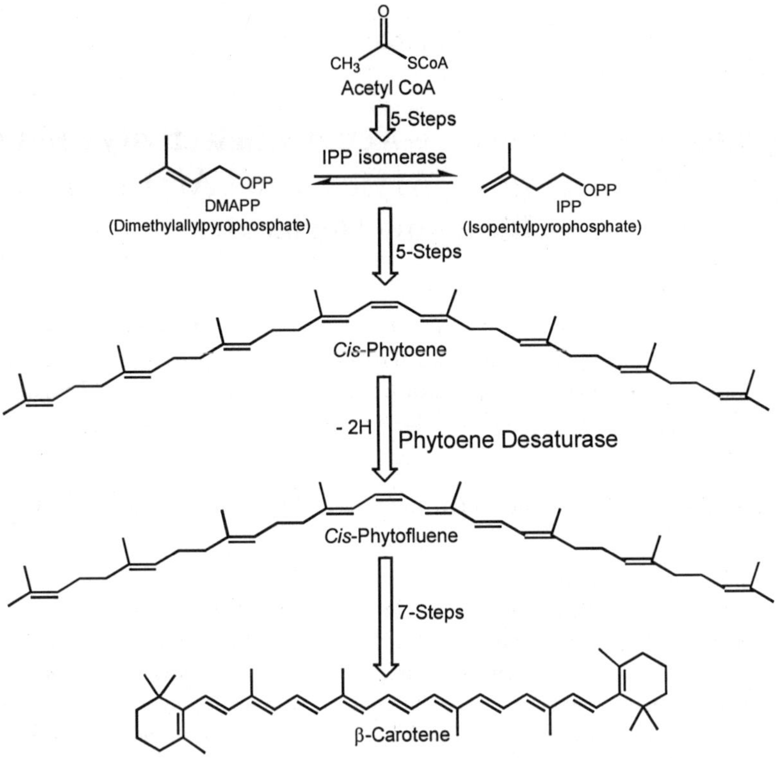

Figure 1. Carotenoid Biosynthetic Pathway.

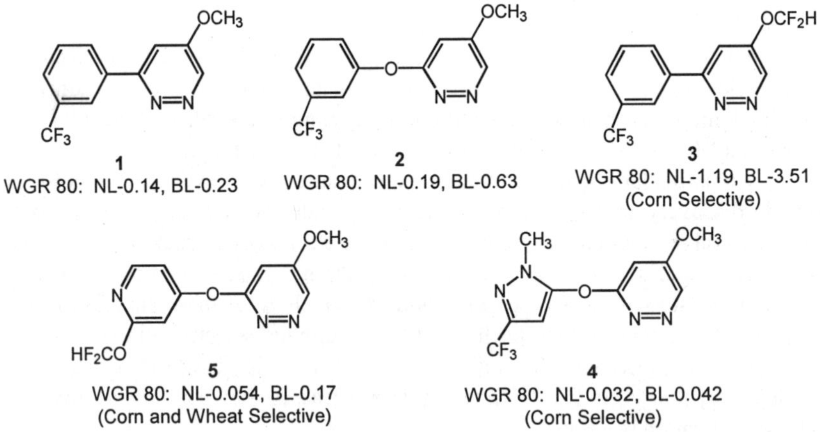

Figure 2. AZOD Field Compounds, WGR 80 in lbs./acre.

Synthesis and Structure-Activity-Relationship (SAR) Studies

3-Phenylpyridazines. Our entry into pyridazine chemistry revolved around the discovery of a novel 4 + 2 cycloaddition reaction of a haloazodiene with electron rich olefins (Figure 3). These azodiene cycloaddition reactions give rise to the synthesis of a number of heterocyclic molecules, which include tetrahydropyridazines, pyridazines, pyridazinones, *N*-aminopyrroles, pyrroles, and triazinones. The details of many of these reactions have been published elsewhere (*13-15*). The initial bleaching herbicide lead in this area was discovered via screening of the 3-phenylpyridazine analogs derived from these novel azodiene cyclizations.

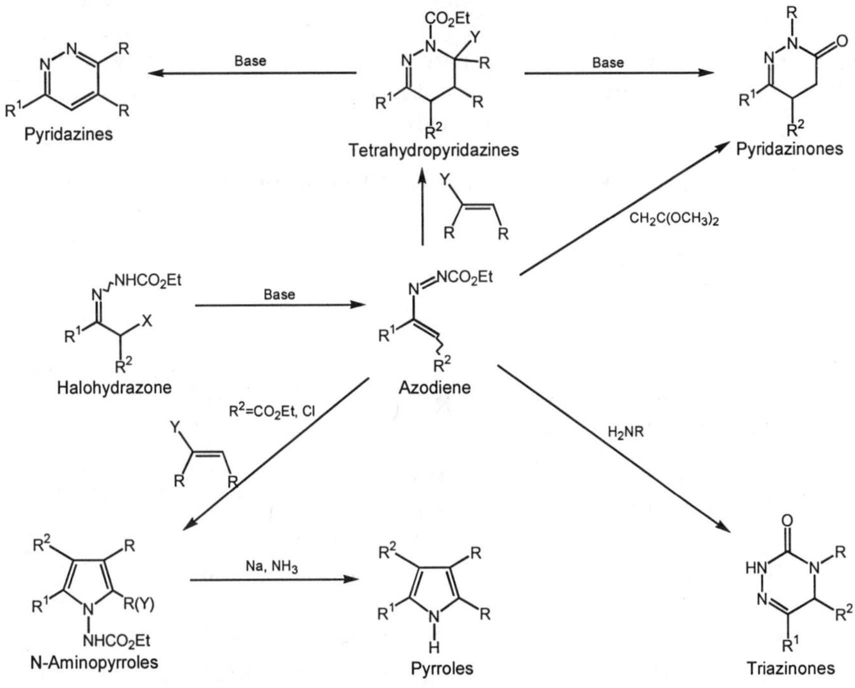

Figure 3. Reaction of Haloazodienes with Electron Rich Olefins.

The synthesis of the 3-phenylpyridazines was accomplished as outlined in Figure 4. Treatment of a dichlorohydrazone with base generates the azodiene *in-situ*. Subsequent reaction with an electron rich olefin gives good yields of the tetrahydropyridazines as separable mixtures of two diastereomers. Aromatization of the tetrahydropyridazine isomers with base leads to a variety of substituted 3-phenylpyridazines in high yield (*13-15*). This reaction is compatible with a number of functional groups on the electron rich olefin and azodiene. The pyridazine derivatives were screened for herbicidal activity and were found to exhibit the bleaching symptomology.

SAR of 3-Phenylpyridazines. The azodiene synthesis provided 3-phenylpyridazines with alkyl and electron withdrawing substituents on the pyridazine ring.

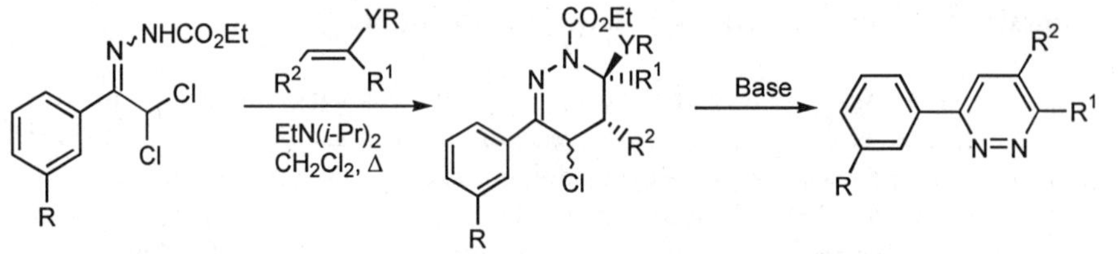

Figure 4. Synthesis of Substituted 3-Phenylpyridazines via Azodiene Cyclizations.

Herbicidal activity is shown in Table I. It was necessary to have the meta-trifluoromethyl substituent on the phenyl ring, as is the case with many phytoene desaturase inhibitors (*1*). Other substitution patterns or differing substituents on this ring lowered the activity (*5*). Substitution at the four-position (*5*) or six-position of the pyridazine also lowered the activity. Interesting activity in this series was obtained with a 5-methyl substituent on the pyridazine ring (compound **6** was described previously, *16*), which warranted further study.

Table I. Herbicidal Activity of 3-Phenylpyridazines.

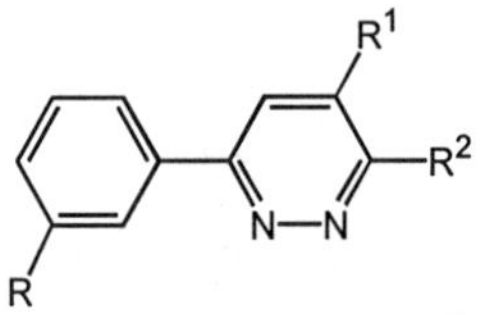

Compound	R	R¹	R²	Primary PE	Primary POE	Secondary-PPI[a] NL	Secondary-PPI[a] BL
6	*m*-CF_3	Me	H	98, 92	62, 82	0.48	0.87
7	*m*-CF_3	Et	H	94, 82	68, 80	3.2	4.6
8	*m*-CF_3	*i*-Pr	H	68, 50	20, 6	4.9	8.1
9	*m*-CF_3	CF_3	H	64, 46	8, 10	5.6	14
10	*m*-CF_3	H	H	84, 62	16, 36	3.1	5.7
11	*m*-CF_3	H	Me	52, 42	22, 44	19	99
12	*m*-CF_3	Me	Et	46, 42	62, 88	8	10
13	H	Me	H	56, 36	10, 28	12	13
14	*m*-CF_3	H	*n*-Bu	12, 4	10, 8		
15	*m*-CF_3	*n*-Pr	Me	4, 2	20, 8		
16	*m*-CF_3	CO_2H	H	0, 0	0, 0		

a) Primary data is percent inhibition at 10 lbs./acre pre-emergent (PE) and post-emergent (POE) for narrow-leaf (NL), broad-leaf (BL) weeds. Secondary data is WGR 80 in lbs./acre, pre-plant incorporated (PPI).

We sought to improve on the activity that was obtained in the initial 3-phenylpyridazine series through further modification of the 5-position of the pyridazine. This was accomplished by functional group transformation of the pyridazine carboxylic acid **16** (Figure 5). Curtius rearrangement of the acid to the amine followed by chlorination via the Sandmeyer reaction gave chloropyridazine **18**. This intermediate was found to react with a variety of nucleophiles giving the analogs shown in Table II. Treatment of **18** with KOH in methanol gave compound **1**, which was the most active pyridazine in the 3-phenyl series. A variety of nitrogen, sulfur, and oxygen containing nucleophiles with varying substitution were incorporated at the 5-position of the pyridazine which lowered the bleaching herbicidal activity relative to the 5-methoxy substitution. Therefore, we chose the 5-methoxy substituent as the optimal pyridazine substitution for our subsequent activity and selectivity studies.

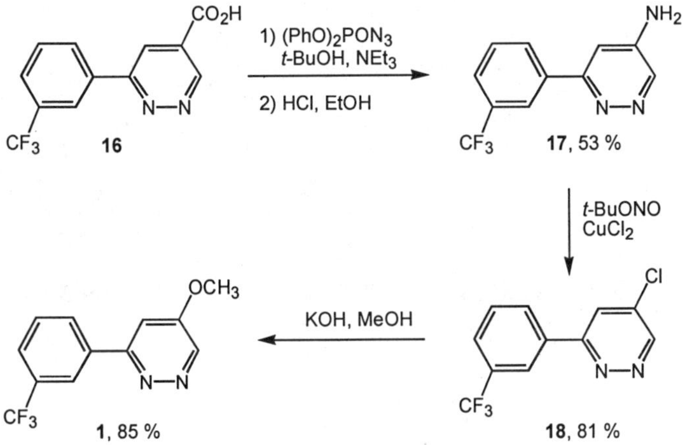

Figure 5. Synthesis of 3-(3-Trifluorometh y)phenyl-5-methoxypyridazine.

Table II. Herbicidal Activity of 3-Phenylpyridazines

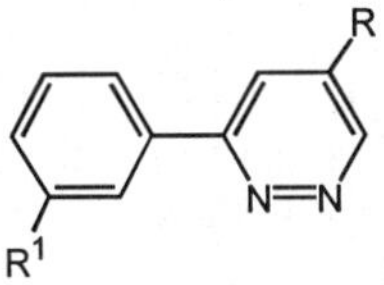

Compound	R^1	R	Primary PE	Primary POE	Secondary-PPI[a] NL	Secondary-PPI[a] BL
1	*m*-CF_3	OCH_3	98, 96	44, 42	0.14	0.23
19	*m*-CF_3	OEt	80, 64	18, 30	0.73	4.1
20	*m*-CF_3	SMe	88, 60	30, 26	1.8	5.0
21	*m*-CF_3	Cl	92, 66	8, 20	3.6	7.8
22	H	OCH_3	60, 64	44, 72	4.9	4.5
23	*m*-CF_3	$N(CH_3)_2$	54, 18	12, 24	5.3	5.4
24	*m*-CF_3	OCH_2CHCH_2	54, 34	16, 34	5.9	9.5
25	*m*-CF_3	OCH_2CCH	42, 12	42, 54	7.3	8.9
26	*m*-CF_3	$NHCH_3$	50, 26	12, 6	7.9	7.3
27	*m*-CF_3	O-*n*-Pr	44, 4	0, 4	9.1	70
28	*m*-CF_3	O-*i*-Pr	32, 8	26, 26	13	16
29	*m*-CF_3	$ONMe_2$	34, 26	24, 38	26	40
30	*m*-CF_3	OH	18, 0	36, 4	32	99
31	*m*-CF_3	NH_2	6, 0	0, 2		

a) Primary data is percent inhibition at 10 lbs./acre for narrow-leaf (NL), broad-leaf (BL) weeds. Secondary data is WGR 80 in lbs./acre.

One additional compound that is noteworthy from the 3-phenylpyridazine series is the 5-difluoromethoxy substituted pyridazine **3** (Figure 6). This compound was somewhat less active than the 5-methoxy analog **1**, but was corn selective and served as a design template for future compounds where further selectivity enhancements with retention of activity were achieved (see selectivity discussion below).

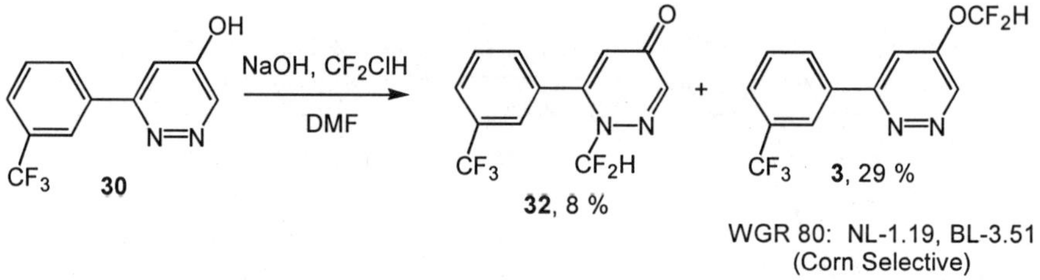

Figure 6. Synthesis of Difluoromethoxypyridazine 3.

A-Link-B Model.

Our studies of the substitution on the pyridazine ring resulted in the optimal group being chosen as the 5-methoxy substituent (compound **1**) on the pyridazine ring (B-ring) for good activity and the 5-difluoromethoxy group (compound **3**) for selectivity. We next turned our attention toward improving the activity and selectivity based on an A-Link-B model, Figure 7. We sought to vary the ring containing the *m*-trifluoromethyl group as well as the linking group between the two rings. This approach required the development of a different route for the construction of the two ring system since the azodiene chemistry described in Figure 4 was not amenable to this strategy. We developed a large scale synthesis of 3-chloro-5-methoxypyridazine **33** (*17*) which served as the B-ring source for coupling with a variety aryl and heteroaryl groups (Figure 7).

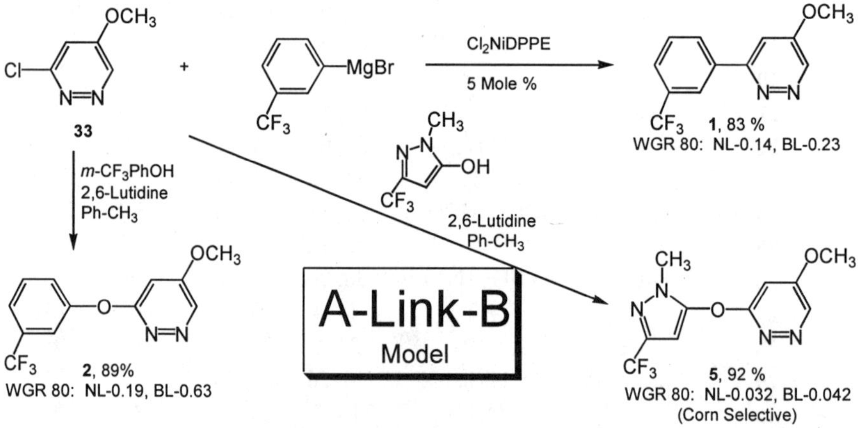

Figure 7. A-Link-B Model.

Chloropyridazine **33** was found to undergo a nickel catalyzed cross coupling reaction with a Grignard to give the previously prepared 3-phenylpyridazine **1** in 83 % yield. Reaction of **33** with *m*-trifluoromethylphenol and base resulted in the formation of oxygen linked compound **2** in 89 % yield. Greenhouse data suggested that this analog retained the activity exhibited by the directly linked pyridazine **1** while gaining some selectivity in corn and small grains. A number of arylphenols were coupled to pyridazine **33** with no improvement in activity or selectivity (*6*).

Large gains in activity were obtained when a pyrazolyloxypyridazine **5** (*7*) was prepared utilizing the same coupling chemistry used for the phenols. This compound proved to be one of the most active bleaching agents that inhibit phytoene desaturase (*4*) developed to date and underwent field trails in corn in the US and small grains in Europe. Compound **5** exhibited selectivity in these crops mainly through placement (soil mobility was limited under normal rainfall) in the top 1 inch of the soil surface.

Due to the superior activity exhibited by **5** and in order to study the SAR around the A-ring we developed a synthesis of fluorinated pyrazoles that relied on the chemistry shown in Figure 8 (*7*). Here the fluorinated group on the pyrazole was varied through conversion of the acid fluoride to a fluorinated acetoacetate. Cyclization with methylhydrazine gave the various substituted pyrazoles which were coupled with 3-chloro-5-methoxypyridazine to give the analogs shown in Table III. Some of these analogs were prepared via functional group transformations on the pyrazole moiety (*7*).

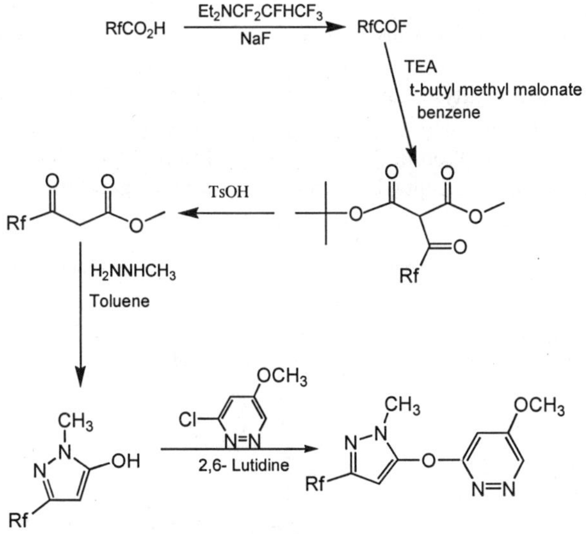

Figure 8. Synthesis of Fluorinated Pyrazoles.

SAR Studies of Pyrazolyloxypyridazines. A variety of substituents were incorporated into the pyrazolyloxypyridazines as shown in Table III. SAR studies of this series is summarized in Figure 9. The best activity was obtained with small fluorinated groups of two carbon atoms or less in the three position of the pyrazole. Larger functionality at the 3-position of the pyrazole lowered the activity. The pentafluoroethyl functionality at this position was about two times more active than the

Table III. Herbicidal Activity of Pyrazolyloxypyridazines.

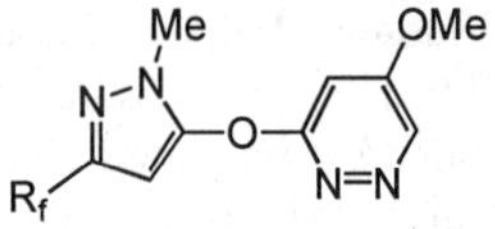

Compound	R_f	NWW-GR80[a]
33	CF_2CF_3	0.015
34	CH_3CF_2	0.029
5	CF_3	0.032
35	HCF_2CF_2	0.130
36	$CH_2{=}C(CF_3)$	0.150
37	CF_3CH_2	0.170
38	$CF_3CH(OCH_3)$	0.540
39	$CF_3CH(OH)$	0.730
40	$CF_2{=}CH$	0.920
41	$CF_2{=}CF$	1.100
42	$CH_2{=}C(CH_3)$	1.300
43	$CF_3CF_2C(OCH_2CH_2O)$	1.500
44	$CF_3(CF_2)_5CO$	1.800
45	CF_3CO	2.200
46	HCF_2O	2.300

a) Narrow-leaf weed warm season GR80 in lbs./acre.

trifluoromethyl functionality. Larger substitution than ethyl on the nitrogen of the pyrazole ring lowered the activity. Placing the methyl group on the other pyrazole nitrogen as well as substitution in the four position lowered the activity. Changing the linking atoms to anything except CF_2 or oxygen lowered the activity.

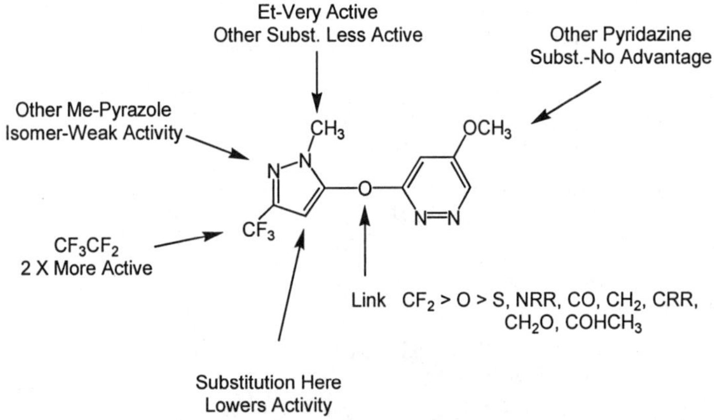

Figure 9. SAR Studies of Pyrazolyloxypyridazines.

SAR Studies of Alternate Heterocycles. Using the chemistry described above (Figure 7) a variety of heterocycles were coupled with the 3-chloro-5-methoxy pyridazine **33** (Figure 10). Pyridine and triazole heterocycles substituted with small fluorinated groups were all found to be quite active as compared to the pyrazole substitution. The pyridine moieties were also found to impart some metabolic selectivity that was superior to the other analogs synthesized and tested so far. The pyridine A-rings were advanced for further synthesis and testing due to their enhanced selectivity profiles and are discussed in the next section.

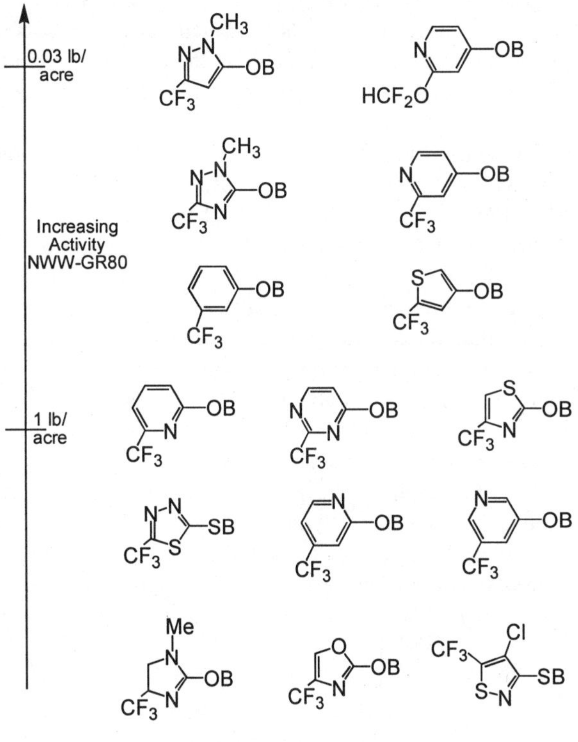

Figure 10. SAR Studies of Alternate Heterocycles.

Synthesis and Structure-Selectivity-Studies (SSR).

Building in Metabolic Selectivity with Retention of Activity. Large gains in efficacy were obtained in the pyridazine area of chemistry via the development of new compounds using the A-Link-B model described above. The activity of the directly linked 3-phenylpyridazine **1** was improved upon by introducing an oxygen linking atom between the two rings to give 3-phenoxypyridazine **2** followed by optimization of the A-ring heterocycle to give the pyrazolyloxypyridazine **4**. These compounds performed well in greenhouse and field testing, however, much of their selectivity was derived from what is termed placement selectivity and not from metabolic selectivity. Studies indicated that most of active ingredient was found in the top one inch of soil in greenhouse and field situations after normal rainfall amounts. Most weeds germinate in this zone containing the herbicide while the crop in question is planted below this 1 in. level. Greenhouse and field crop injury was noted when compounds **1**, **2**, and **4**

migrated into the zone of germination of the crop (2-4 in.) due to heavy rainfall conditions or soil surface disturbance.

Metabolic selectivity is defined as the ability of the crop species to metabolize the herbicidal compound in question to an inactive form via a pathway that is not present in the weed species to be controlled. One such metabolic pathway is conjugation with glutathione catalyzed by glutathione S-transferaces. Corn and to a lesser extent small grains have high levels of glutathione. We sought to utilize the glutathione metabolic pathway to our advantage in the pyridazine area to design a corn and/or wheat selective compound while maintaining the activity against important weed species.

We knew that compound **3** had exhibited (Figure 11) a high level of selectivity in corn from our previous studies with a selectivity factor of 7.12 (selectivity factor = Corn GR20/NWW GR80, larger numbers equal better selectivity) and a somewhat moderate narrow-leaf weed warm season GR80 (NWW GR80) of 0.33 lbs./acre. In an effort to understand this selectivity, compound **3** was incubated with glutathione in the presence and absence of glutathione S-transferaces. Both experiments led to conjugation of **3** with glutathione through replacement of the labile OCF_2H group with the sulfhydryl group of glutathione (Molin, B., unpublished results). We also knew that compound **47** exhibited high levels of activity (NWW GR80 0.037 lbs./acre) with poor selectivity in corn (selectivity factor = 0.84). We theorized that replacing the OCF_2H group on the pyridazine with the optimal 5-methoxy group and moving the OCF_2H group to the labile position alpha to the pyridine nitrogen would give an analog with good selectivity and activity profiles. Synthesis (Figure 12) and testing of compound **5** proved that this was indeed the case. Compound **5** was not only highly active (NWW GR80 0.054 lbs./acre), but also selective in corn (selectivity factor of 2.6).

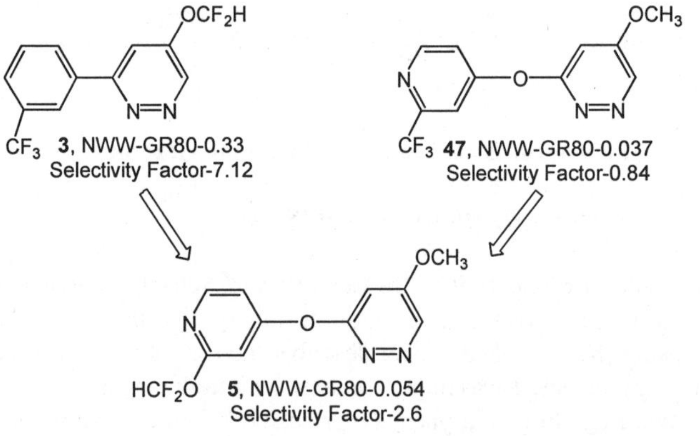

Figure 11. Building in Metabolic Selectivity with Retention of Activity.

Due to the activity and selectivity profiles exhibited by compound **5** we developed a new synthesis of the pyridyloxypyridazines (*8*) as outlined in Figure 12 so that further SAR and SSR could be defined. The synthesis relied on the conversion of a protected pyridone to a fluorinated alkoxy group or displacement of an ortho-chloropyridine with a nucleophile.

Activity data for selected pyridyloxypyridazines is shown in Table IV. A number of small fluorinated alkoxy groups were tolerated ortho to the pyridine nitrogen. Compounds **48-51** were as active as compound **5** with similar selectivity profiles. None of the compounds listed in Table IV was superior to compound **5** in

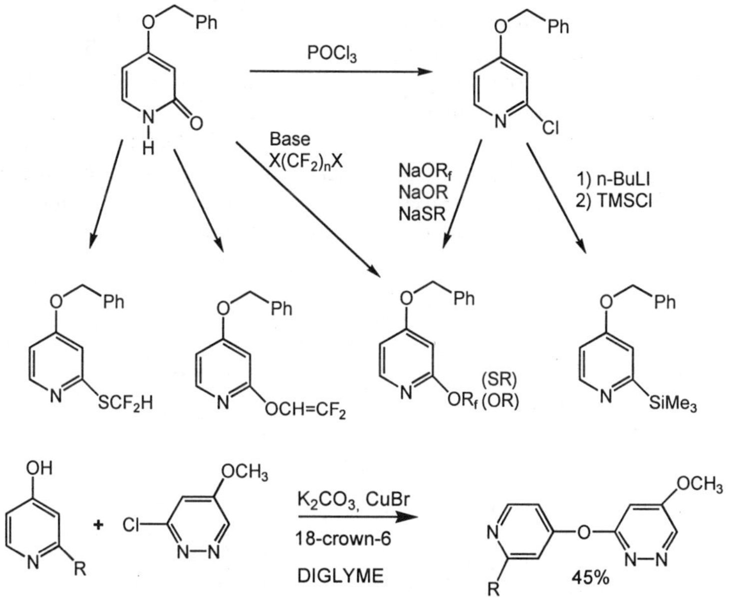

Figure 12. Synthesis of Pyridyloxypyridazines.

field tests in corn and small grains. This series was used however to demonstrate that the soil mobility could be adjusted based on the lipophilicity of the fluorinated side chain (longer fluorinated side chains imparted less soil mobility).

Table IV. SAR of Pyridyloxypyridazines.

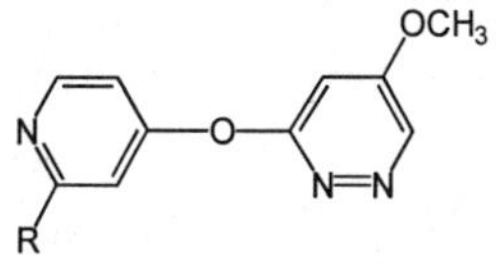

Compound	R	NWW-GR80[a]
5	HCF_2O	0.054
48	CF_3CH_2O	0.032
49	$CF_3(CH_3)CHO$	0.030
50	HCF_2S	0.040
51	$HCF_2CF_2CH_2O$	0.030

a) Narrow-leaf weed warm season GR80 in lbs./acre.

Summary of SAR and SSR. A summary of the structure-activity and structure-selectivity relationships in the pyridazine area are shown in Figure 13. The initial lead (compound **1**) in this area was discovered as a result of the development of the novel cyclization of an azodiene with an electron rich olefin giving a 3-phenylpyridazine. Functional group manipulation resulted in the 5-methoxy group being chosen as the optimal group on the pyridazine ring. Compound **1** was not selective on corn (selectivity factor 1.92). Introduction of an oxygen link in between the two rings (compound **2**) maintained the bleaching herbicidal activity while increasing the selectivity in corn (selectivity factor 4.38). Functionalization at the 5-position of the pyridazine ring with an OCF_2H group (compound **3**) imparted high levels of corn selectivity (selectivity factor 7.12) due to metabolic displacement and inactivation with glutathione. Utilizing a A-Link-B strategy, the optimal heterocyclic A-ring was chosen as the pyrazole in compound **4**. This analog is one of the most active bleaching agents that inhibits phytoene desaturase that has been developed to date. Combining the OCF_2H selectivity element with the pyridine A-ring resulted in compound **5** which proved to be one of the most highly active and corn selective analogs in this series.

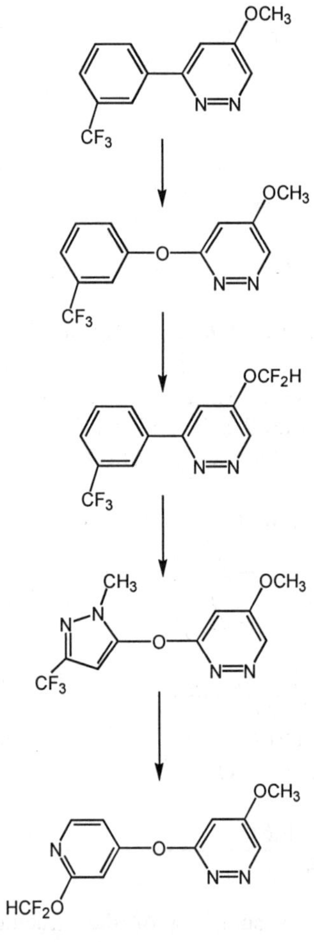

1
NWW-GR80-0.08
Selectivity Factor-1.92

2
NWW-GR80-0.11
Selectivity Factor-4.38

3
NWW-GR80-0.33
Selectivity Factor-7.12

4
NWW-GR80-0.03
Selectivity Factor-1.81

5
NWW-GR80-0.054
Selectivity Factor-2.6

Figure 13. Summary of SAR and SSR.

Literature Cited

1. Sandmann, G.; Boger, P. In *Phytoene Desaturase as a Target for Bleaching Herbicides*; Roe, R. M., Ed.; Herbicide Activity: Toxicology, Biochemistry and Molecular Biology; IOS Press: Amsterdam, Netherlands, 1997, pp. 1-10.
2. Boger, P. *Nippon Noyaku Gakkaishi* **1996**, *21*, 473-478.
3. Mitchell, G. In *Phytoene desaturase. A Model for the Optimization of Inhibitors*; Baker, D. R; Fenyes, J. G; Basarab, G. S., Eds.; Synthesis and Chemistry of Agrochemicals IV, ACS Symposium Series 584; American Chemical Society: Washington, D. C., 1995, pp. 161-170.
4. A buildup of phytoene was observed by HPLC from extracts of plants treated with pyridazine herbicides, South, M. S.; Mayonado, D.; Louseart, D., unpublished results.
5. South, M. S.; Jakuboski, T. L.; Monsanto Co.; U.S. Patent 5,623,072, April 22, 1997.
6. South, M. S.; Miller, M. J.; Monsanto Co.; U. S. Patent 5,559,080, September 24, 1996.
7. Moedritzer, K.; South, M. S.; Monsanto Co.; U. S. Patent 5,536,701, July 16, 1996.
8. South, M. S.; Monsanto Co.; U. S. Patent 5,484,761, January 16, 1996.
9. Jakuboski, T. L.; South, M. S.; Monsanto Co.; U. S. Patent 5,616,789, April 1, 1997.
10. Parlow, J. J.; Normansell, J. E. *Mol. Diversity* **1996**, *1*, 266-269.
11. Parlow, J. J.; Mischke, D. A.; Woodard, S. S. *J. Org. Chem.*, in press.
12. Parlow, J. J. *Tetrahedron*, submitted.
13. South, M. S.; Jakuboski, T. L. *Tetrahedron Lett.* **1995**, *36*, 5703-5706.
14. South, M. S.; Jakuboski, T. L.; Westmeyer, M. D.; Dukesherer, D. R. *Tetrahedron Lett.* **1996**, *37*, 1351-1354.
15. South, M. S.; Jakuboski, T. L.; Westmeyer, M. D.; Dukesherer, D. R. *J. Org. Chem.* **1996**, *61*, 8921-8934.
16. Speltz, L. M.; Walworth, B. L.; American Cyanamid Co.; U. S. Patent 4,623,376, November 18, 1986.
17. Bryant, R. D.; Kunng, F.-A.; South, M. S. *J. Heterocycl. Chem.* **1995**, *32*, 1473-1476.

Chapter 13

Investigation of 5′-Phosphohydantocidin Analogs as Adenylosuccinate Synthase Inhibitors

G. D. Crouse, R. D. Johnston, D. R. Heim, C. T. Cseke, and J. D. Webster

Discovery Research Center, DowElanco, 9330 Zionsville Road, Indianapolis, IN 46268-1053

Hydantocidin **1** is a novel, fermentation derived non-selective herbicide discovered by Sankyo, whose mode of action, inhibition of adenylosuccinate synthase, stems from the 5'-phosphorylated moiety. Phosphate mimics and ribose ring replacements were designed and synthesized. These derivatives show the requirement of a phosphorylated ribose ring, specifically a single epimer, for enzyme recognition.

Resistance to known classes of herbicides eventually results in reduced efficacy and more limited market opportunities. As a result, the identification of new modes of action represents a potentially significant opportunity for the development of new classes of chemistry as weed control agents. In 1990 Sankyo (*1,2*) described the isolation and activity of a novel herbicide called hydantocidin **1**. The activity, as well as the chemistry, were of interest to us, however nothing was reported about the mode of action or the structural requirements for activity. Later, in 1991, Sankyo published a series of papers (*3*) describing the total synthesis of hydantocidin and all of its geometric isomers. Interestingly, of these isomers, only hydantocidin and the anomeric hydantocidin analog showed herbicidal activity. Subsequently, several other syntheses of hydantocidin have been reported (*4*), further evidence of the interest, on the part of both academic and industrial scientists, in this compound and the activity it displays.

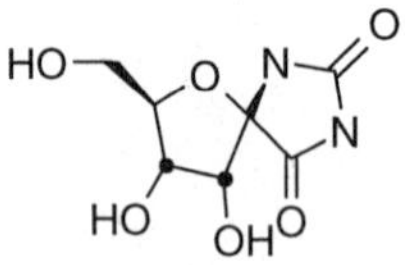

1

It was proposed by Heim, et al (*5*), that hydantocidin was acting as a pro-herbicide, which formed an inhibitor of the enzyme adenylosuccinate synthase (AdSS). This hypothesis was supported by the observation that increasing amounts of an inhibitor of AdSS were formed, over time, in a plant extract following the addition of hydantocidin, ATP, and magnesium chloride. In view of the mechanism of action of AdSS, the endogenous substrate of which is the 5'-monophosphate of inosine, it was further proposed that the actual inhibitor was the 5'-monophosphate of hydantocidin **2**. (In earlier references, hydantocidin nomenclature referred to this position as 2α. We are adopting the more familiar 5' numbering, consistent with other natural nucleosides such as inosine and adenosine.) Hence, hydantocidin is taken up by the organism and phosphorylated by a kinase enzyme which recognizes the ribose component of the molecule (Figure I). This enzyme inhibition results in the immediate cessation of growth and eventual death of the treated plant. While the herbicidal effect is somewhat reminiscent of glyphosate in terms of spectrum and the time of onset of symptoms, no known herbicide acts by this mode of action.

Ultimate proof of this mechanism would depend upon preparation of the phosphate derivative **2**, and determination of its effect on AdSS, and finally by isolation of the actual inhibitor from hydantocidin-treated plants. Papers describing the unequivocal establishment of the mechanism of action of hydantocidin to be through inhibition of the AdSS enzyme by phosphohydantocidin have now been published by us (*6*) as well as others (*7*).

Figure I

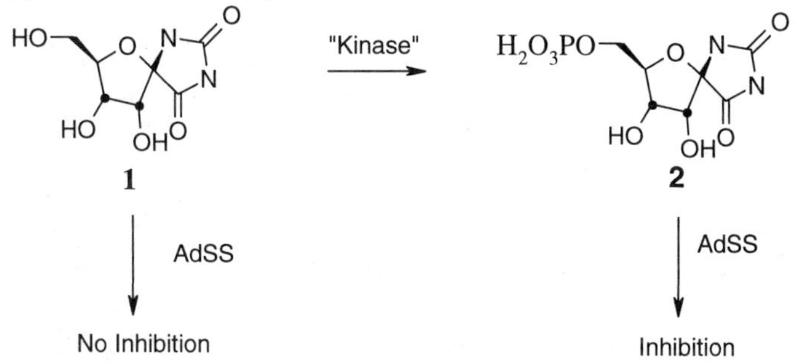

From a synthesis standpoint, using hydantocidin as a starting point for a design effort was appealing. The attributes considered ideal were the novelty of the chemistry and mode of action, and the requirement for phosphorylation. The key negative issues associated with the development of hydantocidin as a herbicide were the inherent cost of a fermentation derived product and the complexity of a total synthesis production.

The development of a lower cost alternative, while retaining the novel mode of action and a significant portion of the activity, would be an important discovery. Reports by Sankyo have argued against the probability of finding such a simple analog. They have shown, through some elegant synthetic work, that all the deoxy (*8*) and isomeric derivatives of hydantocidin were either inactive or much less active than hydantocidin itself. Since our data strongly suggest that two enzymes are required for expression of herbicidal activity, a key synthetic and mechanistic question was to determine whether just one, or both of the enzymes, was responsible for this high regiochemical and stereochemical requirement. It is reasonable, given the high specificity of kinases toward ribose derivatives, that the selectivity might be due, solely or in part, to the kinase enzyme, rather than AdSS itself. Theoretically it should then be possible to design a molecule which did not require activation *via* phosphorylation. In principle, these designed inhibitors may have fewer structural requirements, and thus be simpler to prepare.

This hypothesis took the form of two design questions. First, the high polarity of a phosphate monoester creates a severe limitation in uptake. We needed a more 'mobile' phosphate mimic, or lacking that, a masked phosphate (e.g. a hydrolyzable ester) which could be efficiently converted into an inhibitor. This issue has been resolved in numerous cases, primarily with pharmaceuticals. In EPSP, for example, a malonate group was found to be an effective phosphate mimic (*9*). Use of phosphonates as analogs of natural phosphates has been reviewed (*10*); newer examples include the trifluoromethylphosphonate esters (*11*), fluoromethylene phosphonates (*12*), and a phosphonomethoxy group (*13*). In addition, a sulfamate group is described as a viable phosphate replacement in Nucleocidin (*14*) and in the natural herbicide 5"-O-Sulfamoyltubercidin (*15*). It is important to note that a 'universal' replacement probably does not exist; there is not an optimal phosphate replacement group for every case.

The second question was whether we could demonstrate that a simplified derivative, consisting of a hydantoin linked to a phosphorylated ribose mimic, could inhibit the AdSS enzyme. A considerable body of research exists in which the ribose portion of several purine and pyrimidine nucleosides has been modified in some manner to generate new enzyme inhibitors with therapeutic and economic value. For example, acyclovir (9-(2-hydroxyethoxymethyl)guanine), an antiherpetic and antiviral compound, has what is simply an ethylene glycol unit attached to a guanine residue (*16*). Similar replacements have also been attempted with uracil analogs (*17*). In addition, replacement of the ring oxygen with carbon (*18,19*), or replacement of ribose with other heterocycles (*20*) has been reported to lead to biologically active nucleoside analogs.

With the literature ribose mimics in mind, a series of compounds was designed which linked a hydantoin to a phosphate group via a tether, which would function as the ribose group. The simplest tether would be an ethylene glycol unit, as was demonstrated to be sufficient for compounds such as acyclovir. A cyclic spirotetrahydrofuran analog was also targeted as a conformationally restricted analog. If these showed activity, further embellishments such as additional hydroxy groups or an additional ring could be incorporated at a later stage.

Results

Phosphate Replacements

The phosphate replacement question could be best answered by modification of hydantocidin itself. Hence, several potential phosphate mimics were selected for attachment to the 5' position of hydantocidin and biological evaluation.

The synthetic scheme of Mio, et al (*3*), leading to **1** proceeds through the fully protected intermediate **3** (Scheme 1). The only procedural difference was the modification of reduction conditions leading to **1**. It was found that transfer hydrogenation using cyclohexene and Pearlman's catalyst (wet Palladium hydroxide on carbon) was an effective and much more convenient lab scale debenzylation procedure than the reported high pressure hydrogenolytic removal of the 5'-benzyl group of **3**. Under both sets of conditions, a small quantity of epi-hydantocidin **7** was formed. These isomers are separable by normal phase chromatography. In addition, the acetonides **5** and **6**, required for some modifications, were not described in the literature; debenzylation of **3** under the conditions described above furnished the expected N-acetyl acetonide **5** along with deacylated acetonide **6**. These could be easily separated chromatographically.

Scheme 1

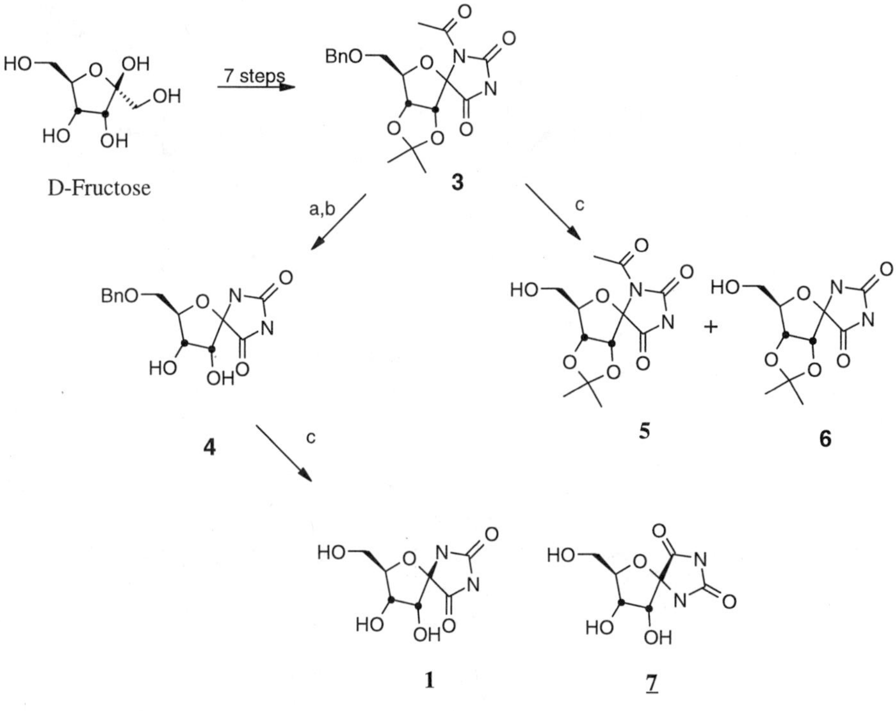

Reagents: a. Dowex, MeOH/H_2O; b. N_2H_4, MeOH; c. Cyclohexene, $Pd(OH)_2$/C

Chemistry leading to the preparation of 5'-phosphohydantocidin is described in Scheme 2. Treatment of **5** with N,N-diethyl dibenzylphosphoramide (DDPA) and tetrazole generated a dibenzyl phosphite ester (*21*), which was treated directly with potassium peroxymonosulfate (Oxone) to form the dibenzyl phosphate ester **8**. Attempted deprotection of **8** with Dowex resin in refluxing aqueous methanol resulted in removal of the acetonide and acetyl groups, as well as loss of one of the benzyl phosphate esters and nearly complete isomerization of the anomeric center to form **9**. Hydrogenolytic removal of the remaining benzyl group gave the epimeric 5'-phospho-N-α hydantocidin **10**. Alternatively, treatment of **1** with phosphoryl chloride in trimethyl phosphite (22) also generated a small quantity of **2** along with much larger quantity of the epimer **10**. Purification of these highly polar materials could be accomplished using ion-exchange chromatography.

Scheme 2

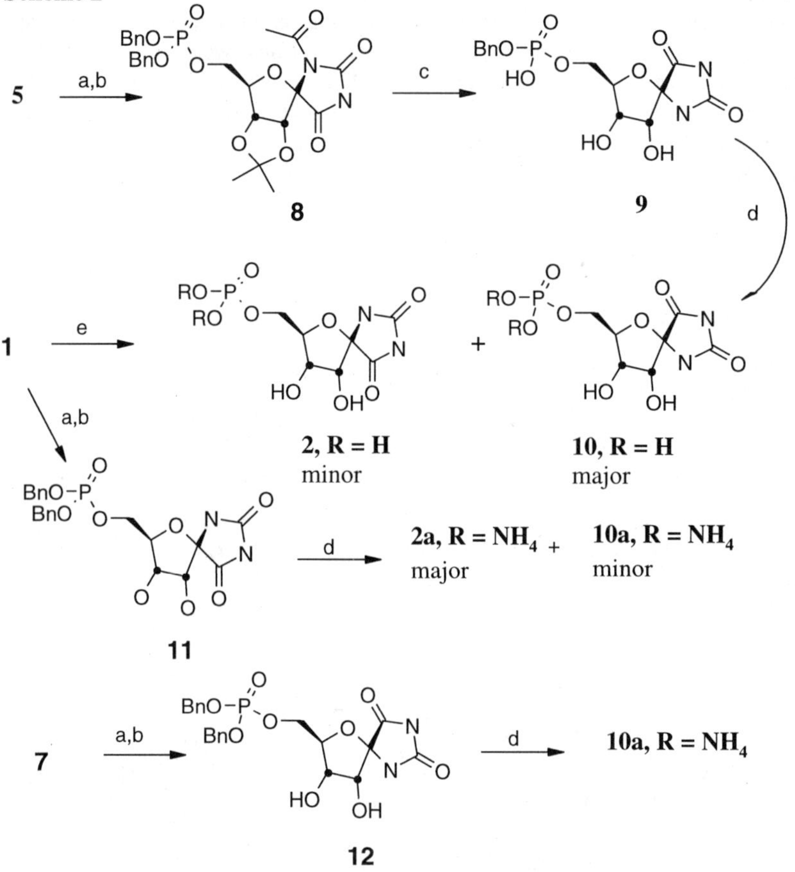

Reagents: a. $(BnO)_2PNEt_2$, Tetrazole; b. Oxone, H_2O; c. Dowex, $MeOH/H_2O$; d. H_2, $Pd(OH)_2/C$, NH_4OH; e. $POCl_3$, $(MeO)_3P$

The above experiments demonstrated that acidic conditions were to be avoided during the synthesis; equilibration at the anomeric center, which occurs slowly in hydantocidin, is apparently more rapid for phosphohydantocidin, and favors (ca. 6:1) the unnatural isomer **10**. When **1** was treated with DDPA as described above, the dibenzyl phosphate **11** was formed. No evidence for epimerization was seen at this stage. Finally, transfer hydrogenation with Pearlman's catalyst and aqueous ammonium carbonate in methanol yielded the ammonium salts of epimeric phosphate monoesters (**2a** and **10a**). Analysis of the ^{13}C spectrum now indicated the ratio of **2a** to **10a** to be ca. 3:1 (see experimental). The signals for the seven carbons of N-β hydantocidin are all between 0.1 and 3 PPM downfield of the corresponding signals of the N-α isomer (see TABLE I). This correlation also holds for 5'-phosphorylated derivatives, and provides a simple way for determining the anomeric configuration and purity of these derivatives. Acidification of the ammonium salt (via elution through acidic Dowex resin) resulted in equilibration to a ca. 4:1 ratio of N-α and N-β isomers. That the rearrangement product was indeed the β isomer was demonstrated by converting **7** to the 5'-phosphate monoester **10a** by the dibenzylphosphate route. The resulting phosphate monoester, stable to acidic conditions, was identical to that formed from isomerization of hydantocidin under acidic conditions.

TABLE I. ^{13}C Data for Hydantocidin and Phosphorylated Derivatives.

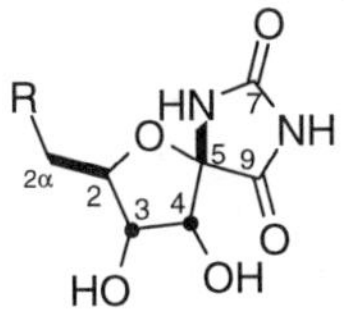

Cmp	R	anomer	C-2	C-2a	C-3	C-4	C-5	C-7	C-9
1	-H	N-β	89.7	63.9	73.7	75.8	97.3	159.5	177.6
7	-H	N-α	87.1	63.6	73.0	74.2	94.4	157.9	177.0
9	BnO(HO)P(=O)—	N-α	86.7	68.0	72.8	73.8	94.6	158.2	174.8
2a	$(H_4N^+\text{-}O)_2$—P(=O)—	N-β	88.8	67	74.3	75.9	97.9	159.0	177.5
10a	$(H_4N^+\text{-}O)_2$—P(=O)—	N-α	86.0	67	73.8	74.1	95.2	159.0	176.0
11	$(BnO)_2$—P(=O)—	N-β	86.3	69.3	72.8	75.5	97.0	159.0	176.0
12	$(BnO)_2$—P(=O)—	N-α	83.7	68.6	72.7	73.9	94.8	158.7	175.0

Preparation of other phosphorus and sulfur analogs of phosphohydantocidin is described in Scheme 3. An attempt to isolate the intermediate dibenzyl phosphite ester by omitting the oxidation step, instead resulted in hydrolysis and generation of diastereomeric (at phosphorus) H-phosphonates **13**. Indicative of this functional group was the large P-H coupling (630 Hz) observed in the ^{1}H NMR. This is the first reported instance of generation of an H-phosphonate from DDPA, although other routes have been described for preparation of these materials (*23,24*).

The phosphonate and sulfamate derivatives were prepared from the 2,3-acetonide **5**. Thus, treatment with sulfamoyl chloride and DBU, followed by removal of the acetonide under standard conditions, afforded the sulfamate **14** in 42% overall yield. Alternatively, **5** was converted to the bromide **15** with carbon tetrabromide and triphenyl phosphite. This bromide was further modified under Arbuzov conditions to generate diethyl phosphonate **16**. Complete deprotection of this ester (to **16a**) was unsuccessful; following removal of the acetonide under standard conditions, attempted hydrolysis of the ester with TMSBr led to intractable mixtures of esterified products. Alternatively, when treated with tris(trimethylsilyl) phosphite (*25*), clean conversion of **15** to a bis-(trimethylsilyl) phosphonate was observed. Treatment of this hydrolytically unstable intermediate with water resulted in removal of both the acetonide and the TMS groups and generation of the desired phosphonic acid **17**. Unfortunately, although ^{1}H NMR clearly indicated that this material was present, it could not be separated from the phosphorus acid generated along with it. The mixture was tested without separation.

Scheme 3

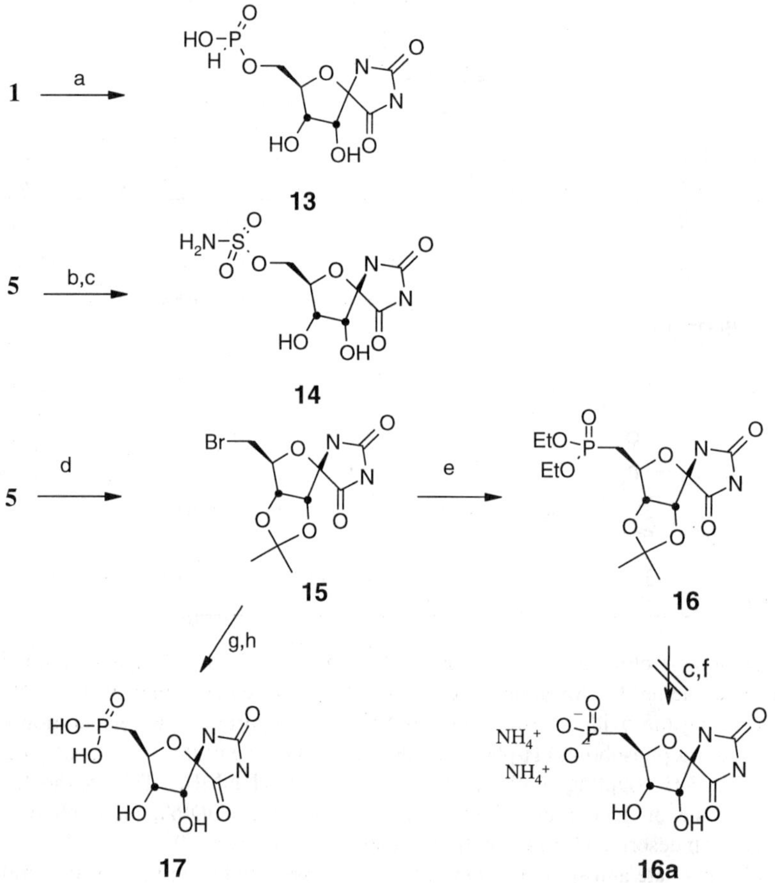

Reagents: a. DDPA, tetrazole, THF; b. $ClSO_2NH_2$, DBU/DMF, RT; c. Dowex, $MeOH/H_2O$; d. CBr_4, Ph_3P; e. $(EtO)_3P$; f. TMSBr, NH_4OH; g. $(TMSO)_3P$, heat; h. H_2O

Interestingly, compound **4**, the 5'-*O*-benzyl derivative prepared initially as a protected form of hydantocidin, showed whole plant activity (vide infra). This was most likely due to metabolism back to hydantocidin; therefore, the preparation of two other 5'-protected analogs was completed as described in Scheme 4. Acylation of **6** with p-chlorobenzoyl chloride and pyridine, followed by deprotection as described previously, generated 5'-benzoyl hydantocidin **18**. In similar fashion, the N-phenyl carbamate **19** was generated from p-chloroisocyanate.

Scheme 4

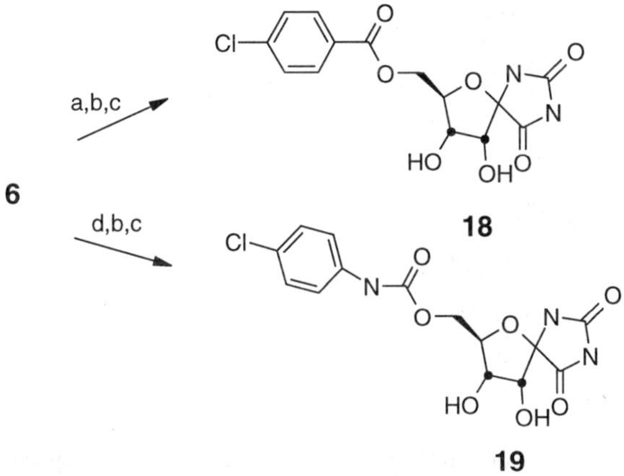

Reagents: a. p-ClPhCOCl, Et_3N; b. H_2NNH_2, MeOH; c. Dowex, $MeOH/H_2O$; d. p-ClPhNCO

Ribose Mimics

The preparation of the acyclic ribose analogs was fairly straightforward. Bis-benzyl hydantoin **22** could be prepared in two steps from bis-(p-methoxybenzyl) urea (26) and oxalyl chloride (Scheme 5). The trione intermediate **21** being reduced with sodium borohydride (*27*) to give the hydroxyhydantoin **22** in 80% overall yield.

The hydroxyl group of **22** is easily displaced with alcohols under conditions of Lewis acid catalysis. With excess ethylene glycol, the protected hydantoin **23** was obtained in 70% yield. Phosphorylation using the dibenzyl phosphate method (*vide supra*), cleanly generated the corresponding dibenzyl phosphate ester **24**. Removal of the PMB groups was accomplished with ceric ammonium nitrate (CAN) in acetonitrile/water in a 60% yield. Catalytic debenzylation then removed the remaining protecting groups, and afforded the phosphate **26** as its ammonium salt.

The acyclic carbon analog **29b** was prepared from 5-benzyloxy-2-pentanone **27** by using the Bucherer-Lieb method for converting carbonyl groups to hydantoins (*28*). Thus, treatment of **27** with potassium cyanide and ammonium carbonate afforded the hydantoin **28a** (Scheme 6). Debenzylation, phosphorylation and deprotection according to the conditions described previously generated target **29b.**

Scheme 5

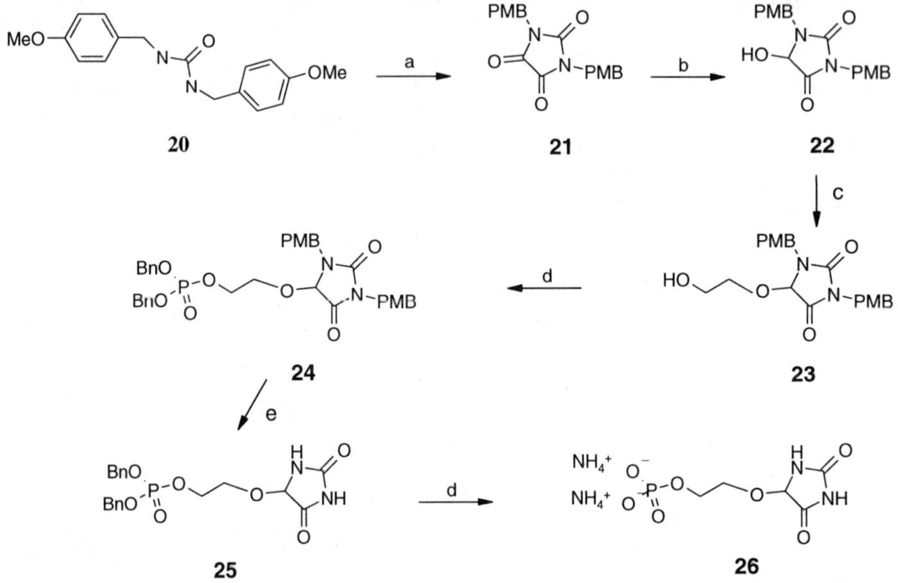

Reagents: a. Oxalyl chloride, dioxane, reflux; b. $NaBH_4$, MeOH; c. Ethylene glycol, TsOH; d. $(BnO)_2PNEt_2$, tetrazole, THF, then Oxone; e. CAN, $MeCN/H_2O$; f. Cyclohexene, $Pd(OH)_2/C$, $NH_4OH/MeOH$

Scheme 6

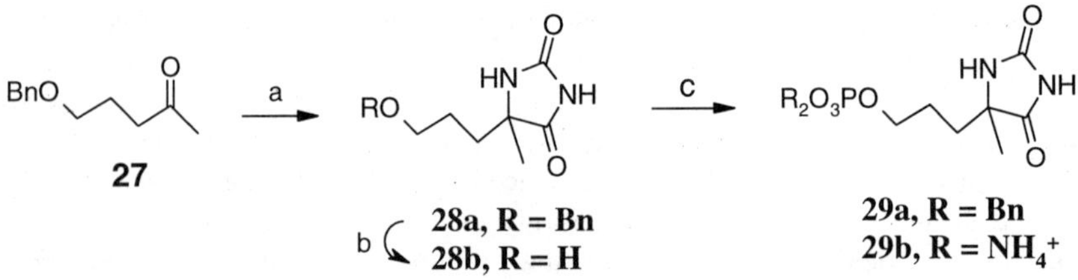

Reagents: a. KCN, $(NH_4)_2CO_3$; b. Cyclohexene, $Pd(OH)_2/C$, MeOH; c. DDPA, tetrazole, THF, then Oxone; d. $Pd(OH)_2/C$, $(NH_4)_2CO_3$, $MeOH/H_2O$

The cyclic example was prepared from the same trione **21** that was used above (Scheme 7). Addition of alkyl Grignard reagents to one of two identical amide carbonyl groups on the trione was found to yield alkylated hydroxy hydantoins. In contrast, alkyllithium reagents were not effective nucleophiles. Addition of 3-butenyl magnesium bromide generated the homoallylic alcohol **30**. Epoxidation of the double bond (to **31**) followed by acid-catalyzed rearrangement (Turner, J. A., DowElanco, unpublished data) generated the dideoxyribose analog **32** as a mixture of diastereomeric isomers. While these could be separated chromatographically, it was found ultimately that a rapid re-equilibration *via* ring-opening of the spiro-hydantoin occurred at a later stage. For our purposes, a mixture of isomers was considered adequate for determining the validity of the model, and thus further chemistry was conducted on the isomeric mixture. The remaining chemistry was essentially as described in Scheme 6; phosphorylation using the DDPA method (to form **33**), followed by sequential removal of the PMB and benzyl protecting groups, gave **35** as a diastereomeric mixture of ammonium salts.

Scheme 7

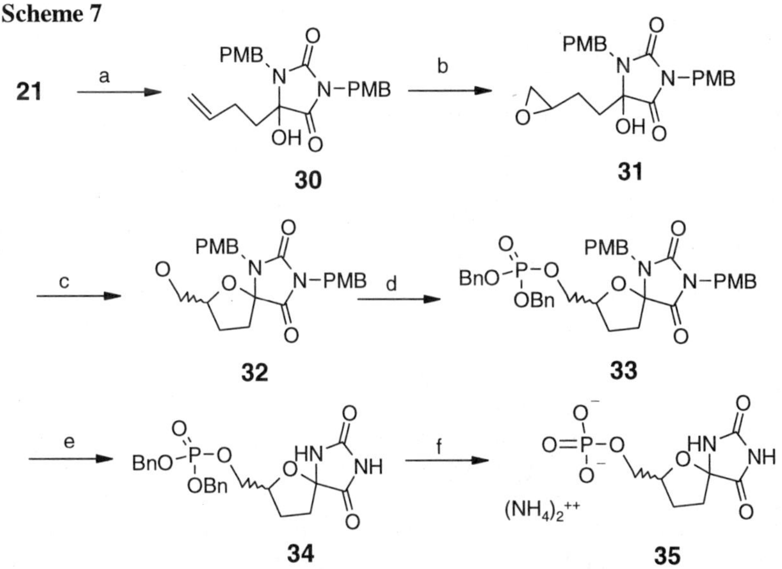

Reagents: a. 4-Bromo-2-butene, Mg, THF; b. MCPBA, CH_2Cl_2; c. TsOH, H_2O, CH_2Cl_2; d. DPPA, tetrazole, THF, then Oxone; e. CAN, $MeCN/H_2O$; f. Cyclohexene, $Pd(OH)_2/C$, $(NH_4)_2CO_3$, $MeOH/H_2O$.

Biological Activity of Hydantocidin Analogs

The aforementioned compounds were evaluated for *in vitro* activity against both plant-derived (maize) and animal-derived (muscle) AdSS enzyme, and in two *in vivo* systems, an *Arabidopsis* plate test (*6*) and whole plant. The *Arabidopsis* assay is a highly sensitive assay in which arabidopsis seeds are grown in an agar medium containing the test compound. In this test hydantocidin and its analogs characteristically display decreased root and shoot length which is reversed by addition of adenine to the agar medium, thus providing a rapid means of confirming the mode of action of new analogs.

Table II lists the structure and activity of the 5'-modified hydantocidin analogs. A large variation in whole plant, as well as enzyme activity, is observed, although the *in vitro* and *in vivo* activity were not at all correlated. Thus, despite a 100x improvement in activity in the *Arabidopsis* assay for the 5'-benzoyl derivative **18** over hydantocidin (**1**), the effects observed in the greenhouse were essentially indistinguishable. Furthermore, no 5'-substituent other than phosphate had any effect on the AdSS enzyme. This suggests that the plant active 5'-substituents are functioning primarily as carriers of the hydantocidin moiety through lipid layers. The carrier group is then metabolically removed and the hydantocidin nucleus is phosphorylated as before to generate the active species. The lack of activity of any of the substituted hydantoins against the enzyme indicates that in this particular system, the AdSS enzyme is highly specific for phosphate the 5' position. Note also that the epimeric (N-α) phosphate **10a** is considerably less active in the enzyme assay than the natural epimer. We have observed that, under laboratory conditions, this epimerization is facile and greatly favors the unnatural and less active isomer. This may be a limitation to activity of these compounds, or, it may provide an opportunity for enhanced activity if a means of preventing epimerization could be found.

One intriguing derivative is the 'phosphite' analog **13**. In spite of a highly polar functional group, which should have considerable difficulty crossing membrane barriers, activity in the *Arabidopsis* assay is quite strong. We have not determined whether this is due to non-specific hydrolysis of the phosphite back to hydantocidin, or whether this compound is in fact taken up and oxidatively converted into an enzyme-active species.

The structures and activity of the ribose replacement models are shown in Table III. As was anticipated based on the polarity of the phosphate group and presumed poor uptake, these compounds showed no whole plant activity. These analogs were all less active than 5'-phosphohydantocidin against the plant enzyme.

Our initial hypothesis suggested that the ribose ring was functioning as a complex tether, and that AdSS might reasonably be expected to recognize simpler phosphorylated hydantoin derivatives. We conclude from this limited set of analogs that enzyme-substrate interaction is more complex and highly specific, and that opportunities for structurally simpler hydantocidin analogs is not straightforward.

Summary

We have shown that simplified, phosphorylated analogs of hydantocidin have considerably diminished (<1%) activity against plant-derived AdSS. Thus, it appears that the AdSS enzyme is also quite specific for a ribose template. In addition, known phosphate mimics or precursors were found to be inactive against AdSS, an observation which further diminishes opportunities for synthetic modification. Even those analogs which appeared to greatly enhance uptake of this highly polar molecule did not, in greenhouse tests, provide a greater level of herbicidal activity than hydantocidin. Finally, the significantly diminished activity of the anomeric 5'-phosphohydantocidin again demonstrates the high degree of structural specificity for this enzyme. This may be one mechanism of deactivation, as the equilibrium between anomeric forms greatly favors the unnatural and relatively inactive epimer.

Table II. 5'-Substituted Hydantocidin Analogs

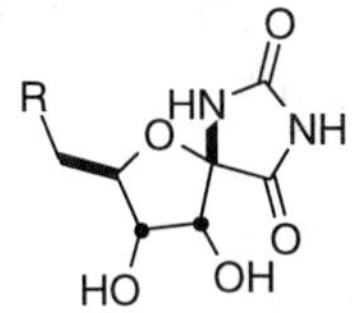

Compound #	R	Arabidopsis I_{50}[a] (PPM)	Greenhouse Post GR_{50}[b] (PPM)	AdSS I_{50} (PPM)[c]
1	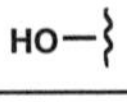	0.3	29	NA
4	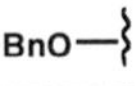	2.0	200	NA
13	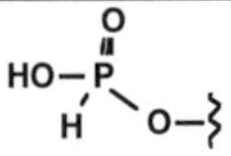	0.04	263	NA
11	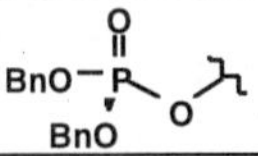	4.0	419	NA
2a	$(NH_4^+)_2$ 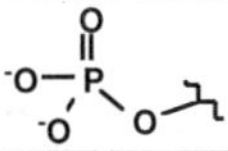	4.0	460	0.5[d] 0.06[e]
10a	$(NH_4^+)_2$ 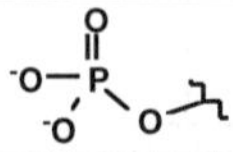 (epimer)	-	-	24[e]
14	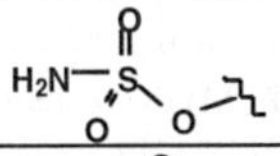	5.0	>4000	NA
16	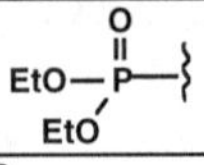	>50	>4000	NA
17	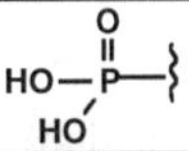 (+H_3PO_3)	>50	>4000	NA
18	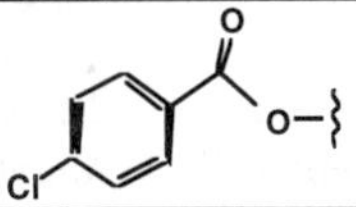	0.005	16.6	NA
19	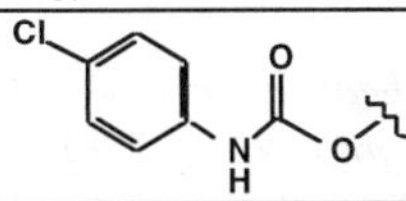	2.0	>4000	NA

[a]Arabidopsis I_{50} is the concentration that is required to cause a 50% reduction in root length. [b]Greenhouse post-emergence GR_{50} is the concentration required to cause an average 50% injury rating to eight grass and broadleaf weed species. [c]AdSS I_{50} is the concentration (in PPM) that is required for 50% inhibition of enzyme activity in an AdSS spectrophotometric assay. [d]Activity using maize-derived AdSS. [e]Activity using muscle-derived AdSS.

TABLE III. Comparison of Phosphorylated Hydantocidin Analogs.

Compound #	Compound Structure	Arabidopsis I_{50}[a] (PPM)	Greenhouse Post GR_{50}[b] (PPM)	AdSS I_{50} (PPM)[c]
2a	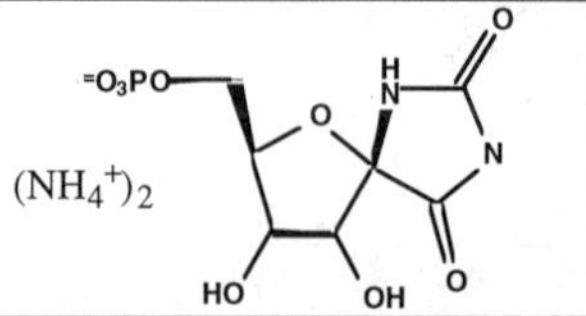 $=O_3PO$—, $(NH_4^+)_2$, HO, OH	4.0	250	0.5[d]
26	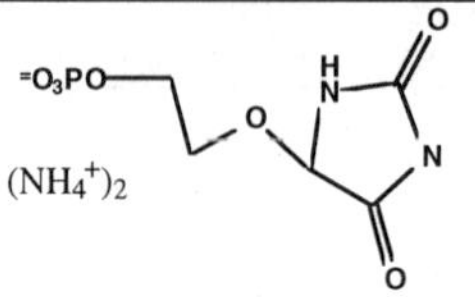 $=O_3PO$—, $(NH_4^+)_2$	>50	>4000	NA
29b	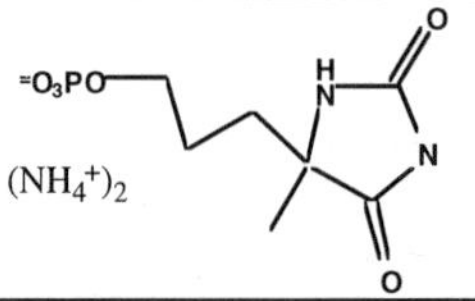 $=O_3PO$—, $(NH_4^+)_2$	>50	>4000	NA
35	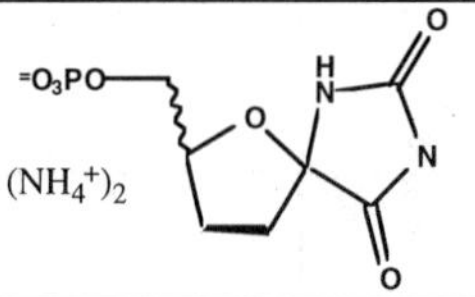 $=O_3PO$—, $(NH_4^+)_2$	>50	>4000	213

[a] Arabidopsis I_{50} is the concentration required to result in a 50% reduction in root length. [b] Greenhouse post-emergence GR_{50} is the concentration required to cause an average 50% injury rating to 8 grass and broadleaf weed species. [c] AdSS I_{50} is the concentration (in PPM) that is required for 50% inhibition of enzyme activity in an AdSS spectrophotometric assay. [d] Activity using maize-derived AdSS.

Literature Cited

1. Harayama, H.; Takayama, T.; Kinoshita, T.; Kondo, M.; Nakajuma, M.; Haneishi, T. *J. Chem. Soc., Perk. Trans. I* **1991**, 1638.
2. Sankyo Patent JP-B 2085 287, 1989. Isolation: Nakajima, M.; Itoi, K.; Takamatsu, Y.; Okazaki, T.; Kawakubo, K.; Shindo, M.; Honma, T.; Tohjigamori, M.; Hneishi, T. *J. Antibiotics*, **1991**, *44*, 293.
3. Mio, S.; Ichinose, R.; Goto, K.; Sugai, S. *Tetrahedron* **1991**, *47*, 2111; Mio, S.; Shiraishi, M.; Harayama, S.; Sato, S.; Sugai, S. *ibid* **1991**, *47*, 2121; Mio, S.; Kumagawa, Y.; Sugai, S. *ibid* **1991**, *47*, 2133; Mio, S.; Ueda, M.; Hamura, M.; Kitagawa, J.; Sugai, S. *ibid* **1991**, *47*, 2145.
4. Chemla, P. *Tetrahedron Letters* **1993**, *34*, (46), 7391.
5. Heim, D. R.; Cseke, C.; Gerwick, B. C.; Murdoch, M.G.; Green, S. B. *Pest. Biochem. Physiol.* **1995**, *53*, 138.
6. Cseke, C.; Gerwick, B. C.; Crouse, G. D.; Murdoch, M. G.; Green, S. B.; Heim, D. R. *Pest. Biochem. Physiol.* **1996**, *55*, 210.

7. Fonne-Pfister, R.; Chemla, P.; Ward, E.; Girardet, M.; Kreuz, K. E.; Honzatko, R. B.; Fromm, H. J.; Schaer, H. P.; Gruetter, M. G.; Cowan-Jacob, S. W. Proc. Natl. Acad. Sci. U.S.A. (1996), 93 (18), 9431.
8. Mio, S.; Hiromi, S.; Shindou, M.; Honma, T.; Sugai, S. *Agric. Biol. Chem.* **1991**, *55(4)*, 1105.
9. Corey, S. D.; Pansegrau, P. D.; Walker, M. C.; Sikorski, J. A. *Bioorganic and Medicinal Letters* **1993**, *3*, (12), 2857.
10. Engel, R. *Chem. Rev.* **1977**, *77*, (3), 349.
11. Blackburn, G. M.; Guo, M. J. *Tetrahedron Letters*, **1993**, *34*, (1), 149.
12. Chambers, R. D.; Jaouhari, R.; O'Hagan, D. *Tetrahedron*, **1989**, *45*, (16), 5101.
13. Harnden, M. R.; Jarvest, R. L.; Parratt, M. J. *J. Chem. Soc., Perk. Trans. I*, **1992**, *18*, 2259.
14. Iwata, M.; Sasaki, T.; Iwamatu, H.; Miyadoh, S.; Tachibana, K.; Matsumoto, K.; Shomuta, Y.; Sezaki, M.; Watanabe, T. *Meiji Seika Kenkyu Nenpo.* **1987**, *26*, 17.
15. Peterson, E. M.; Brownwell, J.; Vince, R. *J. Med. Chem.* **1992**, *35*, 3991 and references therein.
16. Matsumoto, H.; Kaneko, C.; Yamada, K.; Takeuchi, T.; Mori, T.; Mizuno, Y. *Chem. Pharm. Bull.* **1988**, *36*, (3), 1153.
17. Banijamali, A. R.; Foye, W. O. *J. Heterocyclic Chem.* **1986**, *23*, 1613.
18. Nicotra, F.; Panza, L.; Russo, G. *J. Org. Chem.* **1987**, *52*, 5627 and references therein.
19. Dyer, U. C.; Kishi, Y. *J. Org. Chem.* **1988**, *53*, 3384.
20. Coates, A. V.; Cammack, N.; Jenkinson, H. J.; Jowett, A. J.; Jowett, M. I.; Pearson, B. A.; Penn, C. R.; Rouse, P. L.; Viner, K. C.; Cameron, J. M. *Antimicrobial Agents and Chemotherapy* **1992**, *36*, (4), 733.
21. Perich, J. W.; Johns, R. B. *Tetrahedron Letters* **1987**, *28*, (1), 101.
22. Schulz, B. S.; Pfleiderer, W. *Helv. Chim. Acta.* **1970**, *70*, 210.
23. Schofield, J. A.; Todd, A. J. *J. Chem. Soc.* **1961**, 2316.
24. Gibbs, D. E.; Larsen C. *Synthesis* **1984**, *5*, 410.
25. Sekine, M.; Okimoto, K.; Yamada, K.; Hata, T. *J. Org. Chem.* **1981**, *46*, 2097.
26. Ulrich, H.; Sayish, A. *J. Org. Chem.* **1965**, *30*, 2781.
27. Liao, Z.; Kohm. H. *J. Org. Chem.* **1984**, *49*, 4745.
28. Tsang, J.; Schmeid, B.; Nyfeler, R.; Goodman, M. *J. Med. Chem.* **1984**, *27*, 1663.

Control of Insects and Acids

Chapter 14

Synthesis and Insecticidal Activity of *N*-(4-Pyridinyl and Pyrimidinyl)phenylacetamides

Peter L. Johnson, Ronald E. Hackler, Joel J. Sheets, Tom Worden, and James Gifford

Discovery Research Center, DowElanco, 9330 Zionsville Road, Indianapolis, IN 46268-1053

Pyridine (I) and pyrimidine (VII) amides have been found to exhibit broad spectrum insecticidal, acaricidal and nematicidal activity through inhibition of mitochondrial electron transport (MET) at site I. The activity of these compounds against *Heliothis virescens* larvae (tobacco budworm) could be optimized by varying the substituents on both the heterocyclic amine (R^1, R^2) and the phenyl ring (R^3). The discovery, synthesis and structure-activity relationship as it relates to tobacco budworm activity is described.

The control of insects through a novel or unexploited mode of action is a highly desirable goal within insecticide discovery research. Until recently, the control of insects through the inhibition of mitochondrial electron transport had seen very little commercial utilization (*1*). One of the oldest known MET inhibitors is rotenone, which is known for its insecticidal activity (*2*). Other MET inhibitors that have been introduced recently for the control of mites include Mitsubishi's tebufenpyrad (*3*) and DowElanco's fenazaquin (*4*) .

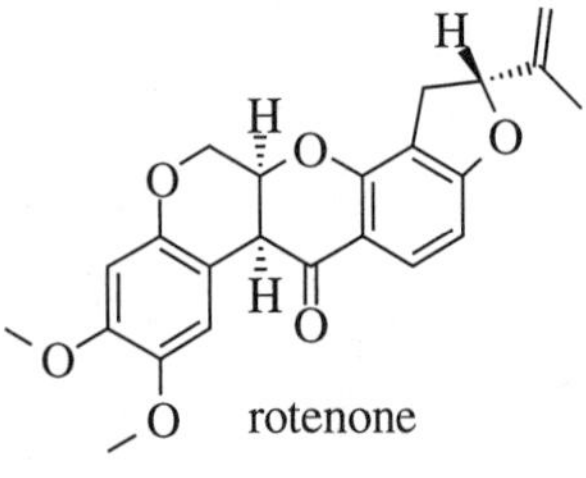

rotenone

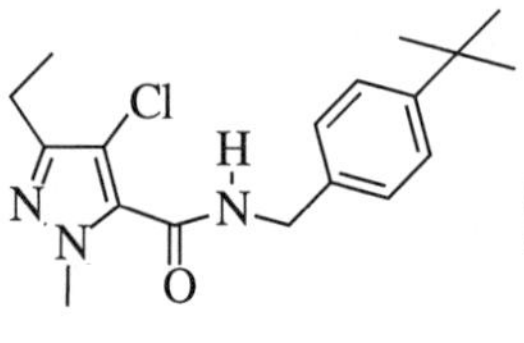

tebufenpyrad

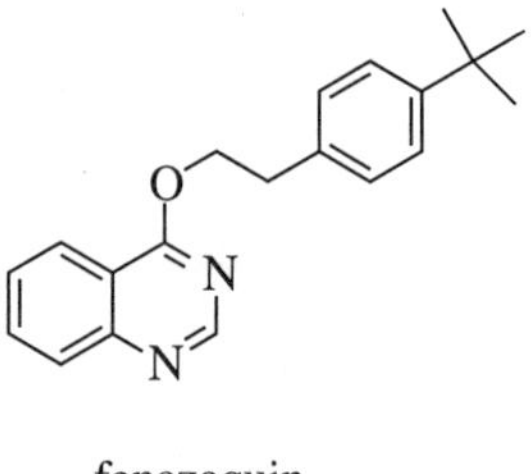

fenazaquin

Within DowElanco, we had been pursuing a series of quinolines (**III**) and quinazolines (**IV**) that showed broad spectrum activity against insects, mites, nematodes and fungi (*5-7*). In an effort to broaden the scope of this series, we began to look at heterocyclic replacements for both the quinoline and quinazoline ring systems. One replacement we looked at was the 2,3-disubstituted pyridine (**II**) as shown in Scheme I. In the synthesis of 2,3-disubstituted phenethylaminopyridine derivatives we were going through an amide intermediate (**I**). It was found that while the phenethylaminopyridines were showing good activity, the amide intermediates were even more active.

Scheme I

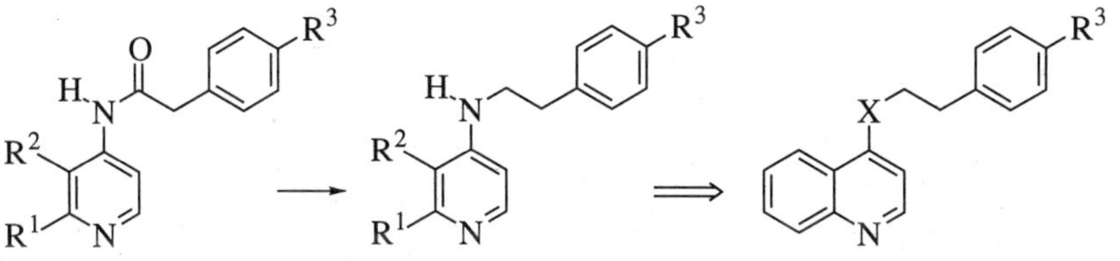

I **II** **III**, X = O or NH

As a replacement for the quinazoline ring (**IV**), we looked at a 5,6-disubstituted pyrimidine ring (Scheme II). While the 5,6-disubstituted phenethylaminopyrimidines (**V**) were in fact quite active as insecticides, they were also the subject of patents filed by both DuPont and Ube and were not pursued by us. Somewhat suprisingly, the 5,6-disubstituted pyrimidine amides (**VI**) were found to be inactive as insecticides.

Scheme II

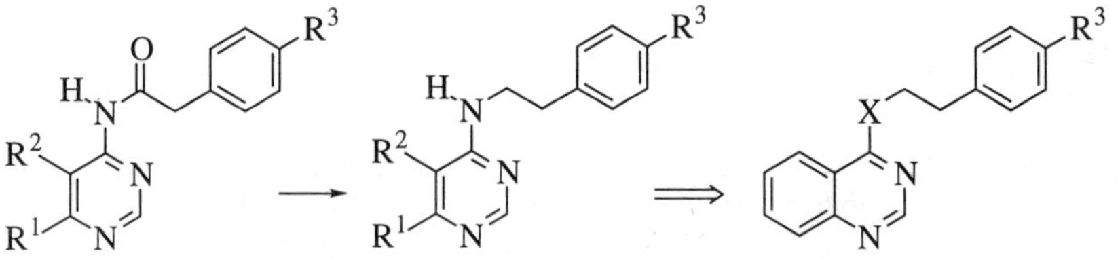

VI **V** **IV**, X = O or NH

We found that pyrimidine amides with substituents in the two position and no substitution in either the five or six position possessed good insecticidal activity (**VII**). In addition to the pyridine, **I** (*8,9*), and pyrimidine, **VII** (*10*) amides, the quinoline, **IX** (*11*), and isothiazole, **VIII** (*12*), amides were also investigated within DowElanco and showed broad spectrum insecticidal activity. This paper will focus only on the pyridine and pyrimidine amides.

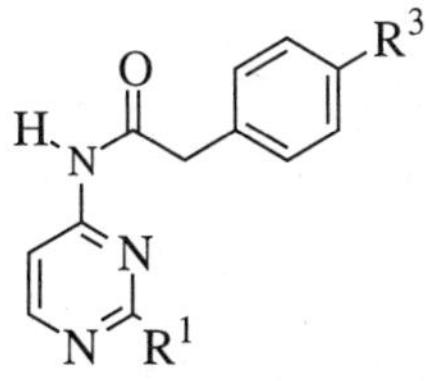

VII

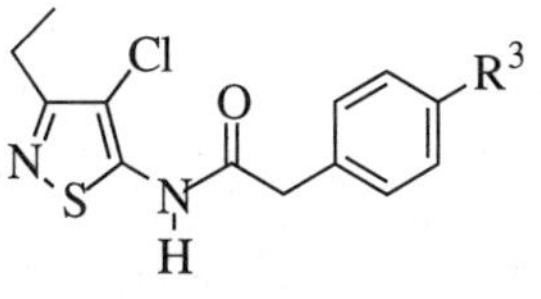

VIII

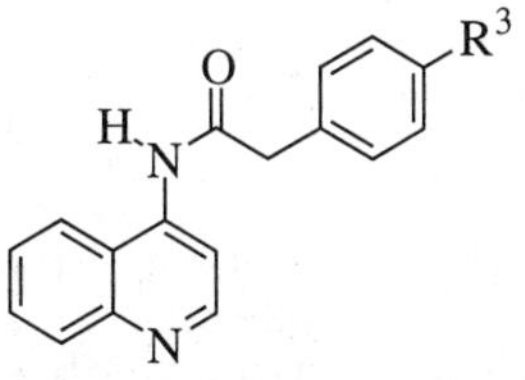

IX

Synthesis

The synthesis of the pyridine and pyrimidine amides can be broken down into two parts: the 4-amino pyridine or pyrimidine head portion of the molecule and the phenylacetic acid tail portion (equation 1).

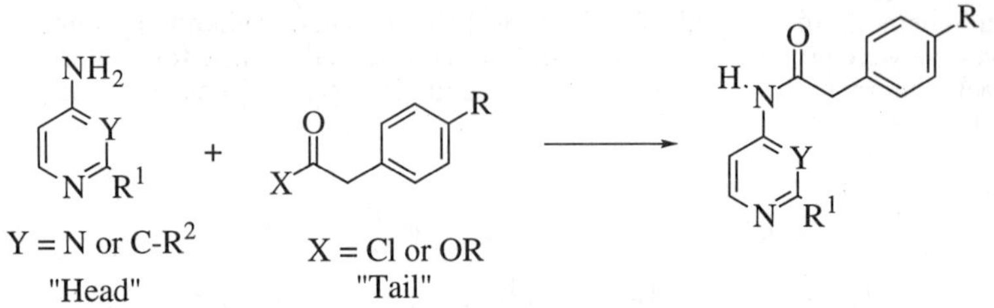

(1)

Phenylacetic Acids. There were three basic synthetic routes used for the synthesis of the phenylacetic acid "tail" portion of the molecule. The first route, and probably the most general, is outlined in Scheme III. This route involves classic chain extension methodology and starts by coupling an appropriate phenol or alcohol with 4-fluorobenzonitrile. The resultant benzonitrile is then hydrolyzed to the benzoic acid which in turn is reduced to the benzyl alcohol with lithium aluminum hydride. The alcohol is converted to the benzyl chloride with thionyl chloride. Displacement of the benzyl chloride with cyanide yields the phenylacetonitrile derivative which can then be hydrolyzed to the desired phenylacetic acid. While this route is quite laborious, each step is high yielding and very general.

Scheme III

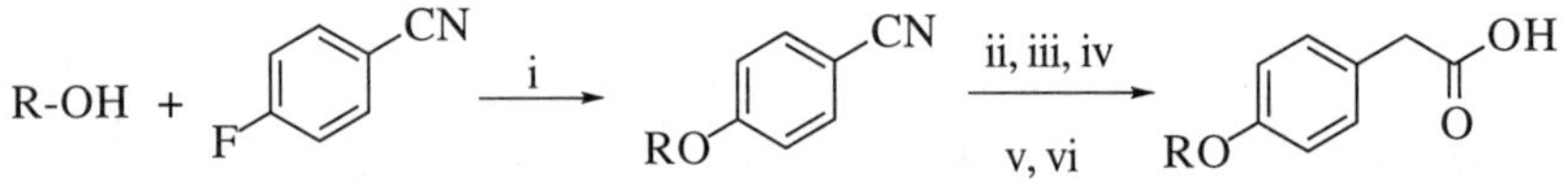

(i) NaH, DMF; (ii) NaOH, dioxane/H_2O; (iii) $LiAlH_4$, Et_2O; (iv) $SOCl_2$, toluene; (v) NaCN, DMSO; (vi) NaOH, dioxane/H_2O

The second route involves coupling either an alkoxide or phenoxide ion with 4-fluoro-acetophenone. Rearrangement of the resultant acetophenone in the presence of boron trifluoride etherate, lead tetraacetate and methanol leads to the corresponding methyl phenylacetate derivative in high yield, as shown in Scheme IV (*13*).

Scheme IV

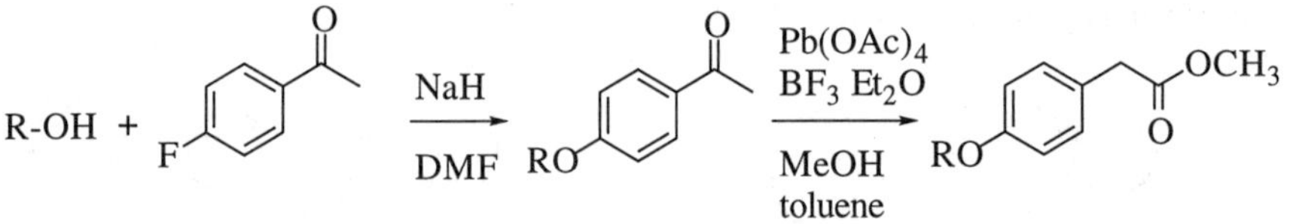

The third, and simplest, route to phenylacetic acids requires generating the dianion of 4-hydroxyphenylacetic acid. The dianion is then reacted with an activated arylfluoride in either DMF or DMSO to give the phenoxyphenylacetic acid in high yield (equation 2). The only drawback to this route is that it is limited to activated arylhalides.

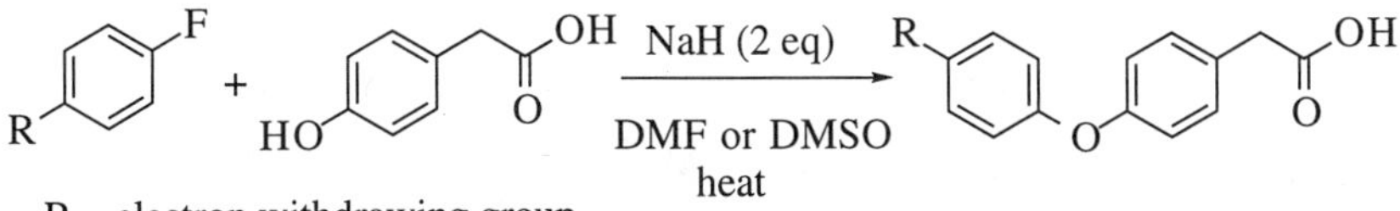

(2)

R = electron withdrawing group

4-Aminopyridines. The synthesis of 2,3-disubstituted-4-aminopyridines started with an appropriately substituted 2-alkylpyridine (Scheme V). The 2-alkylpyridine was oxidized with either hydrogen peroxide in acetic acid (*14,15*) or 3-chloroperoxybenzoic acid in dichloromethane to give the pyridine-*N*-oxide. The pyridine-*N*-oxide was then nitrated with nitric acid in sulfuric acid to give the 4-nitro-pyridine-*N*-oxide (*16*). Reduction of the nitropyridine-*N*-oxide with iron in acetic acid gave high yields of 2-alkyl-4-aminopyridines (*17*). The 4-aminopyridines could then be either chlorinated with chlorine gas in sulfuric acid (*18*) or brominated with hydrobromic acid in hydrogen peroxide (*19*) to give the 2-alkyl-3-halo-4-aminopyridines.

Scheme V

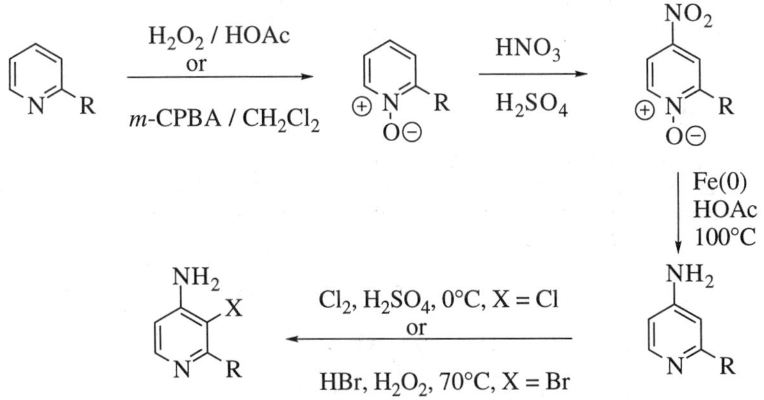

4-Aminopyrimidines. The synthesis of 2-alkyl-4-aminopyrimidines, Scheme VI, started with alkylnitriles which were transformed to imidates with hydrogen chloride in ethanol and then to the amidates by treatment with ammonia gas via the Pinner reaction (*20*). The amidates were then cyclized to 4-aminopyrimidines by first generating the free base with sodium methoxide followed by reaction with 3-ethoxyacrylonitrile (*21*).

Scheme VI

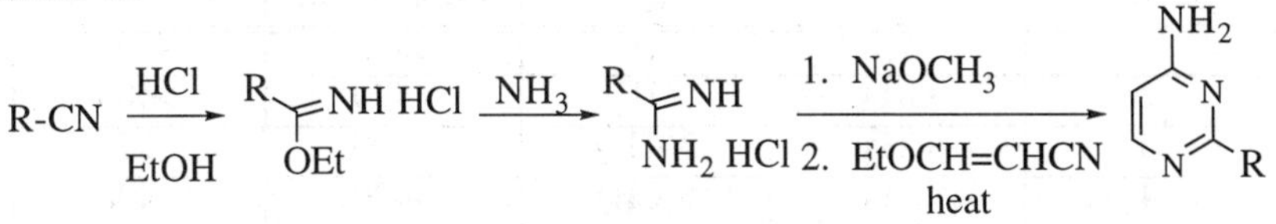

Amides. The reaction of 4-aminopyridines with acid chlorides or with carboxylic acids and DCC gave very poor yields of the desired amides. However, treatment of the 4-aminopyridines with trimethylaluminum followed by reaction with methyl phenylacetates gave good yields of the desired amides, equation 3 (22).

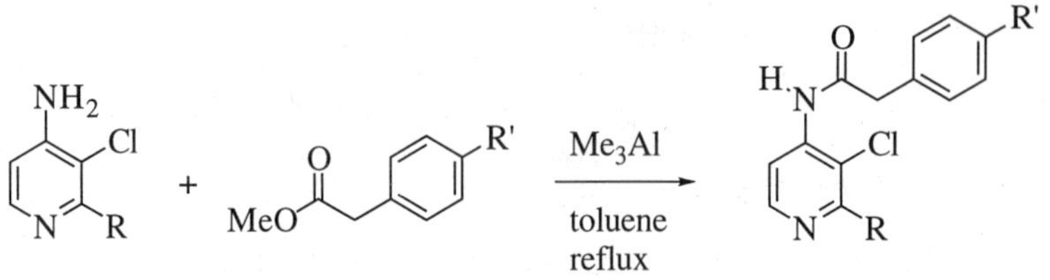

(3)

The reaction of 4-aminopyrimidines with acid chlorides in refluxing toluene gave moderate yields of the desired amides (equation 4). Attempts to improve the yield by treating the 4-aminopyrimidine with trimethylaluminum and a phenylacetate gave very poor yields of the amides.

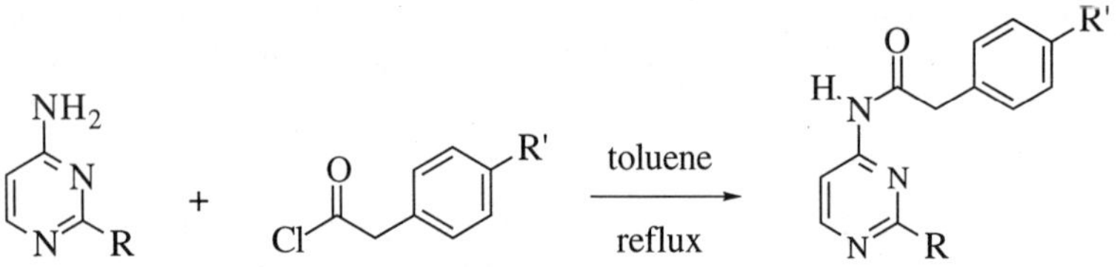

(4)

Structure-Activity Relationships

The structure-activity relationship (SAR) was driven by lepidoptera activity, in particular tobacco budworm. However, some of the early leads showed very weak or no activity against tobacco budworm and in these cases the activity against cotton aphids was used to guide the SAR.

Table I. Effect of Variations of the Pyridine Ring

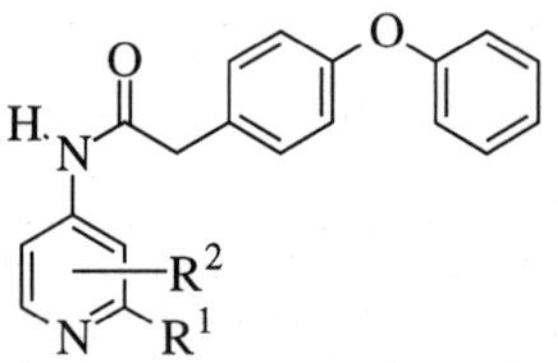

Entry	R^1	R^2	Tobacco Budworm LC_{50} (ppm)	Cotton Aphid LC_{50} (ppm)
1	H	H	>400	>400
2	CH_3	H	>400	>400
3	CH_3	3-CH_3	>400	27
4	CH_3	5-CH_3	>400	>400
5	CH_3	6-CH_3	>400	>400
6	CH_3	3-Cl	6.3	0.9
7	H	3-CH_3	>400	3.7
8	H	3-Cl	>400	0.2
9	CH_3	3,5-di-CH_3	>400	>400

The effects of variations on the pyridine ring on both tobacco budworm and cotton aphid activity are shown in Table I. Both the unsubstituted and the 2-methylpyridine derivatives were inactive (entries 1 and 2). The 2,3-dimethyl derivative had weak activity against cotton aphids whereas the 2,5- and 2,6-dimethyl derivatives were inactive (entries 3-5). With the 2-methyl-3-chloro- derivative we started to see good activity against both tobacco budworm and cotton aphids (entry 6). Substitution in only the three position with either a chlorine or methyl gave compounds that were active against aphids but lost activity against tobacco budworm (entries 7 and 8). A 2,3,5-trisubstituted analog, entry 9, proved to be inactive. From these initial results, it appeared that optimum activity could be achieved with 2-alkyl-3-halo substitution on the pyridine ring.

The SAR about the pyridine was probed further, this time looking at variations in only the 2 and 3 positions and with the 4-(4-chlorophenoxy)phenyl tail as shown in Table II. In going from a 2-methyl to a 2-ethyl, with a chlorine in the 3-position, there was about a six fold increase in the activity against tobacco budworm (entries 1 and 2). Extending the length to propyl resulted in a slight decrease in activity (entry 3). A 2-methoxy derivative, entry 4, was much less active. Variations in the 3-position were now looked at while holding the 2- position constant as an ethyl group. In comparing 3-chloro versus 3- bromo (entries 2 and 5) we saw very similar activity. As mentioned earlier, with a hydrogen in the 3-position the activity began to drop off (entry 6). A 2-ethyl-3-methyl derivative was prepared, entry 7, and found to be less active than the 2-ethyl-3-halo analogs. From the results in Tables I and II we felt that optimum activity was realized when the 2-position of the pyridine ring was substituted with an ethyl group and the 3-position was substituted with either a chlorine or bromine.

Table II. Variations in the 2- and 3-Position of the Pyridine Ring

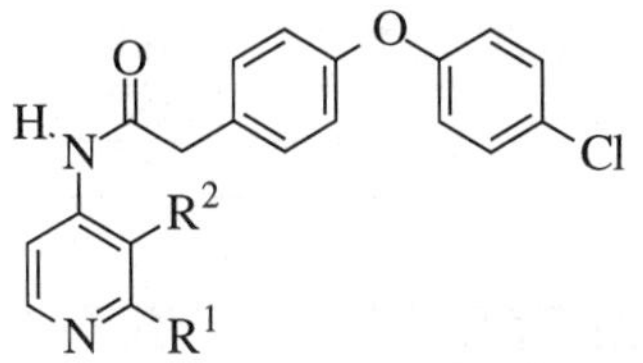

Entry	R^1	R^2	Tobacco Budworm LC_{50} (ppm)	Cotton Aphid LC_{50} (ppm)
1	CH_3	Cl	3.0	<0.1
2	CH_2CH_3	Cl	0.5	0.3
3	$CH_2CH_2CH_3$	Cl	3.1	<0.2
4	CH_3O	Cl	33	<50
5	CH_2CH_3	Br	<0.2	<0.2
6	CH_2CH_3	H	7.2	>400
7	CH_2CH_3	CH_3	3.0	40

A similar SAR was developed about the pyrimidine ring. In this case the 2-substituent was varied while holding the tail portion of the molecule constant as a 4-(4-fluorophenoxy)-phenyl acetamide (Table III). Also shown in Table III is the MET IC_{50} value, which is the concentration at which 50% reduction in electron transport activity was observed using a bovine heart mitochondria preparation. The IC_{50} values shown are in nanomolar concentrations. It was observed that an IC_{50} value of less than 10 nM was needed to see broad spectrum insect control in either of the amide series.

As was seen in the pyridine series, there was an increase in activity in going from hydrogen to methyl to ethyl (entries 1, 2 and 3). The insecticidal activity also began to fall off as the size of two substituent was increased to isopropyl and *n*-propyl, but note that the MET activity remained less that one nanomolar (entries 4 and 5). Somewhat surprisingly, the 2-*t*-butyl derivative, entry 6, was only about two fold less active than the 2-ethyl derivative. Substitution in the 2-position with either a *n*-butyl, methoxymethyl or phenyl group resulted in a substantial loss of activity (entries 7-9). From the results in Table III it appeared that optimum activity in the pyrimidine series was achieved when the 2-substituent was an ethyl group.

Table III. Effect of Variations on the Pyrimidine Ring

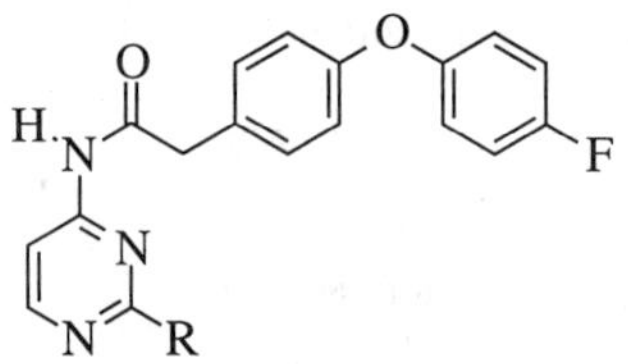

Entry	R	MET IC_{50} (nM)	Tobacco Budworm LC_{50} (ppm)
1	H	18	>400
2	Methyl	2.1	2.5
3	Ethyl	0.7	1.6
4	isoPropyl	0.7	9.8
5	*n*-Propyl	0.8	14
6	*t*-Butyl	1.0	3.8
7	*n*-Butyl	1.4	44
8	CH_2OCH_3	2.8	41
9	Phenyl	3.2	50

As far as optimizing the tail portion of the molecule, much of the SAR followed what had already been established in the quinoline/quinazoline and other amide MET inhibitor series. The basic requirements were a three atom spacer between the phenyl ring and the heterocyclic ring and *para*-substitution on the phenyl ring. Table IV shows what effect various substituents on the phenyl ring had on the tobacco budworm activity for the pyridine amide series.

Alkyl, alkoxy, fluoroalkoxy and phenyl substitution on the phenyl ring showed only moderate to weak activity against tobacco budworm (entries 1-5). There was a large increase in activity when the substituent was changed to 4-phenoxy (entry 6). Furthermore, substituting the phenoxy ring with an electron withdrawing group in the 4-position gave an additional increase in activity (entries 7 and 8), whereas electron donating groups resulted in a slight loss in activity (entry 9).

The pyrimidine amide series showed a very similar SAR in the tail portion of the molecule, with optimum activity being achieved with 4-phenoxyphenyl acetamides in which the phenoxy ring was substituted in the 4-position with an electron withdrawing group.

Table IV. Effect of Variations on the Phenyl Ring

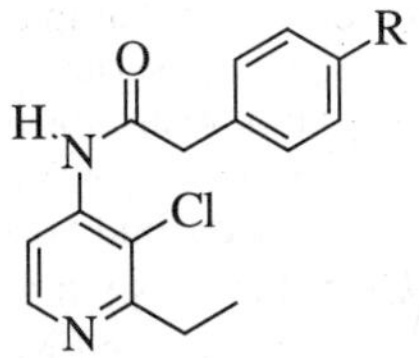

Entry	R	Tobacco Budworm LC_{50} (ppm)
1	*n*-Pentyl	15
2	*t*-Butyl	8.6
3	*n*-Bu-O-	50
4	CF_2HCF_2-O-	13
5	Ph	19
6	Ph-O-	0.4
7	4-Cl-Ph-O-	0.5
8	4-NO_2-Ph-O-	<0.2
9	4-CH_3O-Ph-O-	3.6

The linkage, or connectivity, between the heterocyclic ring and the phenyl ring also played an important role in the intrinsic activity. Table V shows what effect various linkages had on the tobacco budworm activity for the pyridine amide series. The most active linkage is the one in which the amide nitrogen is connected to the pyridine ring (entry 1). This compound showed very good activity against tobacco budworm and also very potent *in vivo* activity. If this amide linkage is reversed, that is the carbonyl carbon is attached to the pyridine ring, there is a dramatic loss in both the *in vitro* and *in vivo* activity (entry 2).

If the amide is replaced by an ester linkage as shown in entry 3, the compound is essentially inactive both *in vivo* and *in vitro*. Reduction of the amide linkage to give a phenethylamino analog resulted in a compound that maintained much of the activity of the parent amide, both *in vivo* and *in vitro* (entry 4).

Table V. Effect of the "Linkage"

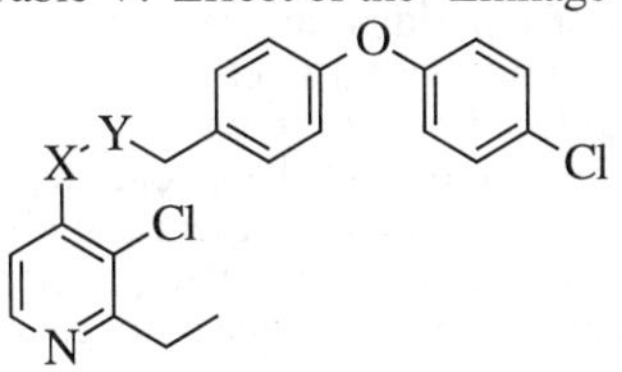

Entry	X	Y	MET IC_{50} (nM)	Tobacco Budworm LC_{50} (ppm)
1	N-H	C=O	1.4	0.5
2	C=O	N-H	40	>400
3	O	C=O	38	>400
4	N-H	CH_2	2.8	2.0

The pyridine and pyrimidine amides were not only active against tobacco budworm, but were also active against a broad spectrum of insects as well as mites and nematodes. Table VI lists representative examples, showing the broad spectrum activity as well as a comparison of the two series. The following abbreviations are used in Table VI: MET is the nanomolar concentration for 50% inhibition of mitochondrial electron transport; BAW is the LC_{50} in ppm against beet armyworm; TBW is the LC_{50} in ppm against tobacco budworm; TSSM is the LC_{50} in ppm against two spotted spider mite; CA is the LC_{50} in ppm against cotton aphid; ALH is the LC_{50} in ppm against aster leafhopper and NEM is the LC_{50} in ppm against root knot nematode. As can be seen from Table VI, all four analogs are very potent MET inhibitors with IC_{50}'s in the one nanomolar range. Against the Lepidoptera species, beet armyworm and tobacco budworm, the pyridine amides are more active than the corresponding pyrimidine amides. Against two spotted spider mites, cotton aphids and aster leafhopper the two series show similar activity. The pyridine amides series is once again more active than the corresponding pyrimidines against nematodes. While the pyridine amides were slightly better than the pyrimidine amides as far as overall activity, both series proved to be highly active and broad spectrum.

Table VI. Broad Spectrum Activity: Pyridine versus Pyrimidine

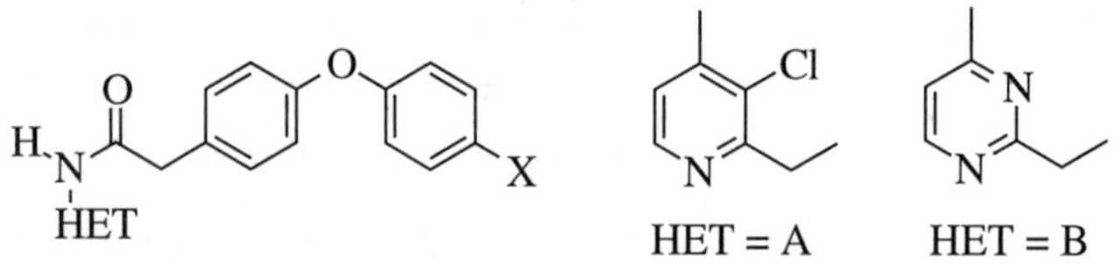

Entry	HET	X	MET	BAW	TBW	TSSM	CA	ALH	NEM
1	A	Cl	1.4	<0.2	0.5	<50	0.28	<0.8	0.5
2	B	Cl	0.9	50	5.1	<50	<0.2	0.7	6.2
3	A	CN	0.4	0.4	<0.2	<3.1	<0.2	<0.8	0.2
4	B	CN	1.0	1.6	0.7	<05	<0.2	<50	50

Metabolism Studies

Metabolism studies were performed on both the pyridine and pyrimidine amides. An example of the metabolism of a pyridine amide using rat liver microsomal protein is shown in equation 5. We found that rat liver microsomal proteins cleaved the amide linkage very rapidly (with a rate of two to three nmoles per minute per mg protein) to give the amino heterocycle and the phenylacetic acid. It was also of interest to find that this metabolism, or cleavage, occurs in the presence or absence of NADPH. When the same metabolism studies were carried out using either fish or tobacco budworm microsomes, very little cleavage of the amide linkage was observed. This lack of metabolism in fish and tobacco budworm and more importantly the lack of differentiation between the two may also be the reason that we observe very high rates of fish toxicity with the most active compounds in both of these series.

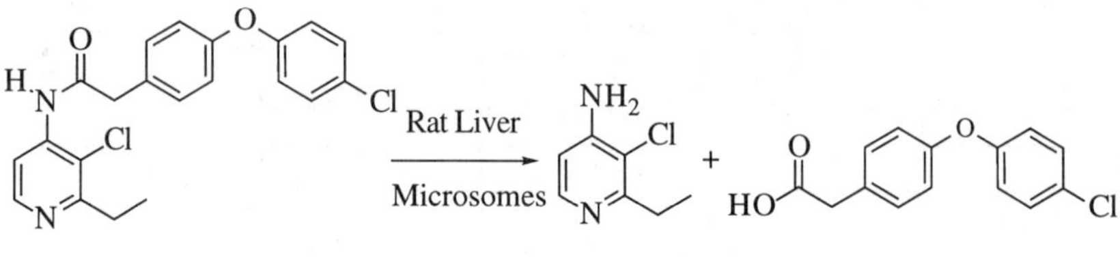

(5)

Conclusions

The pyridine and pyrimidine amides are highly active, broad spectrum insecticides, acaricides and nematicides. They exhibit their activity through the inhibition of mitochondrial electron transport at site I. The activity was optimized through variations of both the heterocyclic head portion of the molecule and the phenyl tail portion of the molecule. Optimum activity was achieved in the pyridine series when the pyridine ring was substituted with 2-ethyl-3-bromo or 3-chloro. Likewise the pyrimidine series was optimized when the substituent in the 2-position of the pyrimidine ring was an ethyl group.

Both series were further optimized when the tail portion of the molecule was a 4-substituted phenoxy derivative in which the phenoxy ring was additionally substituted in the 4-position with an electron withdrawing group.

Acknowledgments

The authors would like to acknowledge the following individuals: Bob Suhr, Bill Johnson, Jack Samaritoni and Brain Thoreen whose expertise in the chemistry of quinolines, quinazolines and amide MET inhibitors helped drive the SAR for both the pyridine and pyrimidine amides; Art Schmidt for generating the *in vitro* MET data; Mark Hertlein, Rod Herman, Larry Larson, Jon Babcock, Michelle Schlenz, Chris Hatton, Joe Schoonover, Jim Dripps, Lena Arndt, Bill Hendrix and Tim Bruce for generating the biological data.

Literature Cited

1. Fukami, J.- I. In *Approaches to New Leads for Insecticides*; von Keyserlingk, H. C., Jager, A. and von Szczepanski, Ch., Eds.; Springer-Verlag: Berlin, 1985; pp 47-69.
2. Fukami, H. and Nakajima, M. In *Naturally Occurring Insecticides*, Jacobson, M and Crosby, D. G., Eds.; Marcel Decker: New York, 1971; Chapter 2.
3. Okada, I., Okui, S., Takahashi, Y. and Fukuchi. T. *J. Pesticide Science*, **1991**, *16*(4), 623.
4. Dreikorn, B. A., Jourdan, G. P. and Suhr, R. G. U.S. Patent 5 411 963, 1995.
5. Dreikorn, B. A., Jourdan, G. P. and Suhr, R. G. U.S. Patent 5 294 622, 1994.
6. Coghlan, M. J., Dreikorn, B. A., Jourdan, G. P. and Suhr, R. G. U.S. Patent 5 296 484, 1994.
7. Dreikorn, B. A., Kaster, S. V., Kirby, N. V., Suhr, R. G. and Thoreen, B. R. U.S. Patent 5 326 766, 1994.
8. Hackler, R. E., Jourdan, G. P., Johnson, P. L., Thoreen, B. R. and Samaritoni, J.G. U.S. Patent 5 399 564, 1995.
9. Hackler, R. E., Jourdan, G. P., Johnson, P. L., Thoreen, B. R. and Samaritoni, J. G. U.S. Patent 5 597 836, 1997.
10. Johnson, P. L. U.S. Patent 5 556 859, 1996.
11. Thoreen, B. R., Samaritoni, J. G., Johnson, G. W., Davis, L. N., Gifford, J. M. and Sheets, J. J. "Synthesis and Insecticidal Activity of *N*-(4-Quinolinyl)phenyl-acetamides". Presented at the 212th National Meeting of the American Chemical Society, Orlando, FL, August 1996; paper AGRO 74.
12. Hackler, R. E., Johnson G. W. and Samaritoni, J. G. WO 95/31448, 1995.
13. Myrboh, B., Ila, H. and Junijappa, H. *Synthesis*, **1981**, (2), 126.
14. Taylor, E. C. and Crovetti, A. J. *Org. Synthesis Coll. Vol. IV*, **1963**, 704.
15. Ochiai, E. and Sai, *J. Pharm. Soc. Japan*, **1945**, *65*(B), 18.
16. Ochiai, E. *J. Org. Chem.*, **1953**, *18*, 534.
17. Hirayama, F., Konno, K., Shirahama, H. and Matsumoto, T. *Phytochemistry*, **1989**, *28*(4), 1133.

18. Ife, R. J., Dyke, C. A., Keeling, D. J., Meenan, E., Meeson, M. L., Parsons, M. E., Price, C. A., Theobald, C. J. and Underwood, A. H. *J. Med. Chem.*, **1989**, *32*, 1970.
19. Dunn, A. D., Currie, A. and Hayes, L. E. *Journal für Pratische Chemie,* **1989**, *331*(3), 369.
20. Dox, A. W. *Org. Synthesis, Coll. Vol. I,* **1941**, 5.
21. Singh, B. and Lesher, G. Y. *J. Heterocyclic Chem.*, **1977**, *14*(8), 1413.
22. Weinreb, S. M., Lipton, M. and Basha, A. *Tetrahedron Letters,* **1977**, (48), 4171.

Chapter 15

Development of Broad-Spectrum Insecticide Activity from a Miticide

Ronald E. Hackler, C. J. Hatton, M. B. Hertlein, Peter L. Johnson, J. M. Owen, J. M. Renga, Joel J. Sheets, T. C. Sparks, and R. G. Suhr

Discovery Research Center, DowElanco, 9330 Zionsville Road, Indianapolis, IN 46268-1053

A series of quinazolines which are insecticidal by virtue of their inhibition of mitochondrial electron transport were examined for their spectrum of activity against mites and several insects. It was shown that several factors could account for the differences between compounds which were primarily acaricidal and those which had broad-spectrum insecticidal activity. Although mitochondria could not be tested from each insect, there is some evidence of differences in the target sites between insects. A major factor in insect selectivity is shown to be differences in metabolism. Examination of penetration of the insect cuticle did not reveal useful data. Additional pharmocokinetic factors such as internal tissue partitioning, protein binding, or spiracle entry and insect behaviour were not considered in this analysis.

For synthesis chemists who seek to understand structure-activity relationships for *in vivo* data, selectivity among biological species can be a great mystery. Depending upon the biological targets, a wide variety of issues may affect the biological results. If the mode of action is understood and *in vitro* data are available, it is not uncommon for the *in vivo* data to diverge widely from the intrinsic data. This paper addresses one such scenario for one series of insecticides, and it is possible that it may have application to other series.

This publication is not intended to be a comprehensive review of the subject, but rather it is the results of our examination of selected issues, notably intrinsic activity, metabolism, and penetration as they affected the *in vivo* results against mites and tobacco budworm. We know that insect behavior also has a profound effect on selectivity. Whether a compound is active on sucking insects or chewing insects may depend upon distribution of the compound upon and within the plant. Where the insect feeds may make the difference in whether or not it is exposed to a toxic dose or even to no compound at all. We do not attempt to address these behavioral issues in this paper, but have restricted our discussion to limited observations which we feel shed some light on some common reasons for selectivity.

Fenazaquin is a commercial miticide marketed by DowElanco. References to the synthesis of fenazaquin may be found elsewhere (1). At about the same time as fenazaquin was being developed, several other miticides were reported that were subsequently determined to operate by the same mode of action as fenazaquin (2). This mode of action is the same as rotenone - inhibition at what is usually referred to as site I of the electron transport chain (NADH: ubiquinone oxidoreductase). All of the compounds shown (Figure 1) also exhibit a similar potency for inhibition of this metabolic process. Fenazaquin exhibits target site binding characteristics similar to rotenone as well.

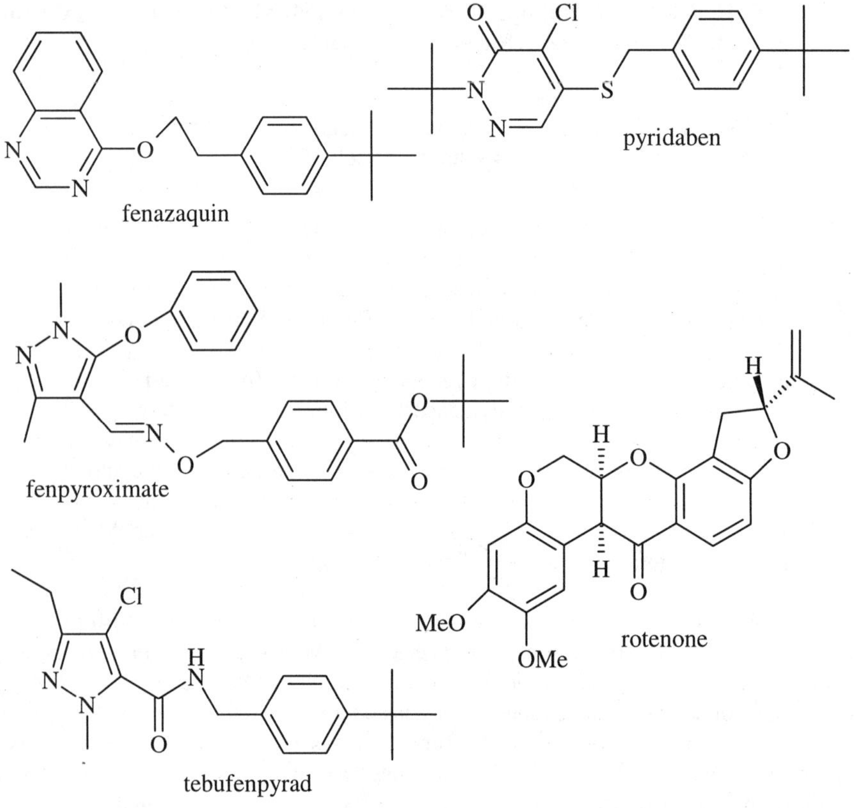

Figure 1. Inhibitors at Site I of the Electron Transport Chain.

Rotenone is as well known for its level of toxicity to fish as for its toxicity to insects. In our in-house tests it had an LC50 of about 10 ppb against Japanese carp, and exhibited even higher toxicity to trout. Fenazaquin is more selective against carp than rotenone, but fish toxicity was a serious issue with our series of quinazolines. In fact we examined several series of mitochondrial electron transport inhibitors and found many derivatives which demonstrated very low LC50's against trout and carp. The compounds which were most toxic to fish were the compounds which demonstrated a broader spectrum against insects. This paper will focus on two such compounds, XR-100 and Compound 11.

Synthesis

Because the previous paper (1) discusses the fenazaquin synthesis, we shall discuss only some of the unique aspects of the synthesis of XR-100 and 11. As shown in Figure 2, the XR-100 side chain is made by reaction of 4-bromobenzotrifluoride with the sodium salt of 4-bromophenol using cuprous chloride in pyridine. This Ullmann reaction would not work with the 4-chlorobenzotrifluoride, but a displacement in DMSO was successful. The diphenyl ether product is then lithiated and the aryl lithium used to open ethylene oxide to give the phenethyl alcohol, which is coupled with 4-chloroquinazoline to give XR-100.

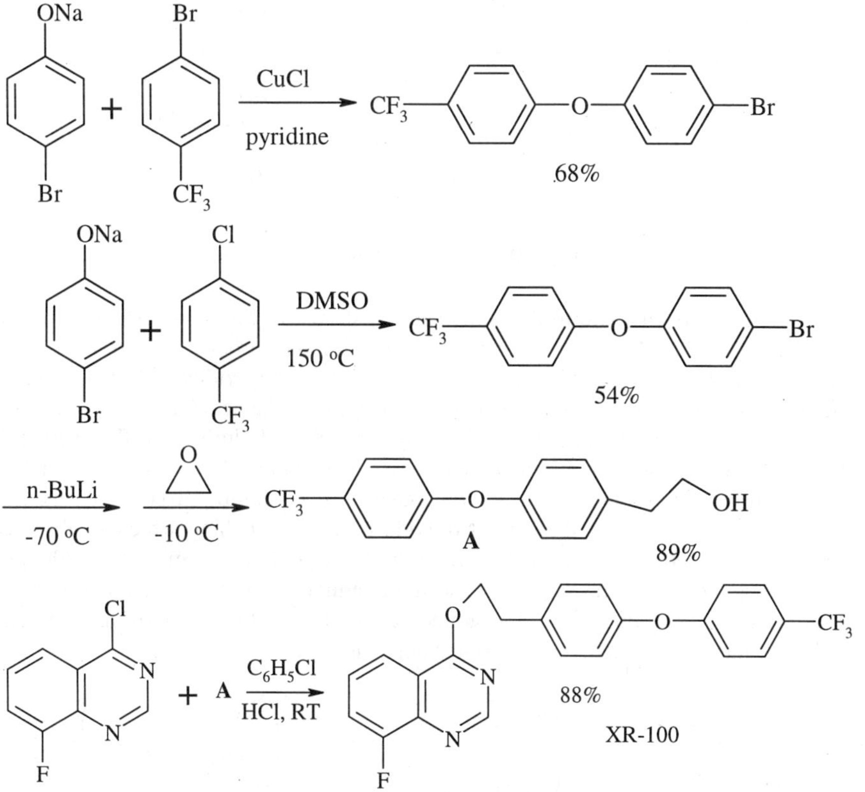

Figure 2. Synthesis of XR-100.

The side chain which is used for Compound 11 was originally made from the hydroxynicotinic acid by a series of classical steps shown in Figure 3. This is typical of the manner in which we made many phenethyl alcohols and phenethyl amines for this series. This chain is elongated through the benzyl alcohol, benzyl chloride, and benzyl cyanide. This was a fairly messy route which gave reasonable yields in each step, but the overall yield was only moderate. This route was used for many compounds starting from various commercial materials. The nitriles may also be hydrolyzed to the acids and these reduced to the phenethyl alcohols.

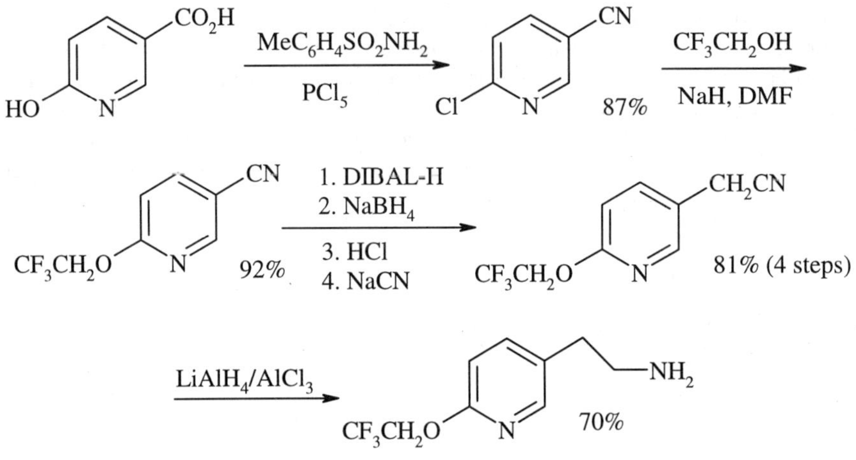

Figure 3. Original Route to Compound 11 Side Chain.

An improved route to this side chain is shown in Figure 4. The key to this route is a modified Heck reaction in which the additional carbons come from the allyl alcohol, and the pyridine unit is derived from 2,5-dibromopyridine. By formation of the oxime of the resulting butanone and rearrangement, the acetamide is made as the major product with a by-product of the propionamide. Hydrolysis of the acetamide gives the desired side chain which can be easily separated from the acid which is derived from the propionamide. We found this separation to be conveniently achieved by formation of the carbon dioxide adduct, something which is observed with many phenethyl amines. Exposure to the air for only a few minutes results in a solid being formed on the surface of the liquid amines. Compound 11 is formed in 90% yield by reaction of this amine with the 4-chloroquinazoline in 1,2-dichloroethane using triethylamine to scavenge the acid.

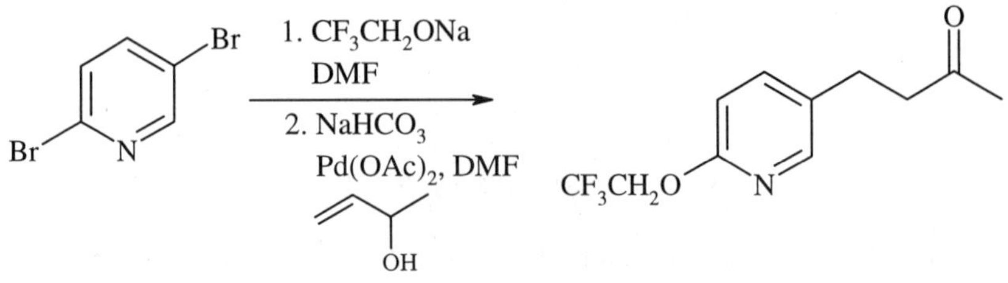

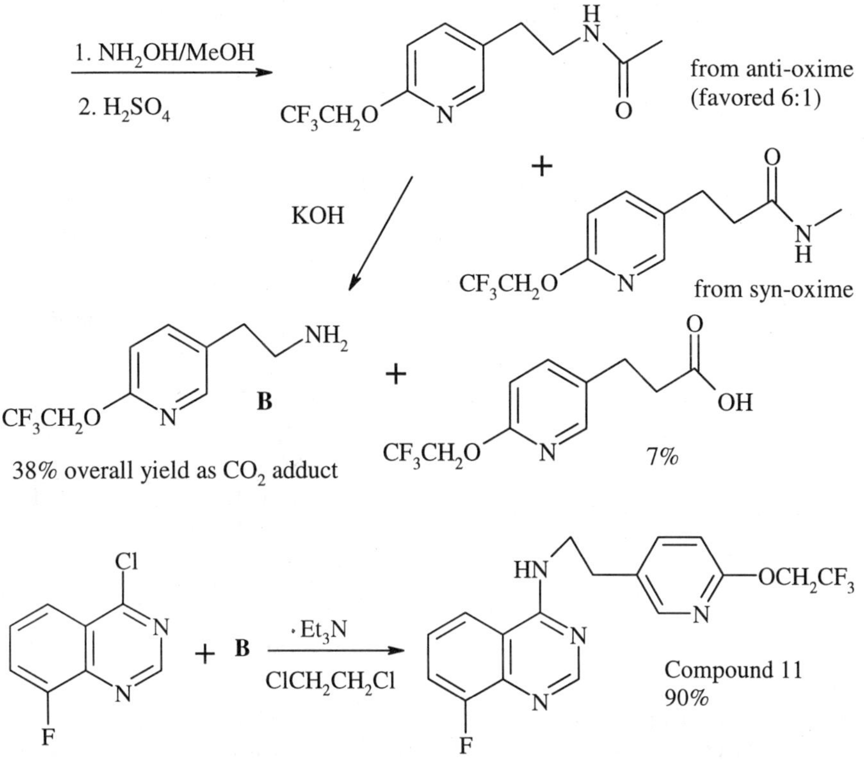

Figure 4. Synthesis of Compound 11.

Biological Activity

The commercial compounds shown previously are all primarily miticides and consist of a basic heterocycle, a bridge, and a phenyl group substituted with either a t-butyl or a t-butoxycarbonyl. Within our series of quinazolines, a t-butyl group on the phenyl ring was one of the best substituents for mite toxicity. In our tests rotenone has very poor mite toxicity with an LC50 more than 800 times that of fenazaquin. Table I shows LC50's for a group of quinazolines against two-spotted spider mite. These results were obtained by spraying squash plants which were previously infested with a mixed age population of two-spotted spider mites until the spray solution ran off the plants. A post-infest spray will give results typically three times greater, and higher LC50 values will be obtained on cotton or bean plants. The table illustrates that a variety of substituents impart miticidal activity to this series of molecules. Para-substitution on the phenyl ring is preferred, with ortho or meta substitution as in compounds 7-9 greatly diminishing activity.

There is an optimal length for alkyl or alkoxyl groups on the phenyl ring of about 5 to 6 atoms, illustrated with compound 5, and consistent with other site I inhibitors of mitochondrial electron transport such as the piericidins (3). A phenyl or

Table I. Two-Spotted Spider Mite LC50's for a series of quinazolines

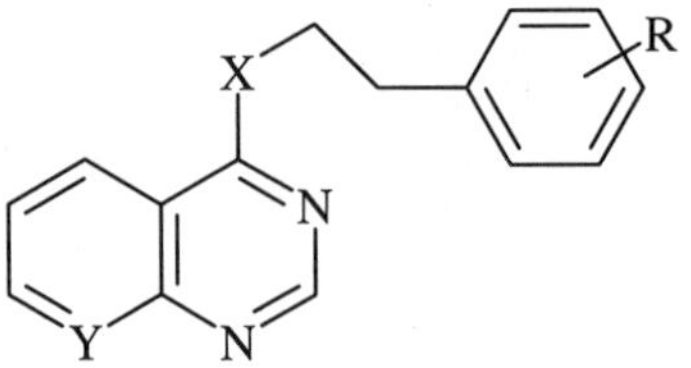

Compound	X	Y	R	TSSM (ppm)
fenazaquin	O	CH	4-t-Bu	1.0
XR-100	O	CF	4- O-C_6H_4-CF_3	< 1
2	O	CH	4-Ph	< 1
3	O	CH	4-cyclohexyl	< 0.5
4	O	CH	4-PhO	1.0
5	O	CF	4-n-pentyl	< 1
6	NH	CF	4-$CHF_2CF_2CH_2O$	1.0
7	O	CH	3-t-Bu	17
8	O	CH	3-Ph	> 400
9	O	CH	2-cyclohexyl	> 400
10	O	N	4- O-C_6H_4-CF_3	1.0

phenoxy in the para position maintains the activity as seen with compounds 2 and 4, and XR-100 emerged as a more potent miticide than fenazaquin. Although the screening data on squash for fenazaquin and XR-100 is similar, there is a larger difference between fenazaquin and XR-100 on other crops such as cotton and beans, and XR-100 has better residual activity than fenazaquin. Although the table does not reflect it, in general an 8-fluorine on the quinazoline results in enhanced activity. Nitrogen as the tether atom as in 6 is often as good as oxygen, and we believe the difference is best explained by log P. Thus at times the nitrogen may be better, and at times the oxygen may be better. The last compound in the table is included as a reminder that other heterocycles may substitute for the quinazoline (1), but to cover the breadth of our studies with various heterocycles and bridges is not feasible in this paper.

The enhanced miticidal activity of XR-100 was accompanied by broader spectrum activity, and that is what prompted our search for an understanding of the issues which contribute to this observation. Table II illustrates the difference in activity between fenazaquin, XR-100, and Compound 11. The latter compound was chosen because of its increased activity against lepidoptera. To simplify our discussion, the remainder of the paper will address these three compounds.

Intrinsic Activity

If the reader will return again to Table II, we shall focus on the *in vivo* activity of fenazaquin, XR-100, and Compound 11 against tobacco budworm (*Heliothis*

Table II. Insect and Nematode Activities for Selected Quinazolines.

Compound	CA	MOSQ	CL	NEM WELL	NEM	SCRW	TBW CON	TBW LF
fenazaquin	2.6	.725	188	$> 10^4$	200	> 400	> 400	> 400
XR-100	< .2	0.02	2.14	0.29	3.1	16.8	93	43
11	< .2	0.0003	0.47	0.28	5.0	200	55	1.9

CA = Cotton Aphid LC50 (ppm)
MOSQ = Mosquito Larvae LC50 in a well test (ppm)
CL = Cabbage Looper topical LC50 (μg/g)
NEM WELL = Free living nematode LC50 in a well test (ppm)
NEM = Root Knot Nematode LC50 in a sand/soil test (ppm)
SCRW = Southern Corn Rootworm LC50 in a soil test (ppm)
TBW CON = Tobacco Budworm LC50 in a petri dish contact test (ppm)
TBW LF = Tobacco Budworm LC50 in a leaf spray test (ppm)

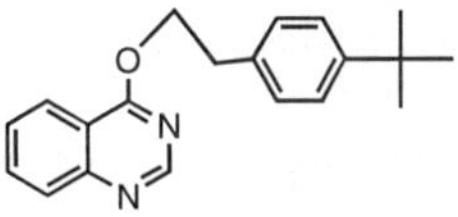

fenazaquin

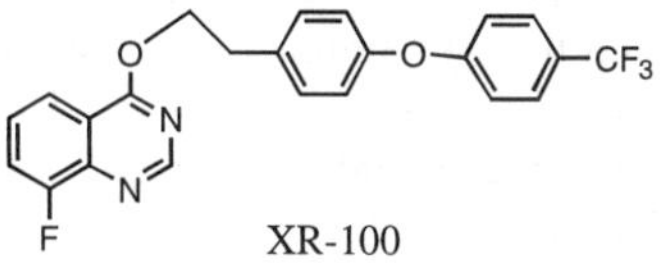

XR-100

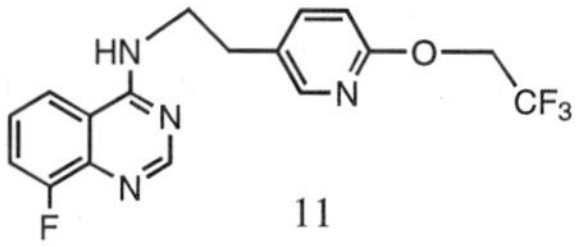

11

virescens). This species is easier to study than many other insects, and it also represents a major commercial market. Fenazaquin is nearly inactive against this organism, XR-100 is much more active, and Compound 11 is one of the most active quinazolines against this species. Let us focus first on the intrinsic activity of these molecules as it may relate to this *in vivo* data. Can there be a difference between the binding sites for different species which explains the selectivity which we observe?

It is not practical to isolate mitochondria from many insects, and in fact our initial measurements of inhibition of electron transport were done with bovine heart tissue. Later assays were developed using tissue from housefly thorax and a cabbage looper cell line. It was hoped that the housefly and cabbage looper assays would be more representative of insects in general, and the cabbage looper assay might better represent binding in lepidoptera.

It is impossible to review all of the data from these tests, so some general observations will be made. Examining data from hundreds of compounds led to the observation that the bovine heart tissue data correlated both with the housefly and the cabbage looper data, but there was no correlation between the housefly and the cabbage looper data. All of the data demonstrated some correlation with toxicity against tobacco budworm as shown in Table II. That is to say that the better the *in vivo* activity against tobacco budworm, the better the level of intrinsic activity in these assays. There was an especially good correlation between *in vivo* tobacco budworm activity and intrinsic data from the cabbage looper cell line. Data for our three selected compounds are shown in Table III.

A broader look including many compounds suggests that the picture is much more complicated than this, and there is not always a simple correlation between intrinsic activity and whole insect activity. High intrinsic activity, usually with an

Table III. Intrinsic assays for inhibition of mitochondrial electron transport as LC50's.

Compound	Bovine	Housefly	Cabbage Looper
fenazaquin	6.9 nM	1.0 nM	1.1 nM
XR-100	2.6	0.6	0.4
11	1.2	0.5	0.012

LC50 less than 10 nM, is necessary for good activity against any insect, but some compounds with high intrinsic activity were inactive in whole insect assays or showed a very narrow spectrum. Many more compounds were strictly acaricidal than were broadly insecticidal. The most active compounds against tobacco budworm or other lepidoptera also had high intrinsic activity as measured in tissue from the cabbage looper cell line. While the cabbage looper data does show some moderate correlation with activity against lepidoptera, it still explains less than 50% of the biological activity. This suggests that intrinsic activity has some ability to predict whole insect activity for specific insects, and thus, by implication differences between active sites in these species. In other words, the activity of Compound 11 is partly explained by a heightened specificity for the tobacco budworm active site, but we must look at additional factors for a more complete explanation.

Metabolism

We believed at a very early stage that metabolism was playing a major role in selectivity, and a series of studies were initiated which confirmed this suspicion. Studies were conducted with tobacco budworm (TBW) midgut microsomes, rat liver microsomes, and trout liver microsomes using radiolabeled fenazaquin, XR-100, and Compound 11. Dramatically different rates of metabolism for these compounds were observed in all three species as shown in Table IV.

Table IV. Rates of metabolism in nmoles/min/mg protein

Compound	TBW	Rat	Trout
fenazaquin	0.91	0.97	0.11
XR-100	0.03	0.14	< 0.01
11	0.07		

Fenazaquin appears to be metabolized at a rate approximately 30 times that of XR-100 using tobacco budworm microsomes. This difference appears to be sufficient to explain the marked difference in toxicity between these two compounds against tobacco budworm. Compound 11 is metabolized at about twice the rate of XR-100 in tobacco budworm, and its greater toxicity is thus better explained by the increased intrinsic activity indicated in Table III.

Selective mammalian toxicity is also indicated by the data in Table IV. Although we do not have rat LD50's for both compounds, there is a large difference in toxicity in mice, with XR-100 having an oral LD50 between 10 and 50 mg/kg, and fenazaquin having an oral LD50 of greater than 1,000 mg/kg. Thus it appears that metabolism at least partially may be responsible for the relative safety of fenazaquin to mammals. Finally, although the metabolism of fenazaquin is much slower in trout than in insects and rats, the relatively faster metabolism as compared with XR-100 is still apparent. In fact XR-100 is highly toxic to fish, having an LC50 nearly 100 times lower than fenazaquin against carp. The major metabolite of fenazaquin in all three systems is the result of hydroxylation of the t-butyl group. This observation is consistent with other analogs in this series. When there is a readily metabolizable group on the phenyl ring such as alkyl or alkoxy the insect spectrum tends to be narrow, but the mammalian and fish toxicity are also much less.

Additional support for diminished insecticidal activity of fenazaquin due to metabolism was obtained through results (Table V) observed on houseflies with and without the addition of piperonyl butoxide (PBO). Piperonyl butoxide is a well known synergist of many insecticides due to an inhibition of oxidative metabolism.

Table V. Effect of piperonyl butoxide (PBO) on activity in houseflies

Compound	Topical LC50 (μg/fly)	Topical LC50 with PBO	Synergism Ratio
fenazaquin	3.13	1.60	2.0
12	1.33	0.35	3.8
XR-100	0.04	0.04	1.0

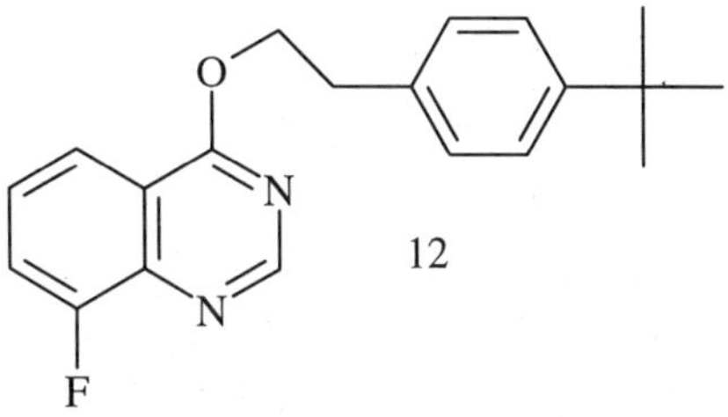

12

When fenazaquin was applied topically to houseflies along with piperonyl butoxide, the LC50 dropped nearly two-fold, apparently because metabolism was inhibited. The same thing was true of the 8-fluoro analog of fenazaquin, Compound 12. No such synergism was seen with XR-100, consistent with the observation of almost no detoxification through metabolism of XR-100 in tobacco budworm.

Penetration

The next thing we examined was penetration. This was done with several insects and using two basic methodologies. Radiolabeled samples were 1) applied topically to insects, or else 2) applied to a glass vial in which the insects were confined. Results are shown in Table VI. In every study performed, the penetration rate was inversely proportional to the activity. In other words, the least active molecule of our set of three, fenazaquin, penetrated to the greatest extent. Apparently the greater intrinsic activity and reduced metabolism of XR-100 and Compound 11 more than make up for this penetration deficiency in terms of overall insecticidal activity. In these tests, the pyrethroid insecticideal standard cypermethrin penetrates to a much greater extent than even fenazaquin, and contact activity remains the weak link for these compounds against *Heliothis*, presumably because of their generally poor penetration.

Table VI. Penetration Studies Given as Percentage of Compound Applied

Compound	% Penetration by topical application to TBW, 3 hr exposure	% Uptake from glass surface, TBW, 24 hr exposure	% Penetration by topical application, *Manduca sexta*, 3 hr exposure
fenazaquin	9.2	1.9	3.7
XR-100	5.7	1.3	2.3
11		0.34	1.1

What about the 8-fluorine substitution? As previously mentioned, the presence of the fluorine on the quinazoline in almost all cases enhances the activity. We have some evidence that the 8-fluorine enhances penetration, but it is not definitive. We have other evidence from QSAR studies that activity in this series of molecules is correlated to the difference between the highest occupied molecular orbital and the lowest unoccupied molecular orbital of the ring system. The 8-fluorine decreases this energy difference. There is also a correlation between activity and the dipole moment of the ring which is co-linear with the tether atom. The 8-fluorine increases this dipole moment, which is also in the right direction. There is no evidence that the 8-fluorine affects metabolism.

Summary

We have examined some of the factors which may contribute to a broadening of the insect spectrum in quinazoline inhibitors of mitochondrial electron transport. We believe that intrinsic activity against the target site of the specific insect has a significant role. However, once a molecule demonstrates a sufficient fit to the target site, metabolism is often the most important consideration for determining the insect spectrum. Metabolism also appears to be the determining factor in observed selectivity to mammals and fish. Because this mode of action is a general one, selective metabolism has played a major role in determining commercial candidates from this class of chemistry. Finally, penetration alone cannot be used to predict *in vivo* toxicity. Additional pharmocokinetic factors such as internal tissue partitioning, protein binding, insect behavior, or spiracle entry were not examined.

Literature Cited

1. R. E. Hackler, R. G. Suhr, J. J. Sheets, C. J. Hatton, P. L. Johnson, L. N. Davis, R. G. Edie, S. V. Kaster, G. P. Jourdan, J. L. Jackson, and E. V. Krumkalns, *Advances in the Chemistry of Insect Control III*, Royal Society of Chemistry, 1994, p 70-84.
2. R. M. Hollingworth, K. I. Ahammadsahib, G. Gadelhak, and J. L. McLaughlin, *Biochemical Society Transactions,* **1994**, *22*, 230-3.
3. K. H. Chung, K. Y. Cho, Y. Asami, N. Takahashi, and S. Yoshida, Z. *Naturforsch*, **1989**, *44c*, 609-616.

Chapter 16

Insecticidal 2-Aryl-5-haloalkylthio-, sulfinyl- and sulfonylpyrroles

K. D. Barnes, Y. Hu, R. E. Diehl, and V. M. Kamhi

Cyanamid Agricultural Research Center, American Cyanamid Company, P.O. Box 400, Princeton, NJ 08543–0400

This report describes the synthesis and insecticidal activity of a series of 2-aryl-5-haloalkylthio-,sulfinyl-and sulfonylpyrroles containing a variety of substituents on the 3- and 4-positions of the pyrrole ring.

Insecticidal pyrroles have been under investigation at American Cyanamid for a number of years (1-4). This work which was based on the insecticidal activity associated with the natural product lead dioxapyrrolomycin provided the broad spectrum insecticide/miticide AC 303630.

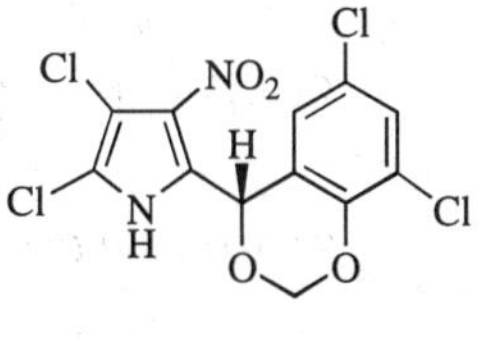

Dioxapyrrolomycin

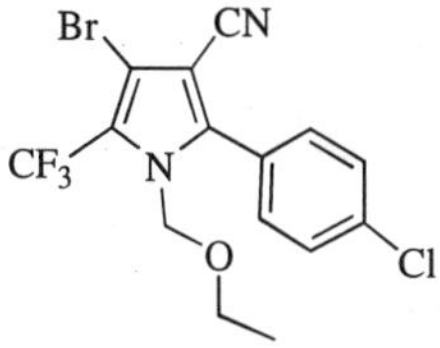

AC 303630

The structure activity relationships that developed during the course of this work suggested that lipophilic, strongly-electron withdrawing groups arrayed around the pyrrole nucleus are requiste for good insecticidal activity. Based on these structure activity relationships, studies were carried out in which the lipophilic strongly electron-withdrawing trifluoromethylsulfonyl functionality was introduced onto the pyrrole nucleus to afford a series of 2-aryl-3-trifluoromethylsulfonylpyrroles **1** (5). Many of the 2-aryl-3-trifluoromethylsulfonylpyrroles that were prepared demonstrated good insecticidal activity across several species of insects. As an extension of this work, the synthesis of a series of the isomeric 2-aryl-5-haloalkylthio-,sulfinyl and sulfonylpyrroles **2** was undertaken.

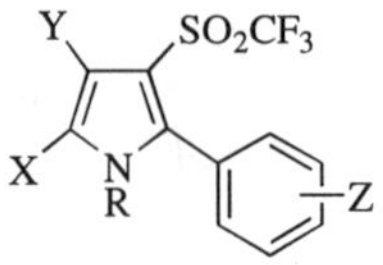

1

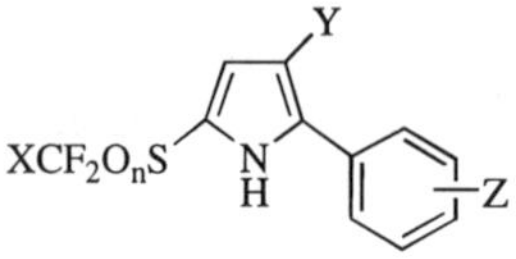

2

Chemistry

The synthetic approaches (Figure 1) toward the targeted molecules involved a variety of methodologies. These methodologies included the introduction of the haloalkythio functionality onto a preconstructed pyrrole nucleus as well as the construction of the pyrrole nucleus from acyclic precursors containing the trifluoromethylthio group. The majority of the 5-haloalkylthiopyrroles were prepared from preconstructed pyrroles. Starting with the appropriate 2-aryl-5-unsubstitued pyrroles **3** a number of targeted compounds were prepared utilizing trifluoromethylsulfenyl chloride or difluoromethylsulfenyl chloride. The use of haloalkylsulfenyl chlorides is a well documented method for the introduction of the haloalkylthio group onto the pyrrole nucleus. 2-Aryl pyrroles could be readily converted to their corresponding 5-thiocyanate derivatives **4**. The *in situ* generation of the 5-thiolate pyrrole *via* either hydrolysis or reduction followed by reaction with difluorochloromethane or a trifluorohaloethylene afforded the difluoromethylthio or 1,1,2-trifluoro-2-haloethylthio pyrroles respectively. Oxidation of the 5-haloalkylthio pyrroles afforded the corresponding 5-haloalkylsulfinyl or 5-haloalkylsulfonyl pyrroles.

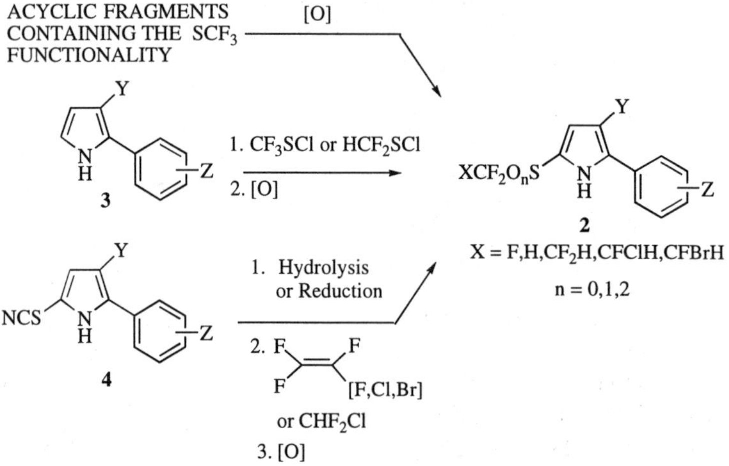

Figure 1. General synthetic approaches utilized.

2-Aryl-5-trifluoromethysulfenylpyrroles. The introduction of a trifluoromethylthio group onto a pyrrole nucleus *via* trifluoromethylsulfenylation with trifluoromethylsulfenyl chloride is a well known method for the preparation of trifluoromethylpyrroles. As shown in Figure 2 this methodology was applied to a variety of 2-aryl-3-substituted pyrroles, **5-8**. The syntheses of **5-8** have been reported previously (1-5).

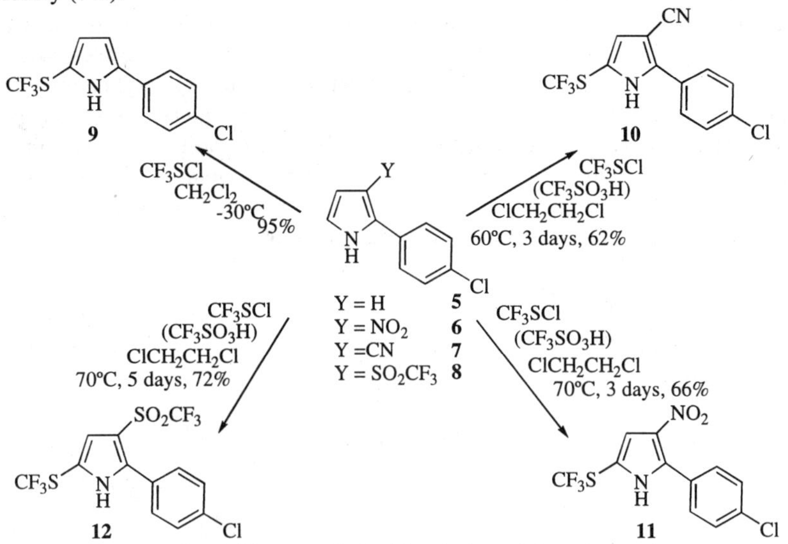

Figure 2. Trifluoromethylsulfenylation of 2-aryl pyrroles.

The ease of the sulfenylation is dependent on the electron withdrawing nature of the 3-substituent. The 3-unsubstituted compound **5** is readily trifluoromethylsulfenylated in high yield at -30°C affording **9**. The trifluoromethylsulfenylations of the 3-nitro **6**, 3-cyano **7** and 3-trifluoromethylsulfonyl **8** pyrroles all require reaction with the trifluoromethylsulfenyl chloride at elevated temperatures in a pressure tube with a catalytic amount of triflic acid, to afford the desired 5-trifluoromethylthiopyrroles. Without triflic acid the reactions proceeded sluggishly or failed. As expected, compound **8** being the most electron-deficient in this series because of the strong electron-withdrawing nature of the trifluoromethylsulfonyl group required the most forcing conditions to effect the trifluoromethylsulfenylation.

Bromination of pyrroles **10-12** occurred readily to afford the 4-bromo derivatives **13-15** as illustrated in Figure 3. Bromination of the 2-aryl-5-trifluoromethylthiopyrrole **9** could be controlled to afford either the 3-bromo or the 3,4-dibromo derivatives **16** and **17** respectively.

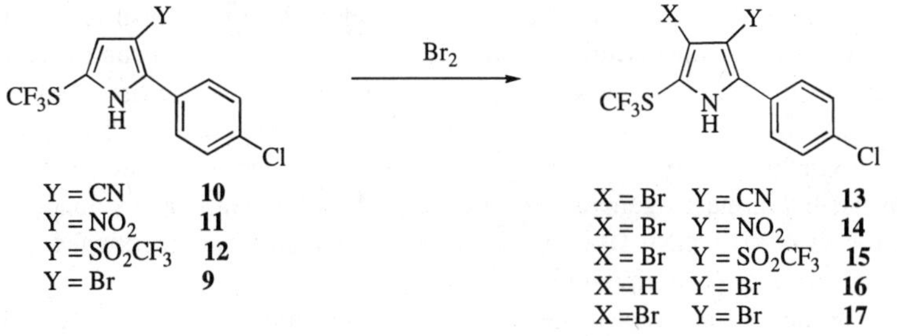

Figure 3. Bromination of 2-aryl-5-trifluoromethylsulfenylpyrroles.

As shown in Figure 4, compound **9** which was readily prepared by trifluoromethylsulfenylation of **5** in excellent yield could be converted to the 3,5-di-trifluoromethylthio derivative **18** by treatment with trifluoromethylsulfenyl chloride in the presence of triflic acid. Alternatively this material could be prepared in 94% yield in one-step from the 2-aryl pyrrole **5** by treatment with excess sulfenyl chloride in the presence of triflic acid. Conversion of **18** to the 4-chloro, bromo and nitro analogs was accomplished in moderate yields with sulfuryl chloride in acetic acid, bromine and nitric acid/acetic anhydride respectively. Whereas mono bromination and trifluoromethylsulfenylation of **9** afforded exclusively the 3-derivatives, nitration afforded the 3-nitro **11** and 4-nitro **22** analogs in 39% and 15% isolated yields respectively.

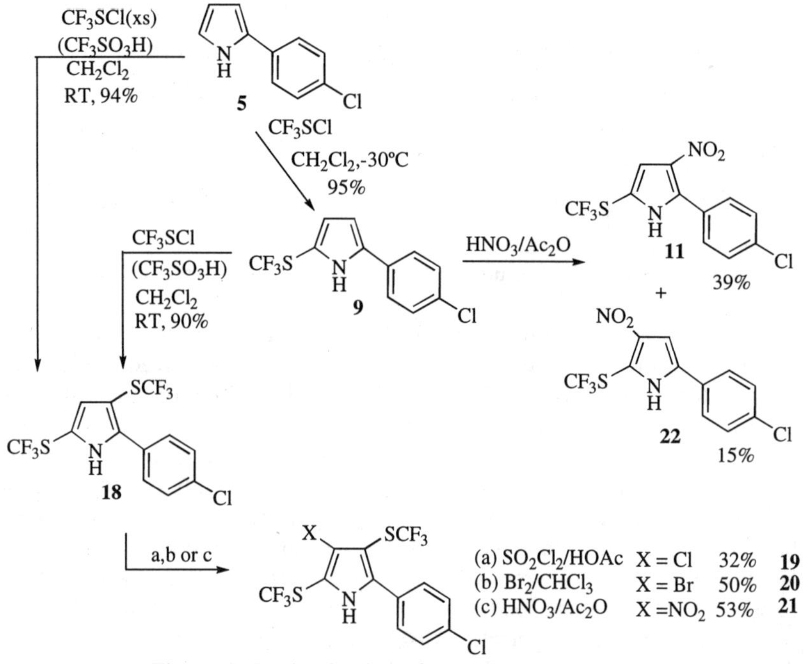

Figure 4. Derivatization of 2-(4-chlorophenyl)-5-trifluoromethylsulfenylpyrrole.

Figure 5 illustrates a novel method developed for the construction of the 2-aryl-3-cyano-5-trifluoromethylthiopyrrole **10** using an acyclic precursor containing the trifluoromethylthio functionality for construction of the pyrrole ring. Conversion of 4-chlorobenzyl amine **23** to the isonitrile **24** followed by reaction with trifluoromethylsulfenyl chloride gave the isothiocarbamoyl chloride **25**. Treatment with triethylamine to generate the nitrile ylid and 1,3-dipolar cycloaddition with 2-chloroacrylonitrile gave **10** in good yield. This material was identical to that obtained *via* trifluoromethylsulfenylation of **7**

Reaction of the nitrile ylid generated from **25** with 1-phenylsulfonyl-2-trifluoromethyl ethene **26** (Figure 6) afforded **27** and **28** as a 70:30 mixture. The 1,3-

dipolar cycloaddition of the nitrile ylid with 1-bromo-trifluoromethyl ethene **30** which was generated *in situ* from 1,2-dibromo-1-trifluoromethyl ethane, was regioselective affording exclusively **28**. As shown in Figure 7, **27** could also be prepared *via* trifluoromethylsulfenylation of **29**. The synthesis of **29** has been described elewhere (6).

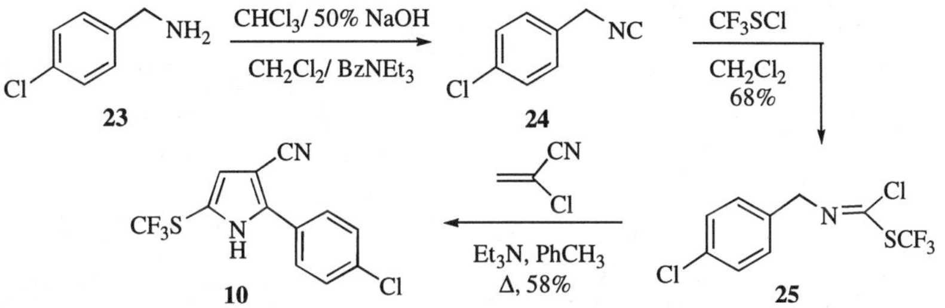

Figure 5. Preparation of **10** *via* 1,3-dipolar cycloaddition chemistry.

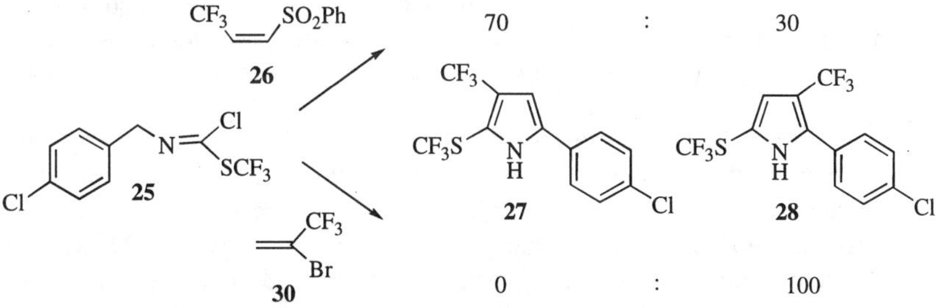

Figure 6. Preparations of **27** and **28** *via* 1,3-dipolar cycloaddition chemistry.

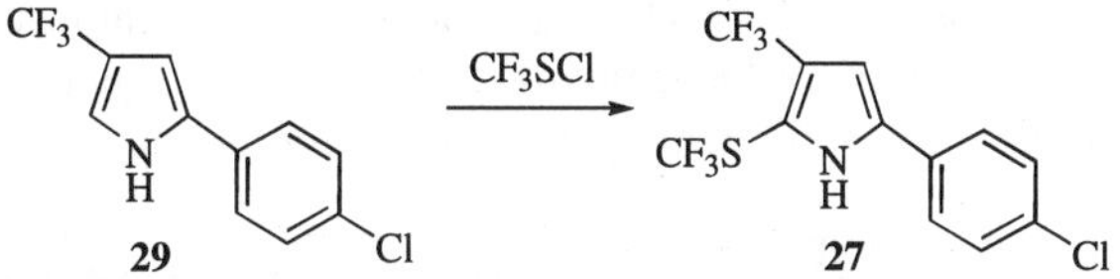

Figure 7. Preparation of **27** *via* trifluoromethylsulfenylation.

2-Aryl-5-trifluorohaloethysulfenylpyrroles. The trifluorohaloethylthio groups possess lipophilic and electronic properties similar to those of the trifluoromethylthio group. The rather straightforward introduction of the trifluorohaloethylthio functionalities onto the pyrrole nucleus *via* reaction of pyrrole thiocyanates with trifluorohaloethylenes as exemplified by examples in the literature prompted us to investigate the synthesis of potentially insecticidally active pyrroles containing these functionalities.

As shown in Figure 8, treatment of 2-(p-chlorophenyl)-3-cyano pyrrole **7** with potassium thiocyanate/bromine in methanol at -60°C afforded the 5-thiocyanate **31** in excellent yield. Hydrolysis of the thiocyanate with potassium hydroxide in a water-alcohol mixture to generate the 5-thiolate, followed by reaction with tetrafluoroethylene afforded the desired 5-tetrafluoroethylthio pyrrole **32** in yields of only 30-45% along with significant amounts of the 5-alkylthio pyrroles **33-35** corresponding to the alkyl alcohol used as the co-solvent. A mechanism for the formation of alkylthiopyrroles *via* reaction of alcohols with thiocyanates under basic conditions has been proposed by Olsen and Snyder (7). We found that using the more hindered 2-propanol as a solvent did not significantly decrease this unwanted side reaction.

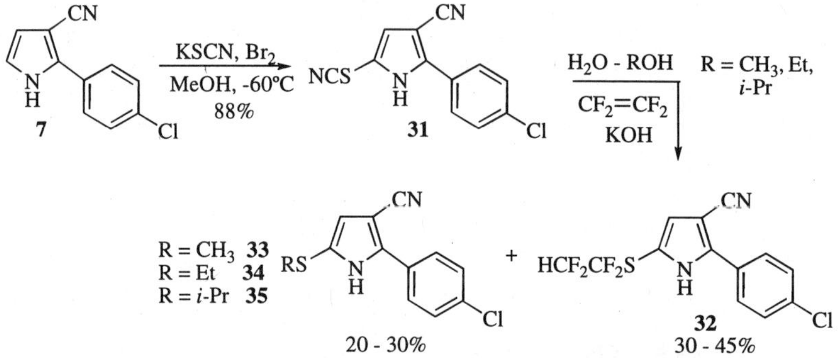

Figure 8. Introduction of the 5- tetrafluoroethylthio group *via* the 5-thiocyanate. Hydrolysis of the 5-thiocyanate and alkylation with tetrafluoroethylene.

The 5-thiolate could also be generated *in situ* as illustrated in Figure 9 by reduction of the thiocyante with sodium borohydride. When the reduction was conducted in methanol followed by treatment with potassium hydroxide/tetrafluoroethylene, the 5-tetrafluoroethylthio pyrrole **32** was isolated in 30% yield along with 8% of the 5-methylthio pyrrole **33**. However when the reduction was conducted in a mixture of tetrahydrofuran/2-propanol followed by reaction with the bromo and chloro trifluoroethylenes, the desired trifluorohaloethylthio pyrroles **36** and **37** were obtained in good yields. The addition of the thiolate occurs at the more highly fluorinated carbon of the trifluorohaloethylenes. Using this latter methodology no formation of the 5-isopropylthio pyrrole **35** was observed.

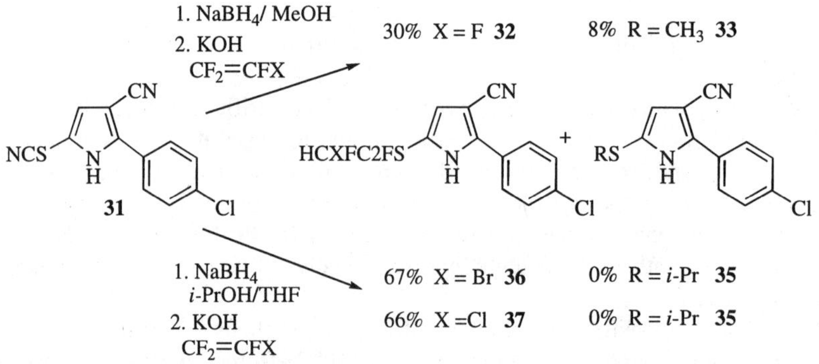

Figure 9.Introduction of the 5-trifluorohaloethylthio group *via* the 5-thiocyanate. Sodium borohydride reduction of the 5-thiocyanate and alkylations with trifluorohaloethylenes.

A number of 5-trifluorohaloethylthio pyrroles were prepared and derivatized as shown in Figure 10. The ease of the thiocyanation is dependent on the electron-withdrawing nature of the 3-substituent. The 3-nitro pyrrole **6** afforded the 5-thiocyanate in only 12% yield whereas the 3-unsubstituted compound **5** was thiocyanated under identical conditions in 99% yield.

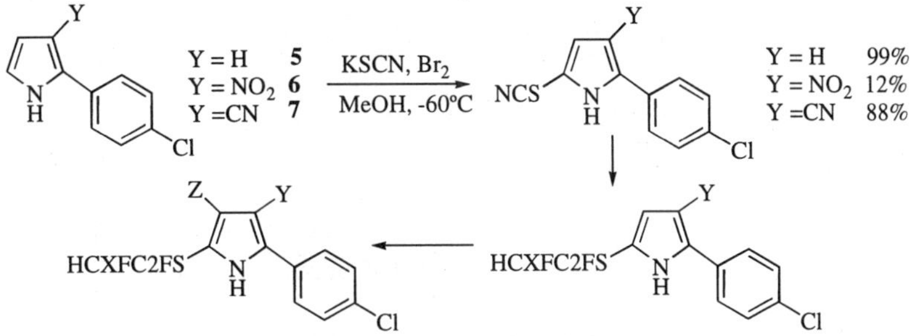

Y = CN, NO_2, Br
Z = Cl, Br

X = F, Cl, Br

Figure 10. Preparation and derivatization of 5-trifluorohaloethylsulfenylpyrroles.

2-Aryl-5-difluoromethysulfenylpyrroles. The preparation of difluoromethylthio analogs were also investigated because of the similarities of the electronic and lipophilic properties of this functionality to those of the trifluoromethylthio group. As illustrated in Figure 11, generation of the 5-thiolate from the thiocyanate **31** as described in the preceding section followed by reaction with difluorochloromethane gave the 5-difluoromethylthio pyrrole **38** in 52% yield.

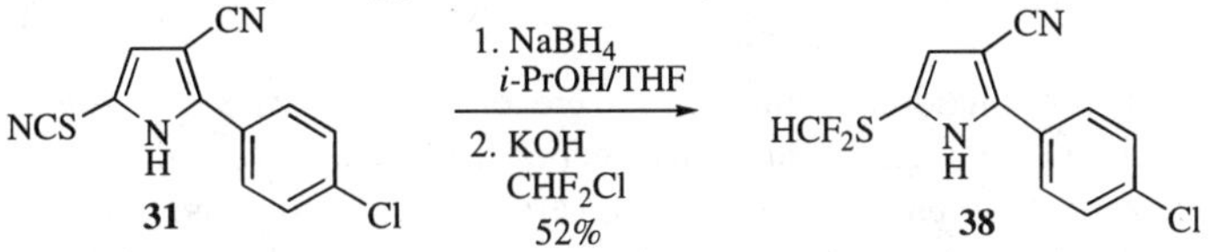

Figure 11. Introduction of the 5-difluoromethylthio group *via* the 5-thiocyanate.

The difluoromethylthio group was also introduced *via* reaction of the 5-unsubstituted pyrroles **5-7** with difluoromethanesulfenyl chloride (Figure 12). Difluoromethanesulfenyl chloride was generated *in situ* by treatment of difluoromethyl benzyl thioether with sulfuryl chloride. The addition of a catalytic amount of triflic acid was necessary for the sulfenylation of the 3-cyano and 3-nitro pyrroles. Bromination gave the 4-bromo analogs.

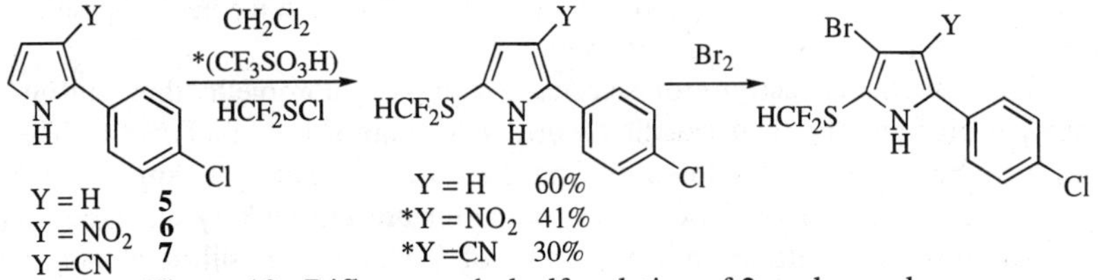

Figure 12. Difluoromethylsulfenylation of 2-aryl pyrroles.

Oxidation. Most of the haloalkylthio pyrroles prepared were converted to their corresponding sulfoxides and sulfones by oxidation with either MCPBA or hydrogen peroxide/acetic acid as illustrated in Figure 13. Treatment with one equivalent of oxidant at temperatures between 0°C and room temperature afforded the sulfoxides and treatment with two or more equivalents of oxidant at room temperature to 50°C gave the sulfones.

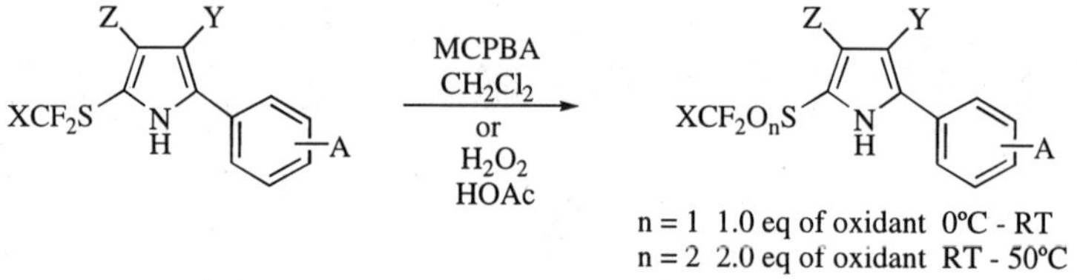

Figure 13. Oxidation of haloalkylthiopyrroles.

N-Derivatization. Earlier work on insecticidal pyrroles at Cyanamid has shown that N-derivatization, especially with N-ethoxymethyl, can result in an increase in the level and spectrum of insecticidal activity. A number of N-ethoxymethyl derivatives were prepared as illustrated in Figure 14.

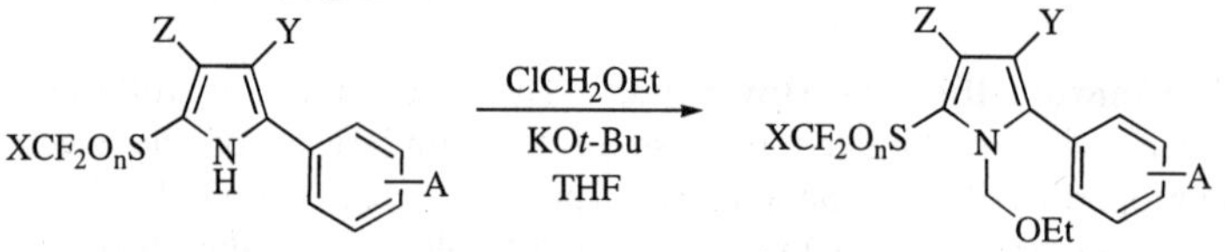

Figure 14. Synthesis of N-ethoxymethyl derivatives.

Insecticidal Activity

Approximately one hundred compounds were prepared by the synthetic procedures described in the preceding section. This section will focus on the insecticidal activity of representative examples in order to illustrate the structure activity relationships demonstrated by the compounds prepared. Insecticidal activity was determined using standard leaf dip assays. The compounds in this study were screened against third instar southern armyworms (SAW, *Spodoptera eridania*) and third instar tobacco budworm (TBW, *Helicoverpa virescens*). The activity is reported as percent mortality at 100 and 10 ppm.

The insecticidal activity of AC 303630 and related pyrroles has been attributed to their ability to function as uncouplers of oxidative phosphorylation. This relationship between uncoupling and insecticidal activity has been described (8). It has been shown that within series of uncouplers there exists an optimum pKa range that is requisite for good activity. Similarly the NH pyrroles have been shown to have an optimum pKa range which falls between 7-8.

The insecticidal activity of selected 2-aryl-5-trifluoromethylthio,-sulfinyl-and sulfonylpyrroles and the their calculated pKa values are shown in Table I. The pKa values were generated as described by Gange *et. al.* (9). This data suggests that the compounds at the lower oxidation states on sulfur are most likely oxidized *in vivo.* The 3,4-unsubstitued sulfone, having a pKa of 9.7 and with sulfur already at its' highest oxidation state, was devoid of insecticidal activity. The 3,4-dibromo-5-trifluoromethylthio and 3-trifluoromethyl-5-trifluoromethylthio analogs with pKa's of 10.6 and 10.3 respectively however demonstrated activity similar to their higher oxidation state analogs. Similarly the 3,4-dibromo-5-difluoromethylthio compound with the highest pKa of 11.3 demonstrated the best tobacco budworm activity of the compounds shown in Table I. The 3-trifluoromethylsulfonyl-4-bromo-5-trifluoromethylsulfinyl and sulfonyl compounds having pKa's of 4.0 and 2.4 respectively show that the low pKa compounds have poor insecticidal activity.

Table I

The Effect of Pka on Insecticidal Activity.

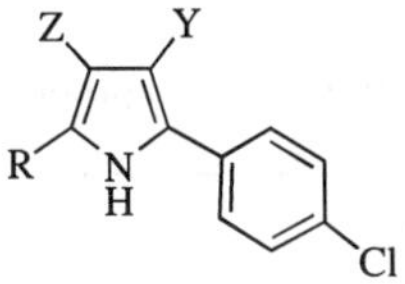

% Mortality at 100 ppm(10 ppm)

R	Y	Z	SAW *Spodoptera eridania* 3rd instar	TBW *Helicoverpa virecscens* 3rd instar	Calcd. pKa
SO_2CF_3	H	H	0	0	9.7
SCF_3	Br	Br	100(100)	100(60)	10.6
$SOCF_3$	Br	Br	100(100)	100(0)	8.3
SO_2CF_3	Br	Br	100(0)	100(0)	6.7
SCF_3	CF_3	H	100(100)	100(90)	10.3
SO_2CF_3	CF_3	H	100(100)	10(60)	6.4
SCF_3	SO_2CF_3	Br	100(100)	100(0)	6.3
$SOCF_3$	SO_2CF_3	Br	0	0	4.0
SO_2CF_3	SO_2CF_3	Br	100(0)	0	2.4
SCF_2H	Br	Br	100(100)	100(100)	11.3
$SOCF_2H$	Br	Br	100(0)	100(0)	9.2
SO_2CF_2H	Br	Br	100(0)	100(0)	7.1

The majority of the compounds prepared were substituted on the aryl ring at the 4-position with chlorine. However a number of 3,4-dichloro, 3,5-dichloro and 4-trifluoromethyl phenyl ring substituted compounds were also prepared. In the aryl ring substitution series that were prepared no major changes in the level and spectrum of activities were observed. Similarly N-ethoxymethylation had little effect on southern armyworm and tobacco budworm activity.

Table II compares the activity of AC 303630, Cyanamid's broad spectrum insecticide to a number of the more active compounds prepared in this study. Also given in Table II are the LC_{50} values for AC 303630. As shown these compounds showed good activites when tested at 100 and 10 ppm, however little activity was observed when these compounds were tested at lower rates.

Table II

Activity Comparison of AC 303630 to Selected 2-Aryl-5-haloalkythio- and sulfonylpyrroles.

% Mortality at 100 ppm(10ppm)

	SAW *Spodoptera eridania* 3rd instar	TBW *Helicoverpa virecscens* 3rd instar
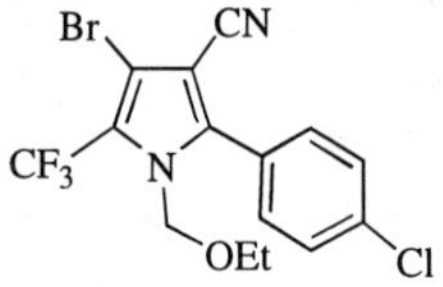	100(100) $LC_{50} = 4.58$	100(100) $LC_{50} = 7.50$
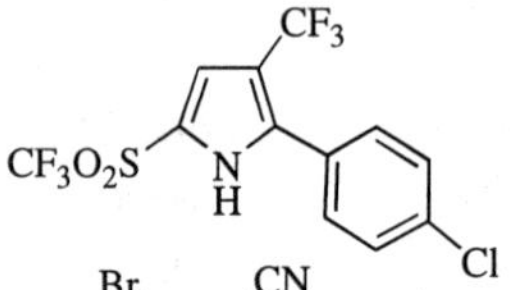	100(100)	100(60)
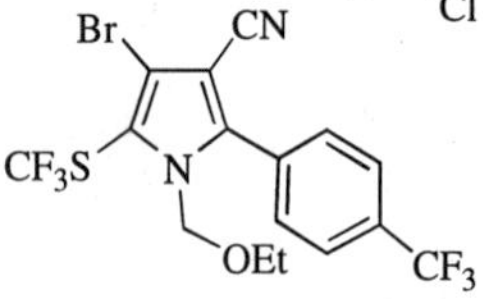	100(100)	100(100)
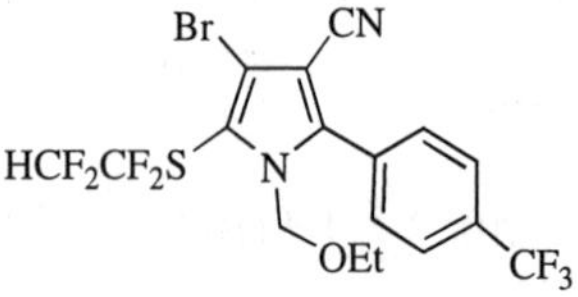	100(100)	100(80)

Conclusions

Although none of the compounds described achieved the benchmark insecticidal activity of AC 303630, good insecticidal activity was observed for a number of the pyrroles prepared, demonstrating the utility of the haloalkylthio-,sulfinyl,-and sulfonyl groups when a lipophilic electron-withdrawing functionality is essential for activity.

The data presented also suggests that the lower oxidation state compounds are oxidized *in vivo*. Additionally, as part of this work a novel method for construction of 2-aryl-5-trifluoromethythio pyrroles was developed utilizing 1,3-dipolar cycloaddition chemistry.

Acknowledgments

The authors wish to thank Drs David Gange and Stephen Donovan for their work in developing an understanding of the structure activity relationships in this area of chemistry and development of the Pka calculations. The authors would also like to thank Dr. Albert Lew, Mr. Barry Engroff and Mr. Mike Rivera for conducting the insecticidal evaluations.

Literature Cited

1. Addor, R. W.; Babcock, T.J. Black, B. C.; Brown, D. G.; Diehl, R.E.; Furch, J. A.; Kameswaran, V.; Kamhi, V. M.; Kremer, K. A.; Kuhn, D. G.; Lovell, J. B.; Lowen, G. T.; Miller, T. P.; Peevey, R. M.; Siddens, J. K.; Treacy, M. F.; Trotto, S. H.; Wright, D. P.; In *Synthesis and Chemistry of Agrochemicals III;* Edited by Baker, D. R.; Fenyes, J. G.; Steffens, J.; American Chemical Society: Washigton, D.C., 1992; pp 281-297.
2. Kuhn, D. G.; Kamhi, V. M.; Furch, J. A.; Diehl, R. E.; Trotto, S. H.; Lowen, G.T.; Babcock, T.J.; In *Synthesis and Chemistry of Agrochemicals III;* Edited by Baker, D. R.; Fenyes, J. G.; Steffens, J.; American Chemical Society: Washigton, D.C., 1992; pp 298-305.
3. Kameswaran, V.; Addor, R. W.; Ward, R.K.; In *Synthesis and Chemistry of Agrochemicals III;* Edited by Baker, D. R.; Fenyes, J. G.; Steffens, J.; American Chemical Society: Washigton, D.C., 1992; pp 306-312.
4. Kuhn, D. G.; Addor, R. W.; Diehl, R. E.; Furch, J. A.; Kamhi, V. M.; Henegar, K. E.; Kremer, K. A.; Lowen, G. T.; Black, B. C.; Miller, T.A.; Treacy. M. F. In *Pest Control with Enhanced Environmental Safety;* Edited by Duke, S. O.; Menn, J. J.; Plimmer, J. R.; American Chemical Society: Washington, D. C., 1993; pp 219-232.
5. Barnes, K. D.; Furch, J. D.; Rivera, M.; Trotto, S.; Ward, R.; Wright, D.; In *Synthesis and Chemistry of Agrochemicals IV;* Edited by Baker, D. R.; Fenyes, J. G.; Basarab, G. S.; American Chemical Society: Washigton, D.C., 1995; pp 300-311.
6. Kuhn, D. G.; Kamhi, V. M.; Furch, J. A.; Diehl, R. E.; Lowen, G. T.; Kameswaran, K. *Pestic. Sci.* **1994**, *41*, 279.
7. Olsen, R. S.; Synder, H. R. *J.O.C.* **1965**, *39*, 3712.
8. Black, B. C.; Hollingsworth, R. M.; Ahammadsahib, K. I.; Kukel, C. D.; Donovan, S. *Pesticide Biochemistry and Physiology.* **1994**, *50*, 115.
9. Gange, D. M.; Donovan, S.; Lopata, R. J.; Henegar, K.; In *Classical and Three-Dimensional QSAR in Agrochemistry;* Edited by Hansch, C.; Fujita, T.; American Chemical Society: Washigton, D.C., 1995; pp 199-212.

Chapter 17

Synthesis and Insecticidal Activity of *N*-(4-Aryloxybenzyl)-pyrazolecarboxamide Derivatives

Itaru Okada, Shuko Okui, Mabuko Wada, Toshiki Fukuchi, Keiko Yoshiya, and Yoji Takahashi

Yokohama Research Center, Mitsubishi Chemical Corporation, Aoba-ku, Yokohama 227, Japan

N-(4-Aryloxybenzyl)pyrazolecarboxamides show high insecticidal activity. Some of them were found to have potent activity against not only hemiptera insects but also lepidoptera insects and spider mites. This paper describes the synthesis and the structure activity relationships of these compounds.

Since 1986, we have been conducting research on N-benzylpyrazole-5-carboxamide derivatives. When our study was started, N-alkyl or N-phenylpyrazolecarboxamide derivatives having fungicidal or herbicidal activity had been reported, but no pyrazolecarboxamide derivatives having insecticidal or acaricidal activity were known (*1-8*). In 1987, we found a potent acaricide, tebufenpyrad (Pyranica, Masai), which was launched in Japan in 1993 and successively in some dozen countries. Tebufenpyrad shows high activity against various mite species (ex. *Tetranychus spp.* and *Panonychus spp.*) due to its rapid action at all growth stages of mites (*9, 10*). Bicyclic compound **1** has shown miticidal activity comparable with tebufenpyrad (*11, 12*).

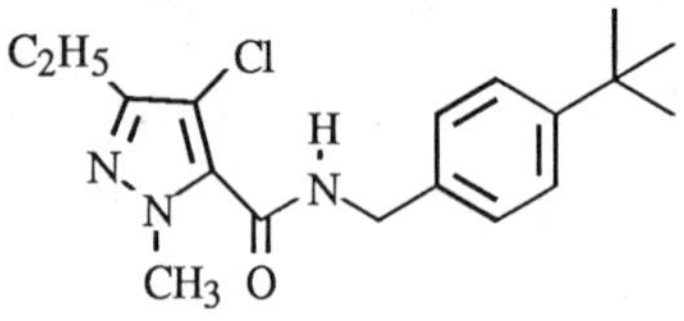

tebufenpyrad

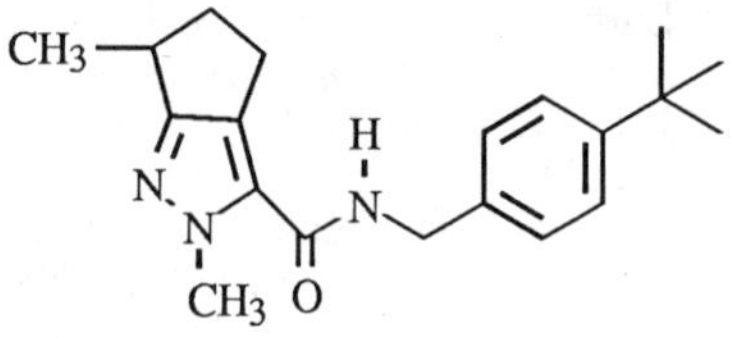

1

Tebufenpyrad and its related compounds having a *tert*-butyl group at the para position on the benzene ring showed excellent activity, but only against spider mites. In the course of our efforts, it was observed that the replacement of the *tert*-butyl group with the *n*-butyl group resulted in activity against a broader spectrum of insect pests, but this activity was weak.

In order to increase insecticidal activity, we synthesized some derivatives having lipophilic substituents on the benzene ring. Among them, 4-substituted phenoxy derivatives **2** showed high activity (*13, 14*).

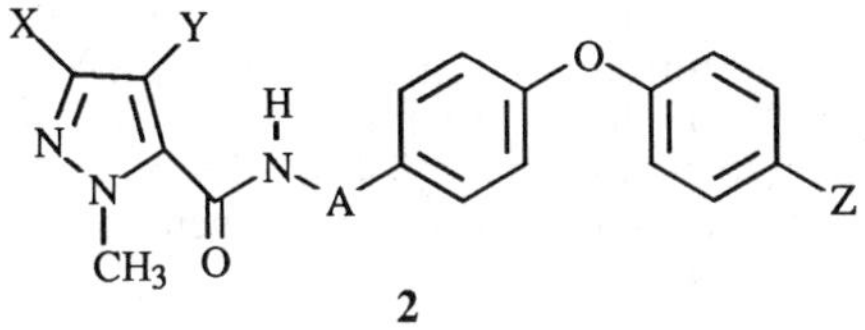

2

Synthesis

Pyrazolecarboxamides. Conversion of pyrazole-5-carboxylic acids **3** to the acid chlorides **4** followed by treatment of phenoxybenzylamines **5** gave pyrazolecarboxamides **2** in excellent yields *(13).*

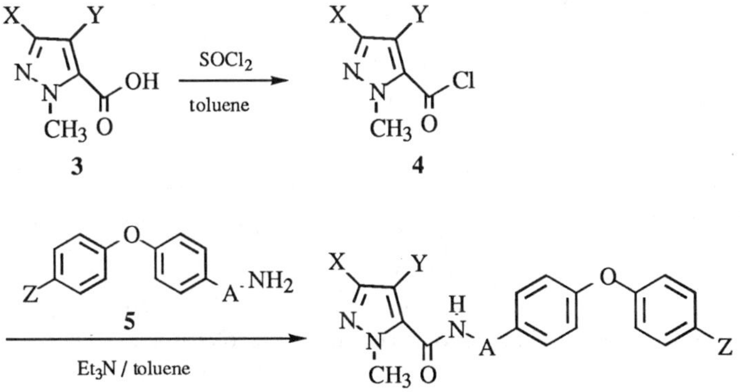

2

Figure1. Synthetic route of pyrazolecarboxamide derivatives.

Ethyl Pyrazolecarboxylate. Claisen condensation of ketones **6** with diethyl oxalate in the presence of sodium ethoxide gave ethyl acylpyruvate **7** (*15-17*). There are two routes to prepare ethyl 1-methyl-3-alkylpyrazole-5-carboxylates (Method A and B). In general, these routes afforded a mixture of ethyl 1-methyl-3-alkylpyrazole-5-carboxylate **9** and ethyl 1-methyl-5-alkylpyrazole-3-carboxylate **10.**

We required ethyl 1-methyl-3-alkylpyrazole-5-carboxylate **9** and we found that new methylation conditions improved the yield of **9** (Method B) (*18, 19*).

Method A. Auwers and Hollmann reported that the reaction of methyl hydrazine with alkyl acylpyruvate **7** gave the desired structural isomer **9** as the minor product (*20*). Structural assignments of the esters were confirmed by comparison of the proton NMR chemical shifts of N-methyl protons of these structural isomers. The N-methyl proton of **9** tended to resonate in a lower magnetic field than **10** (*21*).

Method B. Reaction of ethyl acetopyruvate **7** with hydrazine monohydrate in water led to ethyl 3-alkylpyrazole-5-carboxylate **8** in moderate yield (*22*).

We examined the method of methylation of ethyl 3-methylpyrazolecarboxylate. In the presence of base, alkylation of **8** with methyl iodide or dimethyl sulfate gave **10** equal to twice as much as **9** (Table I run 1, 2). When we treated **8** with dimethyl sulfate at 50-60 ℃ without base, we were able to isolate the desired ester **9** in 87% yield (run 3). Higher temperatures reduced the yields (run 4, 5).

When base was used, the sodium salt of pyrazole ring was formed. We believe that the reactivity of the nitrogen atom in the 2-position on the pyrazole ring was higher than that of the nitrogen atom in the 1-position owing to the electron-withdrawing property of the 5-ethoxycarbonyl group. Without base, the proton in the 1-position on the pyrazole ring was replaced with methyl group selectively and only the compound **9** was obtained because of hydrogen bonding between this proton and the oxygen of the carbonyl group.

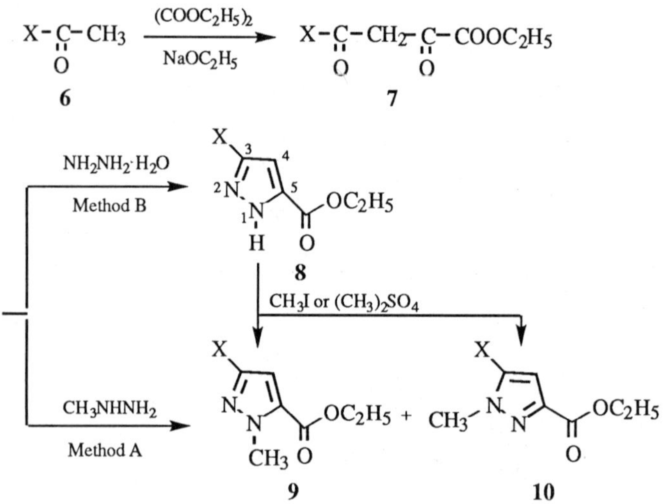

Figure2. Synthetic routes of ethyl pyrazolecarboxylate.

Table I. Methylation of Ethyl 3-methylpyrazole-5-carboxylate

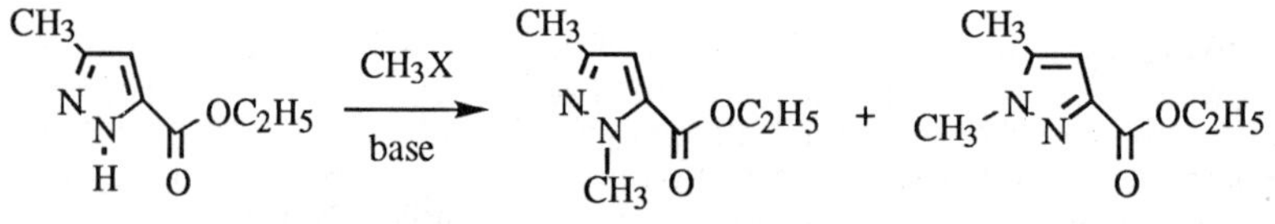

8 9 10

run	CH_3X	(eq)	base	(eq)	solvent	temp(℃)	time(hr)	yield(%) 9	yield(%) 10
1	CH_3I	(1.3)	C_2H_5ONa	(1.1)	C_2H_5OH	r.t.	1.0	27	53
2	$(CH_3)_2SO_4$	(1.5)	K_2CO_3	(1.5)	H_2O, $(CH_3)_2CO$	r.t.	1.0	45	54
3	$(CH_3)_2SO_4$	(1.5)	–		–	50-60	2.0	87	0
4	$(CH_3)_2SO_4$	(1.5)	–		–	100-105	1.5	30	0
5	$(CH_3)_2SO_4$	(1.5)	–		–	150-155	2.0	trace	0

Pyrazolecarboxylic Acid. Compound **9** was halogenated using Fluorine, chlorine, bromine in chloroform in excellent yield and then hydrolyzed to carboxylic acid **3**.

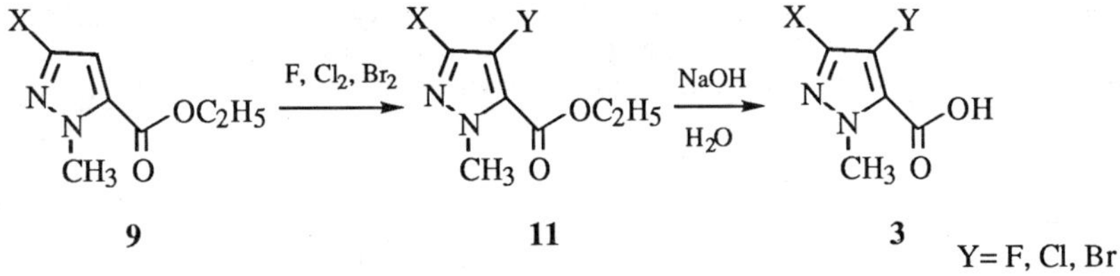

9 **11** **3**

Figure3. Synthetic route of pyrazolecarboxylic acid.

Bicyclic compounds were also prepared as described above starting from the ketones **12**.

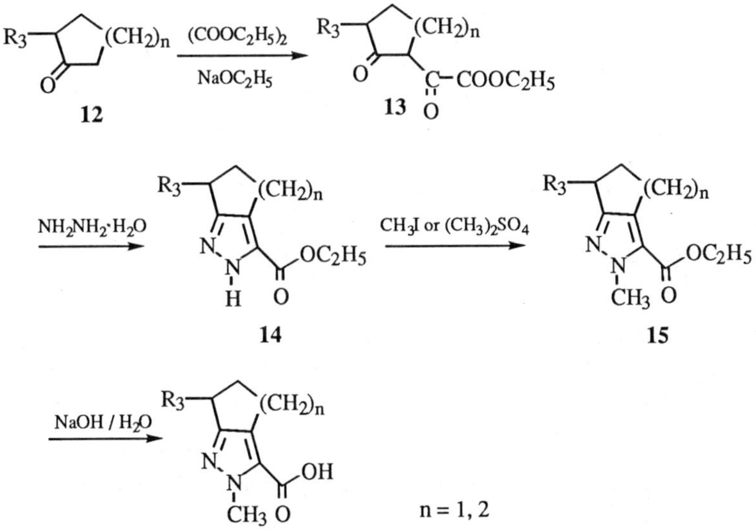

12 **13** **14** **15** **16**

Figure4. Synthetic route of bicyclic pyrazolecarboxylic acid.

Phenoxybenzylamines. When Z was a electron-withdrawing group such as CF_3, NO_2, CN or Cl, the benzylamine **5** was prepared according to a Gabriel reaction in moderate yield. Diphenyl ethers **18** were obtained by the method of G. Soula (*23*). The synthesis of 4-(4-trifluoromethylphenoxy)benzylamine was reported in our previous paper *(13)*.

When Z was alkyl, alkoxy, haloalkoxy or alkylthio group, reaction of the sodium salt of the 4-substituted phenol with p-chlorobenzonitrile in N,N-dimethylformamide gave the phenoxybenzonitrile **21** which yielded the amines on reduction (*24*).

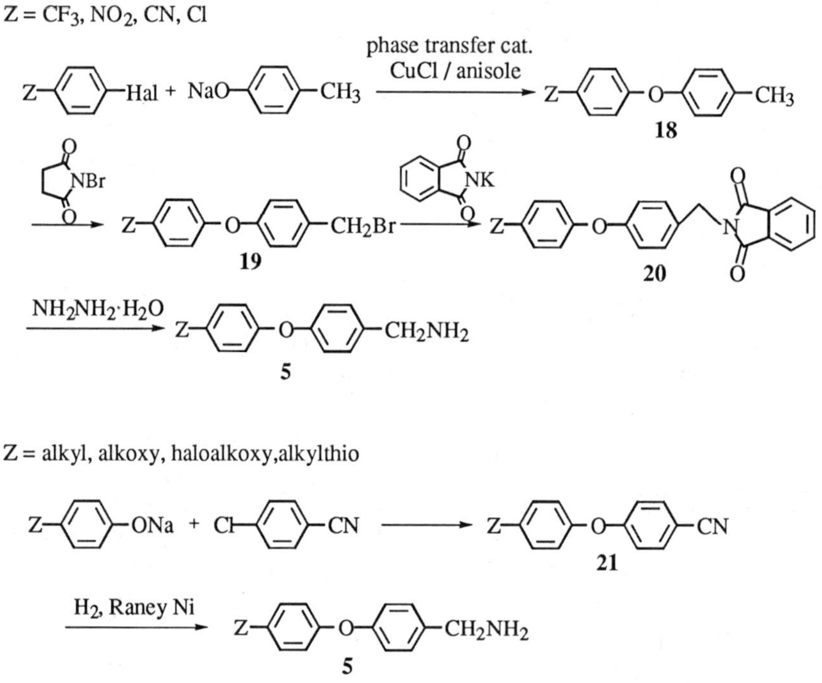

Figure5. Synthetic routes of 4-substituted phenoxybenzylamine.

Biology

Table II-VI show insecticidal activity. The methods used were the same as previously reported. The activity rating was expressed as indexes of 0 to 3, corresponding to 0-29, 30-79, 80-99 and 100% mortality respectively (*13*).

Structure Activity Relationships.

Effects of Substituent Z. When Z was alkyl, alkoxy, haloalkoxy, or alkylthio group, the small substituent such as methyl, methoxy, difluoromethoxy or methylthio group showed high activity. Introduction of strong electron-withdrawing groups such as difluoromethyl, nitro, methylthio, methylsulfinyl, methylsulfonyl or trifluoromethyl group resulted in excellent activity.

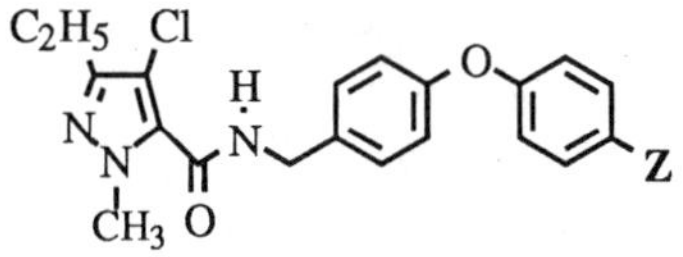

Table II. Effects of substituent Z

Z	activity rating against green rice leafhopper 50	12.5	3.1 ppm
H	3	1	n.t.
CH_3	3	3	1
C_2H_5	3	2	1
i-C_3H_7	2	1	n.t.
t-C_4H_9	0	n.t.	n.t.
CH_3O	3	3	1
i-C_3H_7O	2	1	0
CHF_2O	3	3	2
CF_3CH_2O	3	1	n.t.
Cl	3	2	2
CH_3S	3	3	2
C_2H_5S	3	2	1
i-C_3H_7S	3	1	1
CH_3SO	3	3	1
CH_3SO_2	3	3	1
CHF_2	3	3	3
CF_3	3	3	1
CN	3	3	0
NO_2	3	3	3

n.t. = not tested.

SOURCE: Adapted with permission from reference 13. Copyright 1996.

Effects of Substituent X and Y. When X was alkyl, ethyl derivatives were the most active. In the case of shorter or longer substituents, the activity decreased. Y displayed more latitude. Halogen groups and hydrogen showed almost the same rate of activity. In the case of bicyclic compounds (X and Y were connected), the five-membered ring showed higher activity than the six-membered ring. Introduction of a methyl group into the 6-position of this bicyclic molecule resulted in the highest activity.

Table III. Effects of substituent X, Y

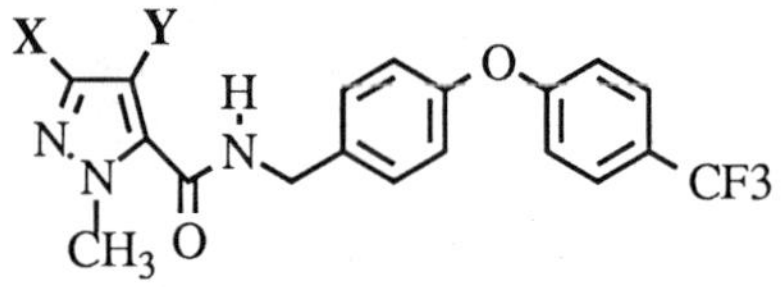

X	Y	activity rating against green rice leafhopper 50	12.5	3.1	0.8 ppm
H	Cl	3	2	1	n.t.
CH_3	Cl	3	2	2	1
C_2H_5	H	3	3	1	n.t.
C_2H_5	F	3	3	1	0
C_2H_5	Cl	3	3	1	1
C_2H_5	Br	3	3	1	1
i-C_3H_7	Cl	2	2	1	n.t.
		3	3	3	0
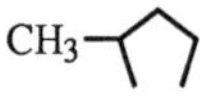		3	3	3	1
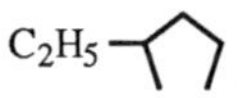		3	3	1	0
		3	2	2	0
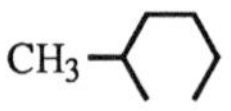		3	3	0	0

n.t. = not tested.

SOURCE: Adapted with permission from reference 13. Copyright 1996.

Effects of Substitution of A. Introduction of a methyl group at the benzylic position reduced the activity. No activity was observed with the dimethyl derivative.

Table IV. Effects of substitution of A

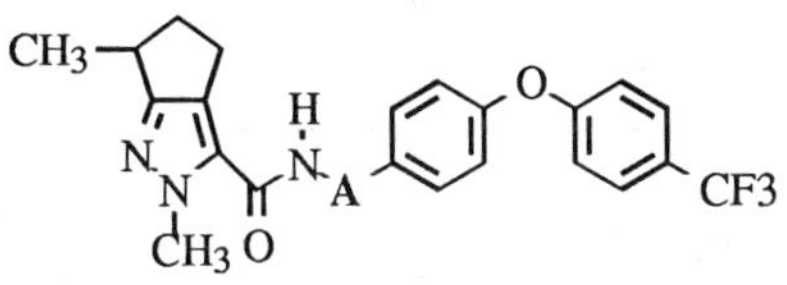

A	activity rating against green rice leafhopper 50	12.5	3.1	0.8 ppm
CH_2	3	3	3	1
$CH(CH_3)^{a)}$	3	3	2	0
$CH(CH_3)^{b)}$	3	3	2	0
$C(CH_3)_2$	0	n.t.	n.t.	n.t.

n.t. = not tested.
a) Diastereomers, low Rf value in TLC
b) Diastereomers, high Rf value in TLC
SOURCE: Adapted with permission from reference 13. Copyright 1996.

Activity of Enantiomers. The (+)-enantiomer was one-fourth as active as the (-)-enantiomer, which was slightly more active than the racemic compound.

Table V. Activity of enantiomers

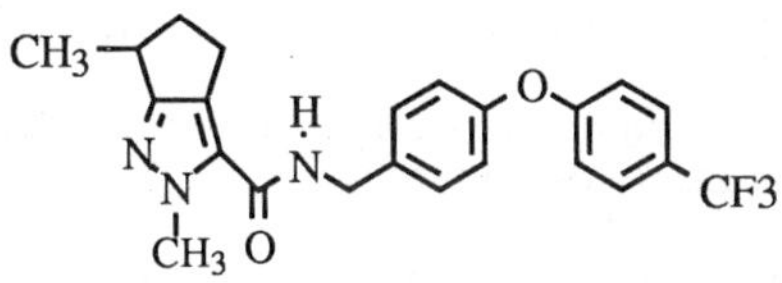

$[\alpha]_D^{20}$	activity rating against green rice leafhopper 50	12.5	3.1	0.8 ppm
—	3	3	3	1
+ 9.7 °	3	3	1	n.t.
– 9.9 °	3	3	3	2

n.t. = not tested.
SOURCE: Adapted with permission from reference 13. Copyright 1996.

Insecticidal Spectrum of Representative Compounds. We have done further bioassays on some highly active compounds of this series. Table VI shows the insecticidal activity against green rice leafhopper (*Nephotettix cincticeps*), brown rice planthopper (*Nilaparvata lugens*), green peach aphid (*Myzus persicae*), diamondback moth (*Plutella xylostella*) and two spotted spider mite (*Tetranychus urticae*).

We have found that when Z is CF_3, SCH_3, CN or NO_2, the compounds are highly active against the green rice leafhopper, the brown rice leafhopper, and the green peach aphid. In addition, not only were they highly active against certain hemiptera insects but also against certain lepidoptera insects such as the diamondback moth. They also showed high activity against the spider mite.

The compound having Z equal to CF_3 in particular was the most active of all.

Table VI. Insecticidal spectrum of representative compounds

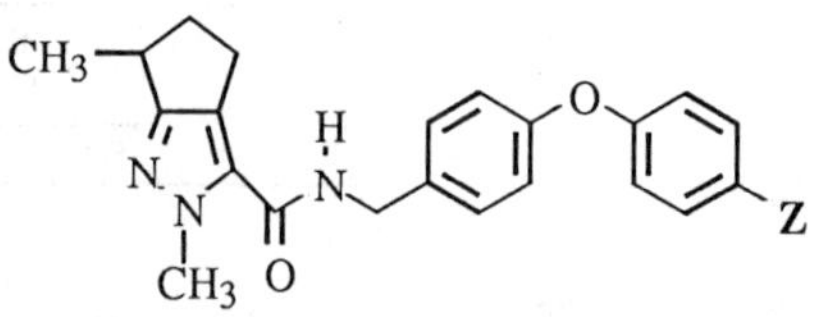

Z	GRL[a) 12.5	3.1	0.8 ppm	BRP[b) 12.5	3.1	0.8 ppm	GPA[c) 12.5	3.1	0.8 ppm	DBM[d) 12.5	3.1	0.8 ppm	TSM[e) 12.5	3.1	0.8 ppm
CF_3	3	3	1	3	3	3	3	3	3	3	2	0	3	3	2
SCH_3	3	2	1	2	1	1	3	3	0	3	2	0	3	2	-
CN	1	0	n.t.	3	2	2	3	2	0	3	3	1	3	3	0
NO_2	3	3	1	2	2	0	3	3	1	3	3	0	2	1	0

(activity rating)

n.t. = not tested.
a) Green rice leafhopper. b)Brown rice planthopper. c) Green peach aphid. d) Diamond back moth. e) Two spotted spider mite.
SOURCE: Adapted with permission from reference 13. Copyright 1996.

Conclusions

1. Many N-(4-aryloxybenzyl)pyrazolecarboxamides show high insecticidal activity against green rice leafhoppper.
2. Both the introduction of small substituents and the introduction of strong electron-withdrawing groups into the benzene ring at the 4-position resulted in high insecticidal activity.
3. Compounds having the CF_3 group as Z in particular have the most potent insecticidal effect on not only the hemiptera insects but also certain lepidoptera insects and the spider mite.

Literature Cited

1. Huppatz, J. L. Jpn. Kokai Tokkyo Koho JP 52-87168 **1977**; Chem. Abstr. **1977**, 87, 147056c
2. Sekinan, A.; Yamaguchi, H.; Nakayama, Y.; Kubo, H. Jpn. Kokai Tokkyo Koho JP 57-106665 **1982**; Chem. Abstr. **1982**, 97, 182409y
3. Huppatz, J. L.; Phillips, J. N.; Witrzens, B. *Agric, Biol. Chem.* **1984**, *48*, 45
4. Yamamoto, S.; Ochiai, Y.; Hanagami, T.; Honda, S. Jpn. Kokai Tokkyo Koho JP 60-34949 **1985**; Chem. Abstr. **1985**, 103, 160502p
5. White, G. A.; Phillips, J. N.; Huppatz, J. L.; Witrzens, B. Grant, S. J. *Pestic. Biochem. Physiol.* **1986**, *25*, 163
6. Hatanaka, T.; Watanabe, J.; Kondo, Y.; Suzuki, K. Jpn. Kokai Tokkyo Koho JP 2-53775 **1990**; Chem. Abstr. **1990**, 113, 78374w
7. Ishii, S.; Yagi, K.; Umehara, T.; Kudo, M. Jpn. Kokai Tokkyo Koho JP 2-129171 **1990**; Chem. Abstr. **1990**, 113, 172014a
8. Beck, J. R. Jpn. Kokai Tokkyo Koho JP 60-172967, 172968 **1985**; Chem. Abstr. **1985**, 103, 141938u, 141948x

9. Okada, I.; Okui, S.; Takahashi, Y.; Fukuchi, T. *J. Pesticide Sci.*(Japan) **1991,** *16*, 623
10. Okada, I.; Okui, S.; Takahashi, Y.; Fukuchi, T. Eur. Pat. Appl. EP 289879 **1988**; Jpn.Kokai Tokkyo Koho JP64-25763 **1989**; U.S. Patent 4950668 **1990**; Jpn. Kokoku Tokkyo Koho JP 5-51582 **1993**; Chem. Abstr. **1989** 110, 95234f
11. Okada, I.; Okui, S.; Sekine, M.; Takahashi, Y.; Fukuchi,T. *J. Pesticide Sci.* (Japan) **1992**, *17*, 69
12. Okada, I.; Okui, S.; Nishimata, M.; Takahashi, Y.; Fukuchi,T. Eur. Pat. Appl. EP 307801 **1989**; Jpn Kokai Tokkyo Koho JP1-156964 **1989**; U.S. Patent 4861777 **1989**; Jpn. Kokoku Tokkyo Koho JP 8-026005 **1996**; Chem. Abstr. **1989** 112, 118814c
13. Okada, I.; Okui, S.; Wada, M.; Takahashi, Y. *J. Pesticide Sci.*(Japan) **1996,** *21*, 305
14. Okada, I.; Wada, M.; Okui, S.; Takahashi, Y. Jpn. Kokai Tokkyo Koho JP 3-81266 **1991**; Chem. Abstr. **1990** 113, 152407c
15. Royals, E. E. *J. Am. Chem. Soc.* **1945**, *67*, 1508
16. Marvel, C. S.; Dreger, E. E. *Org. Syn. Col.Vol.1* **1967**, 238
17. Shinz, H.; Hinder, M. *Helv. Chim. Acta* **1947**, *30,* 1370
18. Okada, I.; Yoshida, K.; Sekine, M. Jpn. Kokai Tokkyo Koho JP 2-292263 **1990**; Chem. Abstr. **1991** 114, 185497f
19. Okada, I. Dr's thesis Tokyo University of Agriculture **1996**
20. Auwers, K.V.; Hollmann, H. *Chem. Ber.* **1926**, *59*, 601, 1282
21. Tensmeyer, L. G.; Ainsworth, C. *J. Org. Chem.* **1966**, *31*, 1878
22. Elguero, J.; Guiraud, G.; Jacquier, R. *Bull. Soc. Chim. France* **1966**, 619
23. Soula, G. *J. Org. Chem.* **1985**, *50*, 3717
24. Yoshida, K.; Okui, S.; Takuma, Y. Jpn. Kokai Tokkyo Koho JP 8-291116 **1996**; Chem. Abstr. **1996** 126, 74570c

Chapter 18

Amidrazones: A New Class of Coleopteran Insecticides

J. A. Furch, D. G. Kuhn, David A. Hunt, M. Asselin, S. P. Baffic, R. E. Diehl, Y. L. Palmer, and S. H. Trotto

Cyanamid Agricultural Research Center, American Cyanamid Corporation, P.O. Box 400, Princeton, NJ 08543–0400

The goal of modern crop protection is becoming increasingly focused on finding agents that control specific agricultural pests while leaving non-target organisms unaffected. At Cyanamid's Agricultural Research Center in Princeton, NJ our search for such agrochemicals has recently uncovered a novel class of amidrazone-based insecticides. More importantly, these novel insecticidally-active compounds appear to target specifically certain coleopteran insects, and in particular the southern corn rootworm *Diabrotica undecimpunctata* while being relatively non toxic to lepidoptera and other insects as well as to acarina, fish, and birds. These compounds also exhibit low toxicity towards mice. A structure-activity profile of these compounds was developed and the details of this work are discussed.

At Princeton our biologists screen a large variety of chemical compounds against a diverse number of insect orders and acarina in order to uncover new agrochemical leads. In that screen the amidrazone **1** (Figure 1) was found to be highly toxic towards the southern corn rootworm *Diabrotica undecimpunctata* with an LC_{50} value of 0.06 ppm, which is equivalent to current commercially applied compounds such as Counter™ (Compound **2**, Figure 1).

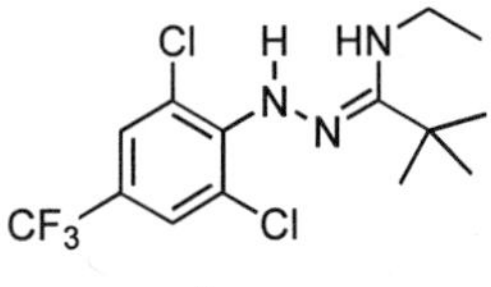

1
0.06 ppm

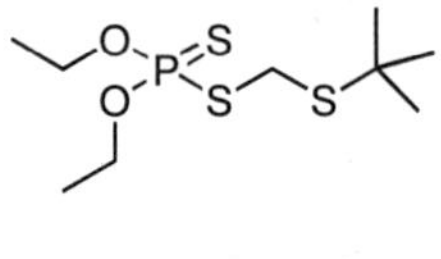

2
Counter®
0.05 ppm

Figure 1. Southern Corn Rootworm LC_{50} (*Diabrotica undecimpunctata*)

More importantly however was that no toxicity was noted when this compound was administered to other insects (i.e. lepidoptera) or acarina. While the amidrazone class of chemistry is well known (*1*), until our discovery their insecticidal properties were not. Because of this potential opportunity to control specifically an important agricultural pest, an analog program was initiated to define further the biological activity of this novel group of compounds.

Synthesis

Owing to their ease of synthesis as well as the large variety and availability of starting materials, an extensive library of amidrazones was conveniently and rapidly prepared by the route shown in Figure 2.

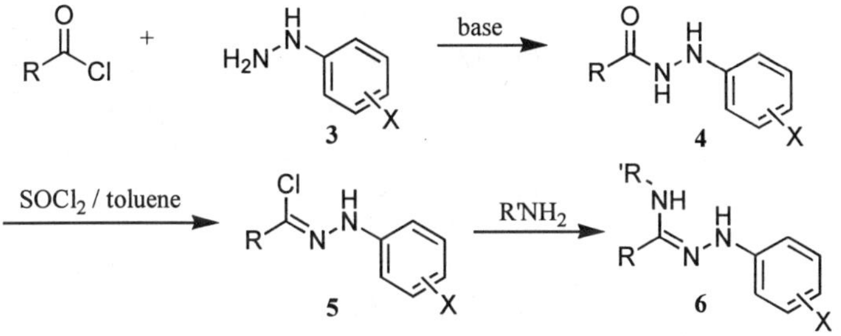

Figure 2. General Synthesis of Amidrazones

The procedure involved reacting a substituted phenylhydrazine **3** with an acyl halide plus base under Schotten-Bauman conditions to generate the hydrazide **4** in greater than 95% yield. Compound **4** was then treated with 2 or more equivalents of thionyl chloride in toluene under reflux for several hours, then the solvent was removed under vacuum to yield the hydrazonyl halide **5** which was used generally without further purification. The reaction of **5** with either two or more equivalents of a nucleophilic amine or one equivalent of such amine plus one or more equivalents of triethylamine in a solvent such as diethyl ether quantitatively yielded the amidrazone **6**.

Biological activity

We quickly discovered that even minor changes to the amidrazone molecule had profound effects on southern corn rootworm (SCR) LC_{50} data.

Effect of Phenyl substitution. Biological activity was especially sensitive to the presence and position of certain functional groups on the phenyl ring. The general trends observed was that compounds containing electron withdrawing groups in the *ortho* and *para* positions of the phenyl ring were more active while *meta* substituents as well as electron-donating groups were less active or inactive. Table I is a summary of the biological data and it is evident that the most active compounds always contained a CF_3 group in the para position of the phenyl ring with either 2,6 dichloro or 2-chloro-6-nitro in the remaining positions (**1**, **7**).

Table I. Effects of Phenyl Substitution on SCR LC_{50} Data

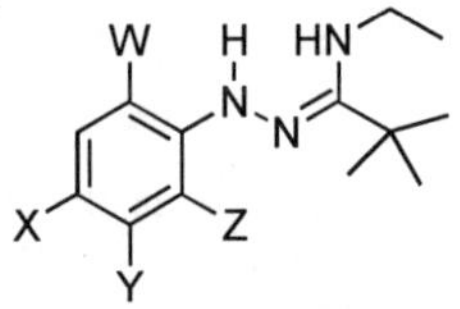

Compound Number	W	X	Y	Z	SCR LC_{50}(ppm)
1	Cl	CF_3	H	Cl	0.06
7	Cl	CF_3	H	NO_2	0.08
8	Br	CF_3	H	Br	1.00
9	Cl	CF_3	H	H	2.10
10	Cl	CF_3	NO_2	H	10.00
11	Cl	Cl	H	Cl	3.30
12	Cl	H	H	Cl	50.00
13	H	CF_3	H	NO_2	0.41
14	H	CF_3	H	H	5.70
15	H	H	CF_3	H	22.00
16	H	H	H	CF_3	22.00
17	Cl	H	CF_3	H	11.00

Replacement of the *ortho* chlorines with bromine **8** or hydrogen (**9**, **13**, **14**) profoundly reduced the activity, as did moving a substituent into a *meta* position (**10**, **15**, **17**). Likewise, replacement of the CF_3 group with a chlorine atom **11** reduced activity 50-fold but **12** clearly shows the need for substitution other than hydrogen at the para position. Diminished or complete loss of activity was also observed when the substituted phenyl group was replaced with other heterocycles and substituted heterocycles (not shown).

tert-Butyl replacements. Table II shows the biological data as related to the R group. Again biological activity was sensitive to substituents at this position and high activity was found to be dependent on the presence of a tertiary carbon center of limited size such as a *tert*-butyl group. Maintenance of the *tert*-carbon center but increasing the groups size (**18**, **19**) as well as removal of the tertiary carbon center (**20**, **22**) significantly and negatively impacted SCR LC_{50} data. Table III even further demonstrates the strict structural requirements within this portion of the amidrazone molecule. For example, replacement of one methyl group of the *tert*-butyl group **1** with an ethyl **23** produced no statistically relevant change in SCR activity but extension to a propyl **24** reduced activity almost two orders of magnitude. Nearly complete loss of activity was observed when one of the methyl groups was replaced with a phenyl or substituted phenyl (**25**, **26**) .

Effects of amine functionality. Moving our attention over to the amino group again we found that there were similar strict structural restraints for high activity. Monosubstituted amino groups were found to be the most active, disubstituted much less so, and the unsubstituted compound **27** was virtually inactive. Even within the class of monosubstituted amines strict structural requirements were necessary for optimal activity (Table IV).

Table II. Effects of Alkyl Substituents on SCR LC_{50} Data

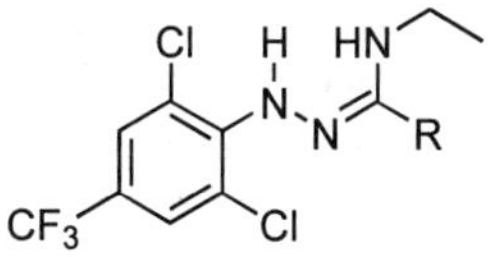

Compound Number	R	SCR LC_{50}(ppm)
1	t-C_4H_9	0.06
18	1-methylcyclohexyl	7
19	1-Adamantyl	11
20	$CH(CH_3)_2$	1.2
21	$CH_2C(CH_3)_3$	3.2
22	$CH_2CH(CH_3)_2$	22

Table III. SCR LC_{50} Data of Tertiary Alkyl Amidrazones

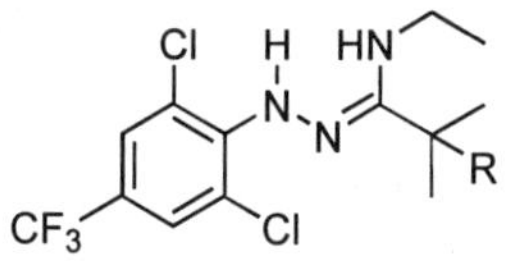

Compound Number	R	SCR LC_{50}(ppm)
1	CH_3	0.06
23	CH_2CH_3	0.08
24	$CH_2CH_2CH_3$	3
25	phenyl	19
26	4-chlorophenyl	>50

Table IV. Effects of Amino Substituents on SCR LC_{50} Data

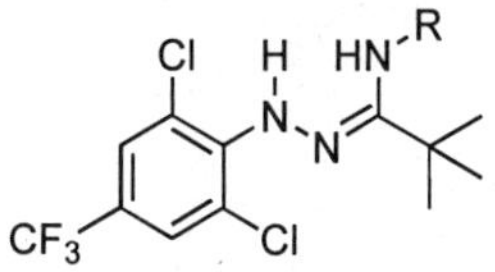

Compound Number	R	SCR LC_{50}(ppm)
27	H	>25
28	CH_3	0.50
1	C_2H_5	0.06
29	C_3H_7	0.46
30	C_4H_9	7
31	$CH_2CH{=}CH_2$	0.04

Compared to the lead compound **1** either shortening or extending the alkyl group (**28**, **29**) by one methylene unit resulted in a magnitude loss of activity and additional alkyl extension further reduced activity as in compound **30**. An allyl group **31** was highly effective however. Phenethyl **34** was half as active as the lead and Table V again clearly illustrates that the optimal alkyl chain attached to the nitrogen is 2 carbon units.

Table V. Biological Effects of Alkyl Chain Length

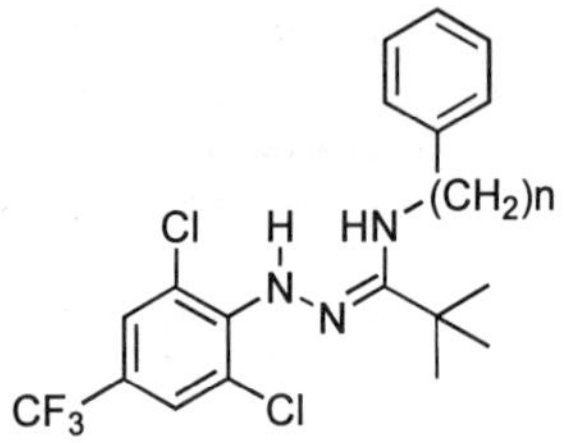

Compound Number	n	SCR LC_{50}(ppm)
32	0	22
33	1	11
34	2	0.13
35	3	11

Effect of alpha amino substituents. Unlike the rest of the molecule where minor changes in substituents had a profound impact on SCR LC_{50} data we found that a wide variety of singly alpha substituted amino groups produced highly and equivalently active insecticidal compounds and a few examples are presented in Table VI. The activity also appeared to be independent of the stereo chemistry (**39** vs. **40**) with the enantiomers having equal LC_{50}s.

Table VI. Biological Effects of Alpha Amino Substituents

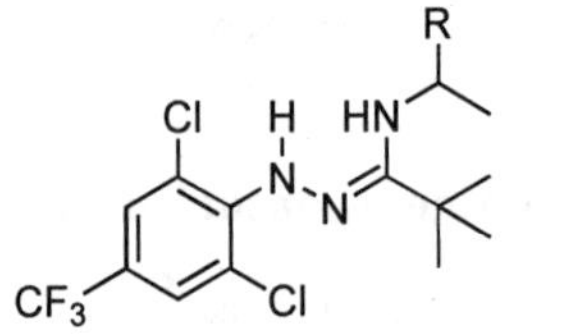

Compound Number	R	SCR LC_{50}(ppm)
1	H	0.06
36	CH_3	0.11
37	i-C_3H_7	0.05
38	C_2H_5 (+)	0.06
39	C_2H_5 (-)	0.06
40	n-C_4H_9	0.06

General Structure-Activity Relationship

Using the biological data generated from several hundred amidrazones of the structures described above, a structure-activity relationship emerged and is presented in Figure 3.

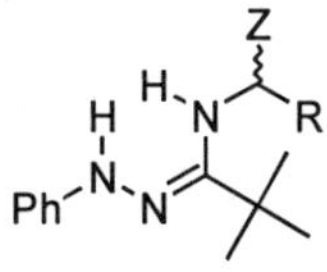

Ph = 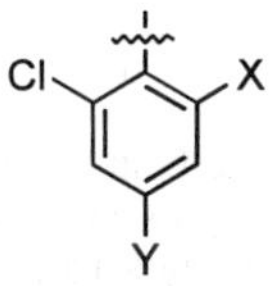

$Y = CF_3$; X = Cl, NO_2

R = CH_3, CH=CH_2

Z = H, C_2 - C_{12} (Branched or straight-chain alkyl) optionally substituted with aromatic, heterocyclic rings, and heteroatoms

Figure 3. STRUCTURE-ACTIVITY RELATIONSHIP SCR $LC_{50} < 0.1$

For *Diabrotica undecimpunctata* LC_{50} data < 0.1 ppm. Both N_1 and N_3 require protons for optimal activity. On the phenyl ring a CF_3 group in the 4 position with either a 2,6-dichloro or 2-chloro-6-nitro substitution is also necessary. Recent work has also suggested that the CF_3 group may be replaced with an OCF_3 or SO_nCF_3 (n=0, 1, 2) groups. The R group can be methyl or allyl. Z substitution on the alpha carbon of the amine function can consist of a wide variety of alkyl groups (C_1 - C_{12} and beyond) optionally substituted with aromatic groups as well as heterocycles and heteroatoms, and independent of stereochemistry.

Ecotox Profile

The ecotox data for **1** are presented in Table VII and show that the compound is non-toxic to bluegill and bobwhite quail and is non-mutagenic in the Ames test. Other favorable ecotox data are also shown.

Conclusion

A new class of insecticides based on the amidrazone molecule has been discovered and the resulting analog program allowed us to optimize the structure for southern corn rootworm activity. The observation that these compounds specifically target certain members of the coleopteran insect order opens up the possibility that selective control of these agricultural pests can be achieved with minimal effects upon non-target organisms. (*2*)

Table VII. ACUTE TOXICOLOGY PROFILE

MOUSE ORAL LD_{50}	----------	386 mg / kg
RAT DERMAL LD_{50}	----------	> 200 mg / kg
AMES TEST	----------	NON MUTAGENIC
BLUEGILL LC_{50}	----------	> 10 PPM (Non-toxic)
DAPHNIA LC_{50}	----------	> 1 PPM; < 10 PPM
BOBWHITE LC_{50}	----------	>2000 mg / kg (Non-toxic)

Acknowledgments

The authors would like to thank the members of the insecticide discovery group for the biological data and the other numerous colleagues at Cyanamids' Agricultural Research Center who have helped make this work possible as well as Dr. Gerald Berkelhammer for his support and interest in this work.

Literature Cited

1. Nelson, D.G.; Roger, R.; Heatles, J.W.M.; Newland, L.R.; Chem. Rev. **1970**, *70*, 151.
2. Furch, J.A.; Kuhn, D.G.; Hunt, D.A.; U.S. Patent 5,420,165, 1995.

Chapter 19

Cycloalkyl-Substituted Amidrazones: A Novel Class of Insect Control Agents

D. G. Kuhn, J. A. Furch, David A. Hunt, M. Asselin, S. P. Baffic, R. E. Diehl, T. P. Miller, Y. L. Palmer, M. F. Treacy, and S. H. Trotto

Cyanamid Agricultural Research Center, American Cyanamid Corporation, P.O. Box 400, Princeton, NJ 08543–0400

A series of cycloalkyl-substituted amidrazones has been prepared. Certain members of this series have shown activity against coleopteran pests such as Southern corn rootworms and boll weevils. Activity against certain lepidopterous species including southern armyworms and tobacco budworms has also been observed. This report details the synthesis and insecticidal activity of these compounds. Preliminary structure-activity relationships are also presented.

The discovery of new chemical moieties possessing insecticidal activity remains one of the most challenging endeavors facing scientists today. In addition, the study of how chemical modification of molecules affects biological activity allows one to define and expand the spectrum of activity within a lead series.

Recent work from our laboratories (*1*) led to the identification of a novel class of amidrazones possessing high *coleopteran* activity and good eco-toxicological properties. However, these compounds had little or no activity on other insect species. These observations prompted us to undertake a synthesis program aimed at expanding the spectrum of insecticidal activity while maintaining the good eco-toxicity profile. We observed, in our initial work, that a tertiary carbon center attached to the central carbon atom of the amidrazone framework was necessary for high *coleopteran* activity. Using this structural requirement, our synthesis efforts turned to the preparation of analogs with cycloalkyl groups attached to the amidiazone framework through a tertiary center as shown in Figure 1.

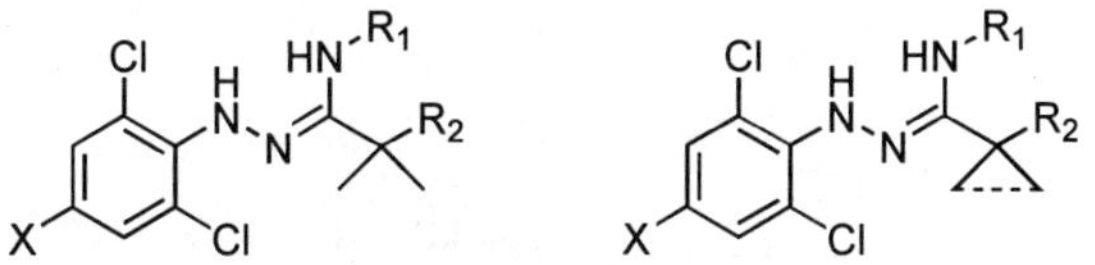

Figure 1. General Structures of the Open-Chain and Cycloalkylamidrazones

Chemistry

Cycloalkyl-Substituted Amidrazones. The synthesis of simple cycloalkyl substituted amidrazones, as exemplified by the 1-substituted cyclopropyl analogs, is shown in Figure 2.

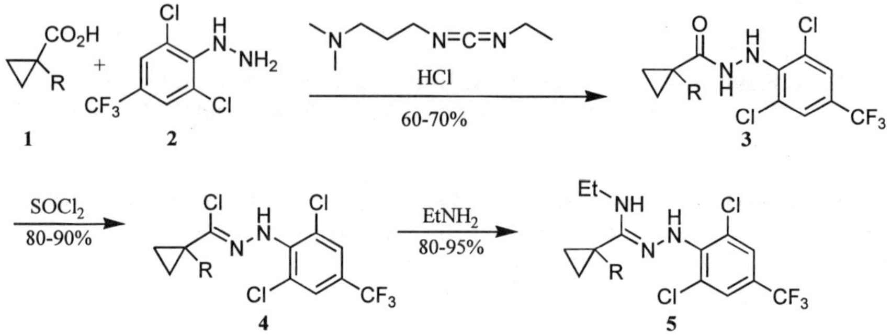

Figure 2. Synthesis of Cyclopropyl Substituted Amidrazones

Coupling of a substituted cyclopropane carboxylic acid **1** with 2,6-dichloro-4-trifluoromethylphenyl hydrazine **2** using a water soluble carbodiimide gave the hydrazide **3** in moderate yield. Treatment of **3** with excess thionyl chloride in warm toluene gave the hydrazonoyl chloride **4**. Reaction of **4** with ethylamine gave the desired amidrazone **5** in good yield. This sequence allowed us to vary, by use of the appropriate amine, carboxylic acid and hydrazine, substitution on the terminal nitrogen of the amidrazone, the one-position of the cycloalkyl substituent or the substitution pattern on the aromatic ring respectively.

Dihalosubstituted cyclopropyl amidrazones. The preparation of amidrazone analogs having substitution on the cyclopropane ring at other than the one position, in particular gem dihalo substitution, was accomplished as shown in Figures 3 and 4.

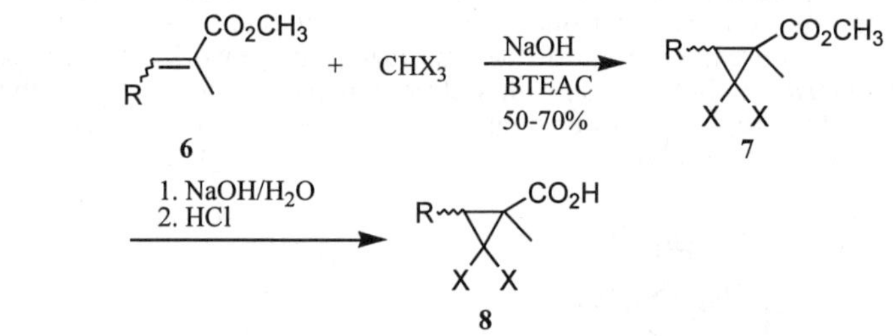

Figure 3. Synthesis of 2,2-Dihalocyclopropane Carboxylic acids

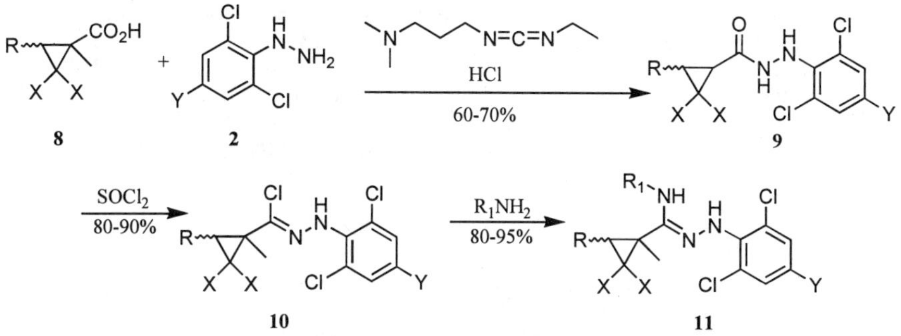

Figure 4. Synthesis of Dihalocyclopropyl Substituted Amidrazones

Treatment of an α,β-saturated ester **6** with chloroform (or bromoform) under basic conditions in the presence of benzyl triethylammonium chloride (BTEAC) gave moderate to good yields of the 2,2-dihaloesters **7** (X=Cl, Br) (*3*). Mild base hydrolysis followed by acidic work-up gave the desired carboxylic acids **8** in 80-90% yield. The final products (**11**) were prepared via the sequence analogous to the simple analogs (Figure 4).

Insecticidal Activity

All compounds prepared in this work were screened against Southern corn rootworm larve (SCR) *Diabrotica undecimpunctata* in a soil assay and third instar southern armyworms (SAW), *Spodoptera eridania* (Cramer) using a standard leaf dip bioassay with technical material. Selected analogs were tested on adult boll weevils, *Anthonomus grandis*, third instar tobacco budworms (TBW), *Heliothis virescens* (F), and adult rice water weevils, *Lissorhoptrus oryzophilus*, all by leaf application of technical or formulated material as indicated.

Cycloalkyl-Substituted Amidrazones. Figure 5 compares the activity of two cyclopropyl analogs.

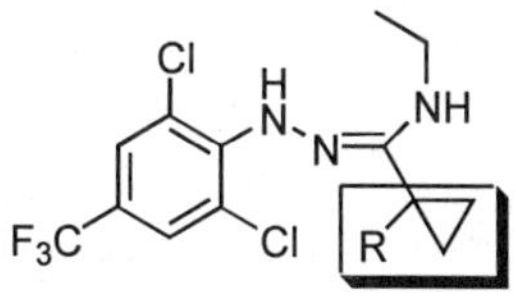

		% CONTROL	
	R	SCR (10 PPM)	SAW (100 PPM)
12	CH_3	50	0
13	C_6H_5	0	80

Figure 5. Insecticidal Activity of Cyclopropylamidrazones

The 1-methyl analog **12** was found to have selectivity on Coleopteran species similar to that of the open chain analogs described previously (*1*). No activity was found on SAW. Replacement of the methyl group at the 1-position of the cyclopropane ring with a phenyl group to give **13** resulted in a loss of SCR activity. However, **13** was found to give 80% control of SAW at 100 ppm. Introduction of one (**14**) or two chlorine atoms (**15**) onto the aromatic ring did increase SAW activity. Both **14** and **15** gave complete control of SAW at 100 ppm. However, none of these compounds showed activity against SCR at 10 ppm (see Figure 6).

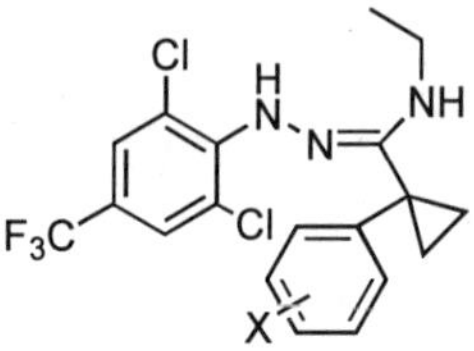

		% CONTROL	
	X	SCR (10 PPM)	SAW (100 PPM)
13	H	0	80
14	4-Cl	0	100
15	2.4-Cl	0	100

Figure 6. Insecticidal Activity of 1-Arylcyclopropylamidrazones

The appearance of activity against lepidoperous insects (i.e. SAW) seen with the 1-aryl analogs prompted us to investigate this series in more detail. Ring size was found to be crucial, as shown in Figure 7. While the cyclopropyl analog **14** gave complete control of SAW at 100 ppm, the cyclobutyl analog was inactive on all species tested.

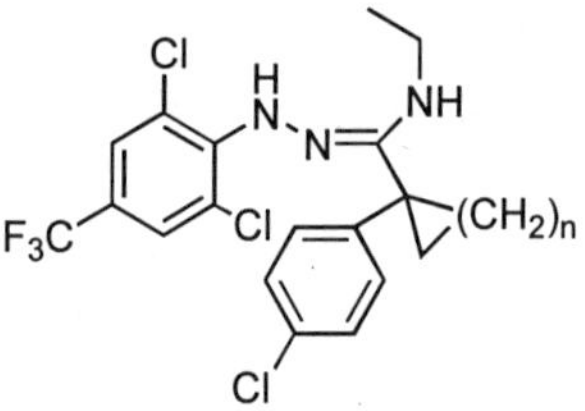

		% CONTROL	
	n	SCR (10 PPM)	SAW (100 PPM)
14	1	0	100
16	2	0	0

Figure 7. Effect of Ring Size on Insecticidal Activity

The position of the aromatic ring on the cyclopropane ring was also important (Figure 8). The 1-phenyl isomer (**13**) gave complete control of SAW at 100 ppm and moderate control of TBW (~30%). The 2-phenyl compound, as a mixture of cis and trans isomers, (**17**) was totally inactive.

	% CONTROL		
	SAW (100 PPM)	TBW (100 PPM)	
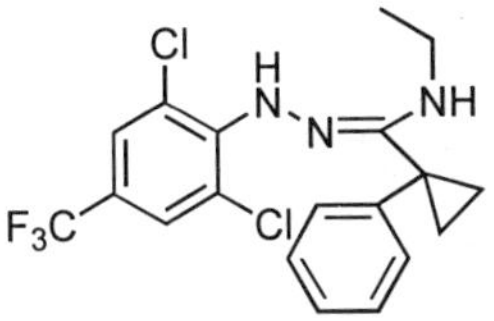	80	30	**13**
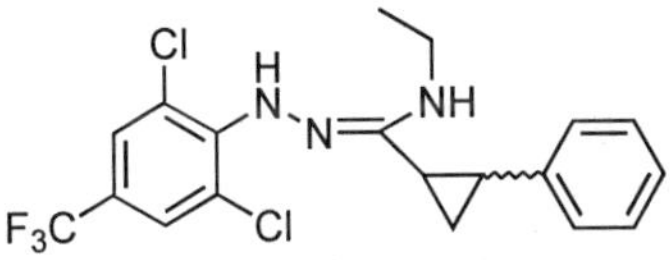	0	0	**17**

Figure 8. Comparison of the Activity of 1- and 2-arylcyclopropylamidrazones

In an effort to improve activity in this series, further substitution on the cycloalkyl ring was studied.

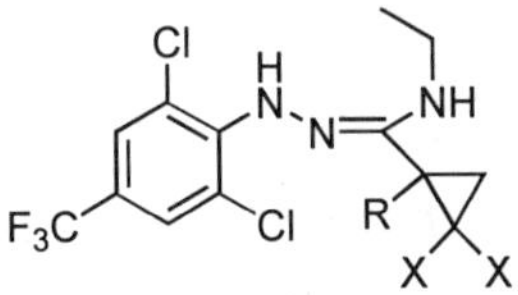

	R	X	% CONTROL SCR (10 PPM)	SAW (100 PPM)
12	CH_3	H	100	0
13	C_6H_5	H	0	80
18	CH_3	Cl	100	80

Figure 9. Insecticidal Activity of 1-Substituted Dihalocyclopropyl Amidrazones

Figure 9 compares the activity of three of these analogs. As was seen earlier, the 1-methyl compound **12** gave moderate control of SCR while the 1-phenyl analog **13** was active on SAW. Replacement of the hydrogens at the 2-position of the cyclopropane ring with chlorine atoms gave a compound (**18**) that was active on both SCR and SAW. Figure 10 summarizes the activity for a series of cyclopropylamidrazones in which the substitution on the terminal nitrogen has been varied.

The unsubstituted analog **19** gave good control of SAW but SCR activity was reduced relative to the N-ethyl compound **18**. Branching at the α-carbon of the nitrogen substitutuent (**20**) or replacement of the methyl group with a phenyl (**21**) resulted in a loss of activity. However addition of a phenyl group to the terminal carbon of **18** to give the phenethyl derivative **22** resulted in high SAW activity but little or no control of SCR. The allyl analog **23** was only moderately active on both species.

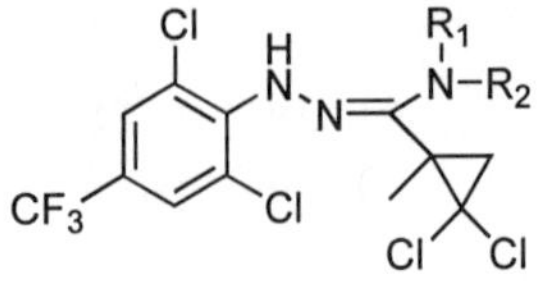

	R_1	R_2	% CONTROL SCR (10 PPM)	SAW (100 PPM)
19	H	H	-	100
18	Et	H	100	80
20	$(CH_3)_2CH$	H	100	40
21	$C_6H_5CH_2$	H	0	0
22	$C_6H_5CH_2CH_2$	H	100	100
23	CH_2=$CHCH_2$	H	100	60

Figure 10. Effect of N-Substitution on Insecticidal Activity

In the dichlorocyclopropane series, as in the alkyl series (*1*) disubstitution on the terminal nitrogen resulted in a loss of activity on both species tested (Figure 11).

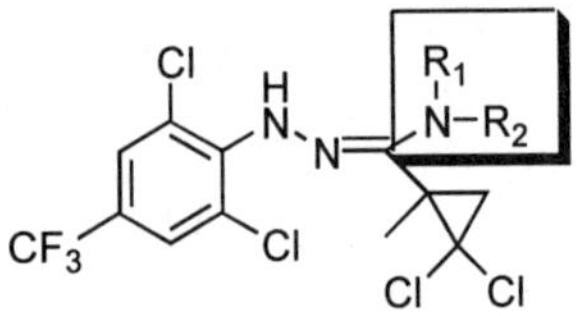

	R_1	R_2	% CONTROL SCR (10 PPM)	SAW (100 PPM)
18	Et	H	100	80
24	Et	Et	100	0
25	Et	Me	0	0

Figure 11. Insecticidal Activity of N,N-disubstituted Amidrazone

While halogen substitution on the cyclopropane ring was necessary for broad spectrum activity, both the dichloro (**18**) or dibromo (**26**) analogs were active on the SCR and SAW (Figure 12.) Alkyl substitution at the remaining position of the cyclopropane ring in the dihalo series resulted in a loss of activity against SCR as the size of the alkyl group increased (Figure 13). The same structural changes did not result in a similar decrease in activity against SAW (**18** vs **27** or **28**). However, branching at the α-carbon (**29**) or replacement of the alkyl group with an aryl group (**27** vs **30**) did decrease SAW activity.

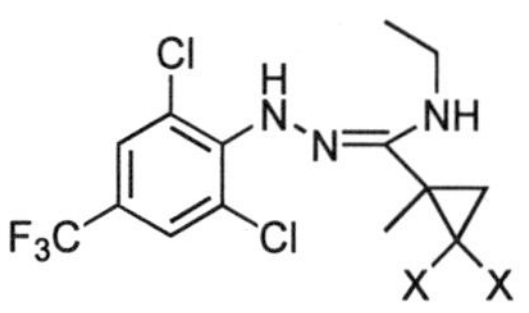

	X	% CONTROL SCR (10 PPM)	SAW (100 PPM)
12	H	100	0
18	Cl	100	80
26	Br	30	100

Figure 12. Insecticidal Activity of 2,2-Dihalocycloalkylamidrazones

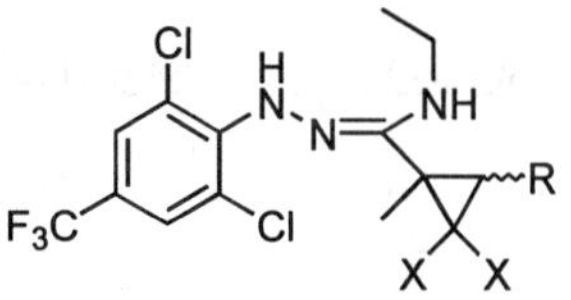

	X	R	% CONTROL SCR (10 PPM)	SAW (100 PPM)
18	Cl	H	100	80
27	Cl	CH_3	30	100
28	Cl	CH_3CH_2	40	90
29	Cl	$(CH_3)_2CH$	10	60
30	Cl	4-ClC_6H_4	40	60
31	Br	CH_3	0	100

Figure 13. Insecticidal Activity of Dihalocyclopropylamidrazones

Attempts to combine the good lepidopteran activity seen in the 1-arylcyclopropane series with the broad spectrum activity observed with the dihalo series were unsuccessful as shown in Figure 14. While the 1-methyl-2,2-dichlorocyclopropyl amidrazone, **18**, was active on both SCR and SAW, the 1-(4-chlorophenyl)-2,2-dichlorocyclopropyl amidrazone, **32**, was only weakly active on SCR and totally inactive on SAW.

The activity seen in the dihalo series prompted us to investigate the activity of these compounds on other coleopteran species. Figure 15 shows the results of a field simulation study using the lead compound, CL 341436, **18,** as an EC formulation, on boll weevils.

The activity seen was superior to the standard, Vydate® at the same rate.

Against adult rice water weevil, CL 341436, formulated as a 20% EC, gave control comparable to the standard, ethofenprox, at the same dose rate (Figure 16).

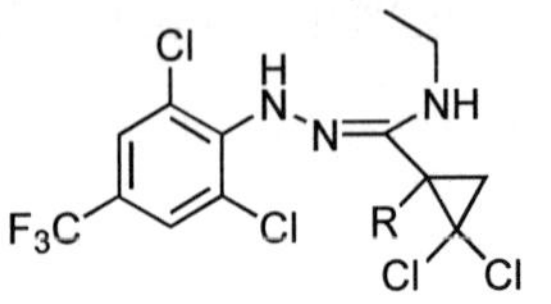

		% CONTROL	
	R	SCR (10 PPM)	SAW (100 PPM)
18	CH_3	100	80
32	4-ClC_6H_4	20	0

Figure 14. Activity Comparison of 1-Methyl vs. 1-Arylcyclopropyl Amidrazones

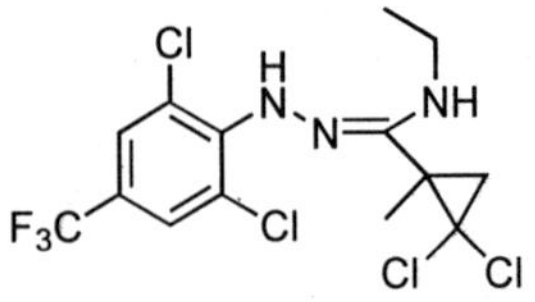

AC 341436, **18**

TREATMENT (LB AI/A)	% CONTROL (0 DAT)
AC 341436 1.67EC (0.125)	70
AC 341436 1.67EC (0.25)	92
AC 341436 1.67EC (0.35)	100
VYDATE 2EC (0.25)	62

Figure 15. Activity of AC 341436 on Adult Boll Weevils, *Anthonomus Grandis*

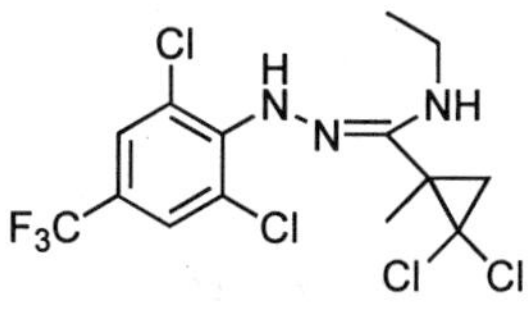

AC 341436, **18**

TREATMENT (G AI/A)	% CONTROL 3 DAT	5 DAT
AC 341436 20%EC (200)	96	100
ETHOFENPROX 20% EC (200)	92	92

Figure 16. Activity of AC 341436 on Rice Water Weevil, *Lissorhoptrus Oryzophilus*

Conclusion

In conclusion, synthetic modification of the alkyl amidrazones, compounds with high coleopteran activity, has resulted in a series of cycloalkyl-substituted amidrazones with a broader spectrum of insecticidal activity including lepidopterous insects. One compound, CL 341436, in limited trials, demonstrated levels of activity comparable to commercial standards on adult boll weevils and adult rice water weevils at both 3 and 5 days after treatment (DAT).

Acknowledgments
The authors wish to express their appreciation to the members of the Insecticide Discovery group for obtaining the biological results presented in this work.

Literature Cited

1. Furch, J.A.; Kuhn, D.G.; Hunt, D.A.; Asselin, M.; Baffic, S.P.; Diehl, R.E.; Pamer, YL.; Trotto, S.H. companion chapter in this volume.
2. Nelson, D.G.; Roger, R.; Heatles, J.W.M.; Newlands, L.R.; *Chem. Rev.* **1970,** *70*, 151.
3. For a review of the preparation of 2,2-dihalocyclopropane carboxylic ester, see: Dehmlow, E.; Angew. Chem. Internat.Edit. **1974,** *13,* 170.

Chapter 20

Selective Probes for Nicotinic Acetylcholine Receptors from Substituted AE-Bicyclic Analogs of Methyllycaconitine

William J. Trigg[1], David J. Hardick[1], Géraldine Grangier[1], Susan Wonnacott[2], Terrence Lewis[3], Michael G. Rowan[1], Barry V. L. Potter[1], and Ian S. Blagbrough[1,4]

[1]Department of Medicinal Chemistry, School of Pharmacy and Pharmacology, and [2]School of Biology and Biochemistry, University of Bath, Bath BAZ 7AY, United Kingdom

[3]Zeneca Agrochemicals, Jealott's Hill Research Station, Bracknell RG12 6EY, United Kingdom

A concise, practical approach to substituted [3.3.1]-AE-bicyclic analogs of methyllycaconitine (MLA) has been developed employing dianion alkylation and double Mannich reactions. Acetylide anion addition to substituted cyclohexanones and manipulation of the acetylide alkoxide dianion generates useful bis-propargyl tertiary alcohols. The design and synthesis of probes for nicotinic acetylcholine receptors (nAChR) affords possible leads for pesticides. MLA also has potential in probing neuronal nAChR which are implicated in some mechanisms of neurodegeneration.

Methyllycaconitine (MLA) **1** is a hexacyclic norditerpenoid alkaloid which occurs in many *Delphinium* species as well as in *Consolida ambigua* and *Inula royaleana* (*1-4*). These plants, especially the *Delphinium* species, are known to be toxic to mammals and to a wide variety of insect species (*5-9*). The insecticidal property of *D. staphisagria* which contains related norditerpenoids, but possibly not MLA, has long been exploited as a herbal treatment for head lice infestations, first reported by Pliny the Elder (*5,9,10*). Each year, ingestion of various wild *Delphinium* spp. is one of the major causes of death for significant numbers of grazing cattle across North America (*7,8*). The toxicity of *Delphinium* plants is in part due to neuromuscular blockade, but a contribution also comes from MLA which acts as a competitive antagonist at insect and mammalian neuronal nicotinic acetylcholine receptors (nAChR) (*11-13*). Thus, MLA has use as a potent, selective ligand for molecular studies of neuronal nAChR implicated in neurodegeneration, but MLA is more potent at neuronal insect nAChR than at the mammalian neuromuscular junction and has potential as a lead compound for the rational design of insecticides acting at nicotine binding sites (*9,14*). We are continuing our structure-activity relationship (SAR) studies of diverse insect and

[4]**Corresponding author.**

mammalian nAChR (*15*). In this *Chapter*, we report our concise synthetic routes to substituted [3.3.1]bicyclic analogs of MLA, templates for the design of selective molecular probes, in order to investigate the structure and function of nAChR.

MLA Isolation and Characterization

MLA was first isolated and identified in 1938 by Manske (*16*) from the aerial parts of *D. brownii*. Goodson determined (*17*) the structure of MLA **1** in the 1940s, and we unambiguously confirmed that the 2-methylsuccinimidobenzoate ester contains an *S*-stereocenter (*18*). Benn (*19,20*), Pelletier (*1,2,21,22*) and their independent research groups have worked extensively on the isolation of alkaloids from *Delphinium* and other plant species. Recent research work in this area of norditerpenoid alkaloid phytochemistry has been comprehensively reviewed by Yunusov (*3,4*). In addition to our studies (*15,18,23-30*), other research groups who have recently reported efficient syntheses of MLA analogs include those of Kraus (*31-33*) and Whiting (*34*).

Nicotinic Receptors and MLA Neurotoxicity

Jennings, Brown and co-workers (*5,6*) first investigated the insecticidal properties of MLA. The insect and mammalian toxicity are a result of MLA **1** acting as a competitive nAChR antagonist. Substituted hexacyclic norditerpenoid alkaloids are the most potent, non-proteinaceous antagonists at nAChR and MLA **1** binds to rat α7-type (α-bungarotoxin sensitive nAChR) with a K_i in the nanomolar region. The closely related norditerpenoid alkaloid lacking the angular methyl group on the succinimide, lycaconitine (LA) **2**, has also been isolated from *Delphinium*. Compared to MLA **1**, LA **2** displays 5-fold reduced binding at nAChR (*35*). Invertebrate and mammalian nAChR have a multisubunit protein structure and are pentameric. There are many different subtypes of nAChR due to the structural heterogeneity of the subunits (e.g. vertebrate α1-9, β1-4, and also insects subunits) (*36*). MLA competes with ligand binding to nAChR in flyheads, locust ganglia, Manduca (tobacco hornworm) (*37*) and cockroach motorneurone with K_i values in the nanomolar range (and at mammalian receptors containing α7 and chick α8 subunits), whereas there is a significantly lower affinity at vertebrate neuromuscular junction nAChR (α1 subtype with nAChR stoichiometry $\alpha 1_2\beta 1\gamma\delta$) (*13*). MLA is therefore an attractive lead compound for the design of insecticides. Furthermore, MLA shows a significantly greater affinity (10,000-fold) for nAChR than that displayed by the parent alcohol lycoctonine **3**. Following SAR comparisons between natural products MLA **1**, LA **2**, and lycoctonine **3**, we conclude that the 2-methylsuccinimidobenzoate ester function is significant for efficient nAChR antagonism within this series of small molecules. Our related studies with the *Aconitum* hexacyclic norditerpenoid alkaloid aconitine **4** confirm this conclusion (*28*). Aconitine **4** is a potent, well-characterized neurotoxin which has essentially no activity at nAChR and acts by maintaining voltage-sensitive sodium channels in an open conformation. In order to test the above hypothesis, we have designed and prepared an MLA-aconitine hybrid, the 2-*S*-methylsuccinimidobenzoate ester of 3-deoxy-18-*O*-demethylaconitine **5**. This semi-synthetic alkaloid was essentially equipotent with MLA **1** and lacked biological activity at voltage-sensitive sodium channels (*28*). Therefore, such an ester functional group appears to be important for nAChR antagonism (*15,25,28*).

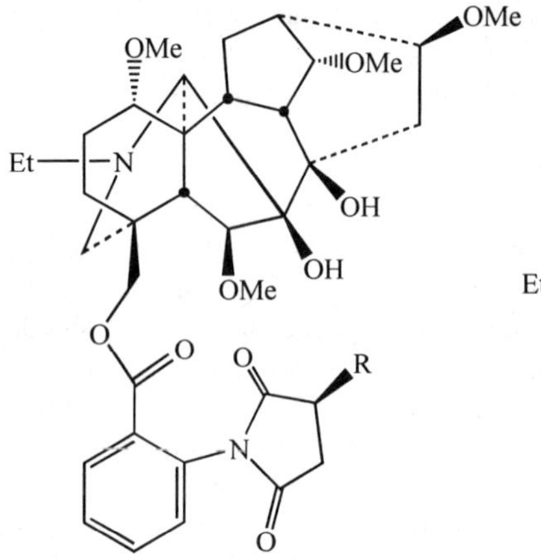

R = Me Methyllycaconitine (MLA) **1**
R = H Lycaconitine (LA) **2**

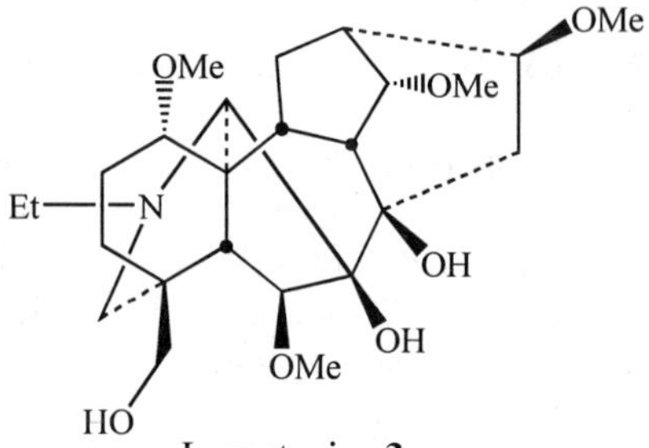

Lycoctonine **3**

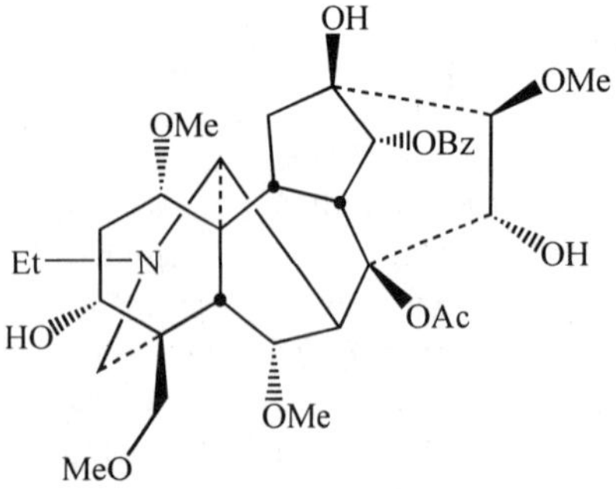

Aconitine **4**

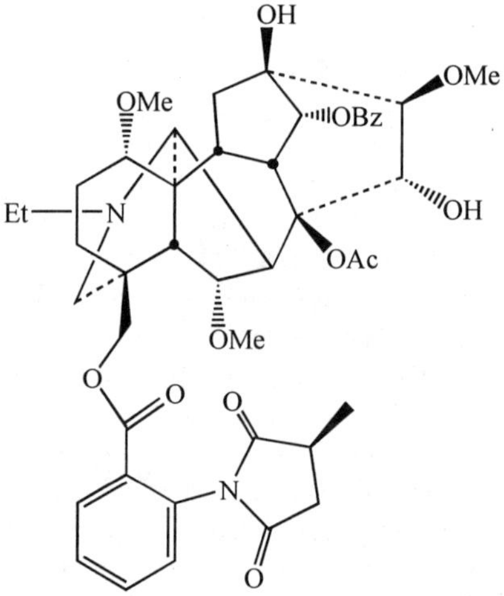

3-Deoxy-18-*O*-demethylaconitine-MLA-hybrid **5**

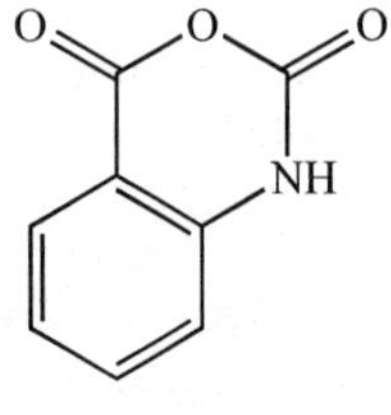

Isatoic anhydride **6**

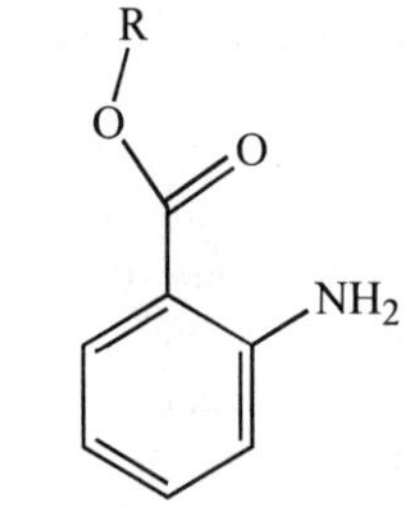

Anthranilate ester **7**

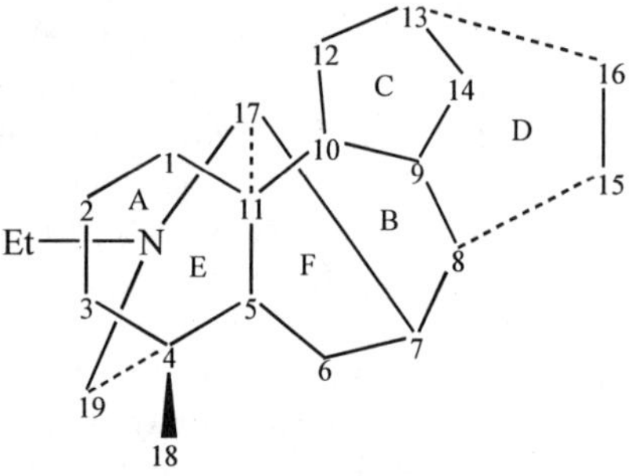

Norditerpenoid labelling **8**

Design of [3.3.1]Bicyclic Analogs of MLA

Practical quantities of alcohols that mimic the AE-bicycle of MLA are therefore required for the next phase of our SAR program. These sterically hindered, neopentyl-like alcohols will be esterified in order to introduce the important 2-methylsuccinimidobenzoate moiety. We can achieve this esterification by reaction of an alcohol (ROH) with isatoic anhydride **6**, under basic catalysis, to afford the desired anthranilate ester **7** together with the liberation of an equivalent of carbon dioxide and no detectable formation of the corresponding isatoate (carbamate functional group). In the penultimate step in the preparation of our MLA analogs, this neopentyl alcohol esterification will need to be performed in the presence of secondary and sometimes tertiary alcohols with regiochemical control. Based upon literature precedent (*38*), we have established a practical protocol for this esterification (*25*). Neopentyl alcohols were esterified in good yield (~60 %) in this reaction, and in significantly better yields than secondary alcohols; not unexpectedly, tertiary alcohols did not react.

We decided to prepare the required AE-bicycles (see: **8**) by a double Mannich reaction between a cyclic β-ketoester, formaldehyde (two equivalents), and ethylamine. In order to add the desired carbon-chain substituents (incorporating C11-C10 of MLA) to these small molecule MLA analogs, a way to substitute efficiently at the γ-position of ethyl cyclohexan-2-onecarboxylate was required. The alkylation of a β-ketoester is relatively easy to achieve at the α-position as has been shown by the extensive use of substituted ethyl acetoacetates in synthesis. In the early 1970s, Huckin and Weiler (*39,40*) showed that ethyl acetoacetate could be alkylated at the γ-position by making a dianion of the β-ketoester using NaH followed by n-butyl lithium as the bases. This is an extension and an improvement to the work of Hauser and Harris (*41*) in the 1950s who employed two equivalents of sodium amide as the base in liquid ammonia. Dianion formation in effect prevents any alkylation at the α-carbon atom making the process regioselective, due to preferential alkylation of the more reactive, least stabilized enolate under kinetic control. If this strategy is successful, dianion alkylation followed by double Mannich reaction will allow access to AE-bicycles further substituted with rings C or D **8**. Appropriately chosen carbon chains will allow subsequent transformations in order to introduce oxygenation at patterns which mimic those found in MLA. Thus, we can incorporate oxygen substituents at C6, C7, C8, C14, and C16 (see: **8**) individually or in selected patterns. Such a flexible approach is highly desirable for SAR studies at selected nAChR.

Preparation of C11-C10 Substituted Analogs

Therefore, we wanted to prepare bicyclic diols represented as **9**. Disconnection to piperidinone **10** and from thence to substituted β-ketoester **11**, gave the requirement for regioselective alkylation at the γ-position of cyclic β-ketoester **12**. The forward synthesis is in Scheme 1. Ethyl cyclohexan-2-onecarboxylate (Aldrich) **12** was alkylated with a range of alkyl halides (allyl, benzyl, cyclohexylmethyl, and cyclopentyl bromides and n-butyl and n-pentyl iodides). Whilst it may be assumed that both the ethyl ester and the newly introduced carbon chain prefer to occupy equatorial conformations (around a cyclohexanone chair), nevertheless, there was always a significant proportion of the corresponding enol tautomer present. This cyclohexenol was detected by ^{1}H NMR spectroscopy in $CDCl_3$, the integrals of the acidic α-proton (methine) and the enol signal combining for one proton. Even with the addition of a carbon chain (or ring) at γ-position in the starting material, this enolization contributes to the similarity in polarity of the starting material and the alkylated product. Although less likely, the ethyl ester could also be in an axial orientation after protonation of the enol tautomer from the less favored face. This tautomerization process will give rise to mixtures of diastereoisomers, rather than to an enantiomeric pair. The dianions were not reactive under the conditions initially investigated (THF as the solvent with the anions formed at -78 °C and then the reaction mixture warmed to room temperature after the electrophile had been added). We thought that this lack of nucleophilicity might be due to a solvent effect (THF solvating the dianion). Therefore, we investigated the use of a co-solvent to increase the reactivity. Following literature precedent (*42*), we used a dimethylpyrimidone urea (1,3-dimethyl-3,4,5,6-tetrahydro-2(1*H*)-pyrimidinone, DMPU) as co-solvent and we established that a 30 % mixture of DMPU in THF (v/v) both increased the reaction rate and improved the alkylation yields. Another way to increase the reactivity and rate of reaction is to use a concentrated reaction mixture, but with simple alkyl halides the use of DMPU as a co-solvent is an efficient protocol. Yields were further improved by carefully controlling the reaction conditions, the temperature was only allowed to rise to 0 °C rather than to room temperature.

Although the dianion alkylation can be performed in good yield (typically 50-80 %), it was neither practical nor essential to separate product **11** from unreacted starting materials at this stage, in part due to enolization modulating the polarity. An immediate double Mannich reaction on the alkylated β-ketoester afforded the desired piperidinone **10**. We were able to effect an efficient separation of this product from the unsubstituted [3.3.1]bicycle by flash silica gel chromatography. Therefore, it was expedient to take the ethereal extracts from the dianion alkylation directly on to the double Mannich reaction, utilizing (in effect) an excess of reagents based upon an equivalent of aqueous ethylamine pretreated with two equivalents of formalin (37 % aqueous formaldehyde) relative to the starting β-ketoester. When R = H (Scheme 1), the unsubstituted case, the double Mannich reaction proceeded in ~40 % yield. When R = alkyl, the yields were always equal to or greater than the unsubstituted case (40-60 % over two steps). The possibility of diastereoisomeric pairs (previously alluded to) was not a problem, as the double Mannich reaction is under equilibrium conditions (acetic acid catalysis) (*43*). The newly formed carbon-carbon bonds, by alkylation of the iminium salt formed *in situ*, firstly intermolecular and then intramolecular alkylations, must both be axial relative to the cyclohexanone, if the piperidinone ring is to be closed. Thus, although the intermolecular Mannich reaction will proceed initially from either face of the cyclic β-ketoester, only those conformers with a newly formed axial aminomethyl substituent can form the piperidinone, the most stable product. Therefore, given that (in the desired product) these aminomethyl substituents must both be axial, the quaternary carbon atoms in **10** must have *R,S*- or *S,R*-chirality. Reduction of the ketone and ester functional groups ($LiAlH_4$, Et_2O, typically 65 %) afforded the corresponding diols **9**.

As piperidinones **10** have been formed as an *R,S/S,R*-enantiomeric pair, the $LiAlH_4$ reduction gives the opportunity to form diastereoisomers with the generation of a new chiral center. The newly formed secondary alcohol **9** was typically isolated (R = n-pentyl) as a 9:1 mixture of diastereoisomers (relative to the *R,R/S,S* pair of enantiomers whose priority follows from ester reduction). That the hydride ion has probably adopted an equatorial position is possibly due to $LiAlH_4$ coordination to the basic nitrogen atom (tertiary amine) with corresponding directing effects. Possibly, there is a less hindered path for attack generating an axial hydroxyl functional group (axial relative to the cyclohexane, but of course equatorial relative to the piperidine in the [3.3.1]bicycle). Diol **9** was then esterified with isatoic anhydride **6**, using *N,N*-dimethylaminopyridine as the nucleophilic base in hot (70 °C) DMF to afford anthranilate ester **13** (~60 %). Treatment of this series of aniline esters **13** with methylsuccinic anhydride afforded the corresponding half-acid amides which were cyclized *in situ* with 1,1'-carbonyldiimidazole as the dehydrating agent to give the desired 2-methylsuccinimidobenzoates **14**.

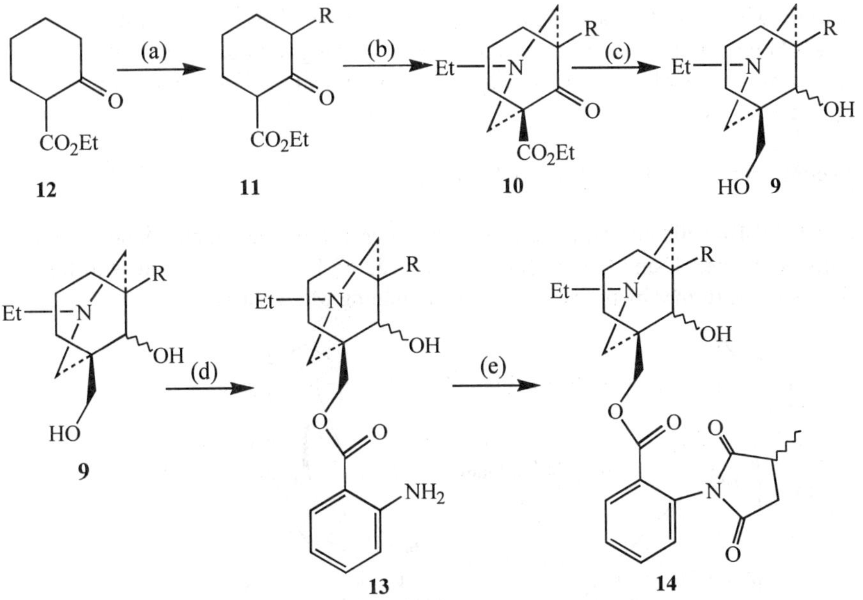

SCHEME 1
(a) NaH, nBuLi, RX, 30 % DMPU/THF (b) 2 CH_2O, $EtNH_2$, EtOH, AcOH
(c) $LiAlH_4$, Et_2O (d) isatoic anhydride **6**, DMAP, DMF
(e) methylsuccinic anhydride, CDI, DCM

This reaction proceeded in good yield (~65 %) and the AE-bicyclic analogs containing a five-membered ring **14a**, mimicking ring C of MLA **1**, a six-membered ring D mimic **14b**, the corresponding planar aromatic analog **14c**, allyl **14d**, n-butyl **14e**, and n-pentyl **14f** analogs were prepared by this concise route.

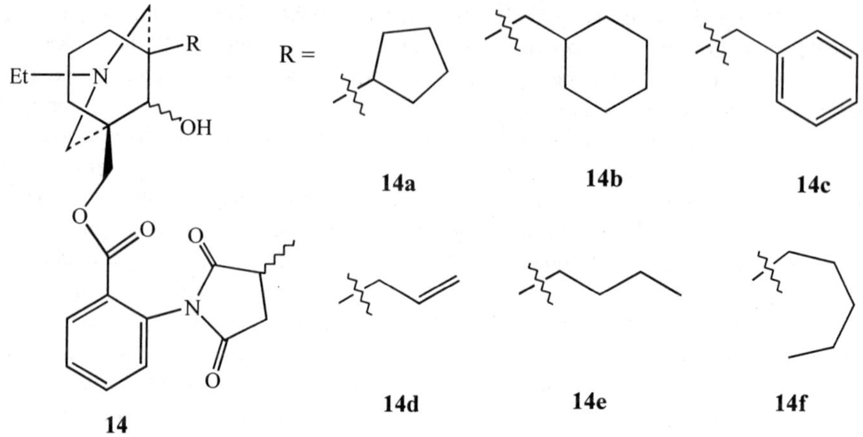

14

Substituted AE-bicyclic analogs **14a-14f** showed no significant insecticidal activity when tested on a range of pest species. They also displayed low affinity for insect brain nAChR in a competition assay, displacing iodo-α-bungarotoxin (*15*). In this assay, where MLA **1** achieved 100 % displacement of [^{125}I]α-bungarotoxin at 0.1 ppm, analogs **14a-14f** reached only 1-20 % at 1 ppm and 20-53 % at 10 ppm.

Preparation of C5-C6 Substituted Analogs

[3.3.1]AE-Bicyclic piperidinone **15** is easily prepared by the double Mannich strategy outlined above. Using this hindered ketone **15**, we have designed routes to take our SAR studies further by preparing C5-C6 containing MLA analogs.

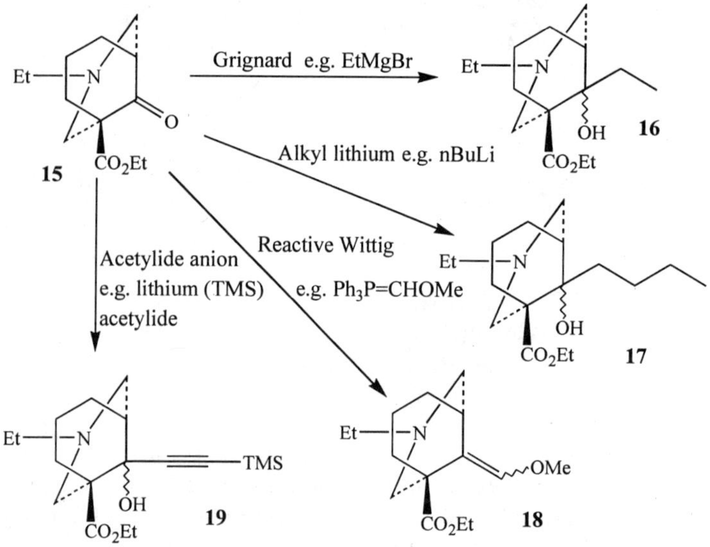

With this synthetic strategy, we can investigate the significance of carbon substituents around ring B and of linking from the AE-bicycle to ring D through C6 and C7. We can now selectively introduce each of the carbon atoms in MLA and regioselectively vary the substitution pattern. The contribution of these substituents to nAChR affinity, within a defined conformational space, can therefore be assessed. Treatment of piperidinone **15** with a Grignard reagent (*44,45*) regioselectively gave tertiary alcohol **16**. Reaction of **15** with an alkyl lithium salt (*46*) afforded tertiary alcohol **17**. Wittig reaction of **15** was only practical with reactive (and small) ylids, yielding enol ether **18**. Stabilized ylids did not react under typical Horner-Emmons conditions. A one-carbon homologation of ketone **15** to enol ether **18** (*47*) is interesting, but of somewhat limited utility. The nucleophilic addition of an acetylide anion (e.g. lithium TMS-acetylide) to ketone **15** afforded propargyl alcohol **19** in good yield (*48-51*). The functionality present in this C5-C6 substituted analog allows us to design practical entries to highly oxygenated, functionalized probes for SAR studies.

Acetylide Alkoxide Dianion Strategy

One such potentially important entry for the design of molecular probes is the highly oxygenated carbon framework containing the three 6-membered MLA rings A, D, and E **20**. We rationalized that such a molecule could be prepared from the corresponding alkyne **21** by reduction (*52,53*). Using acetylenic nucleophiles (*48-51*), we can incorporate C6 and C7 (norditerpenoid numbering) introducing appropriately positioned functional groups. One illustration of the versatility of this strategy is that partial reduction of this alkyne functional group **21** will allow (enantioselective) epoxidation of the resultant allylic alcohol. Vicinal-diol synthesis and potentially regiochemically controlled methyl ether formation can also be designed. Such changes in oxygenation pattern around C6-C7-C8 closely mimic some of the important differences between *Delphinium* (e.g. LA **2** and lycoctonine **3**) and *Aconitium* (e.g. aconitine **4**) alkaloids. Therefore, a practical route to alkyne **21** was devised.

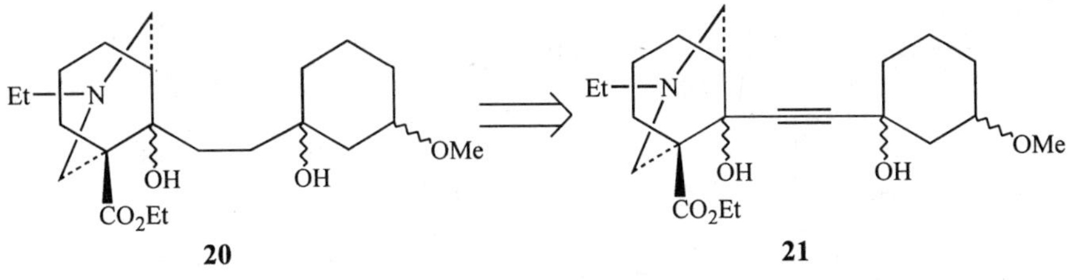

20 **21**

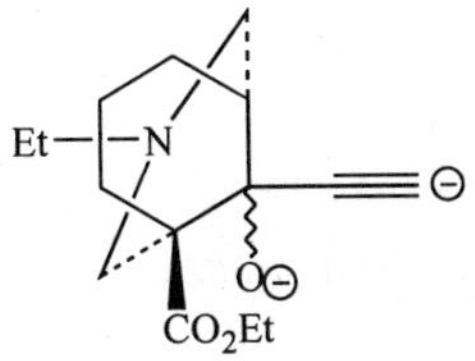

22

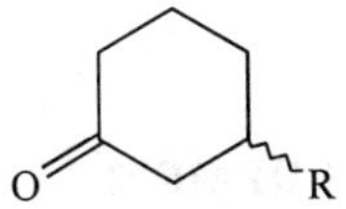

23a R = H
23b R = OMe

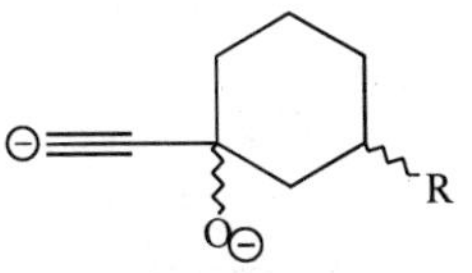

24a R = H
24b R = OMe

Acetylide alkoxide dianion **22** was prepared from TMS-acetylene, but reaction with cyclohexanone **23a** gave only a poor yield (~5 %) of the desired product. Reaction of basic dianion **22** with 3-methoxycyclohexanone **23b** gave only cyclohex-2-enone by elimination of methanol. The alternative order of C-C bond formation, forming C5-C6 after C7-C8, was more rewarding. Acetylide alkoxide dianions prepared by reaction of TMS-acetylene with cyclohexanone **24a**, and from 3-methoxycyclohexanone **24b**, did react in the expected manner with bicyclic β-ketoester **15**.

Preparation of C14/C16-Methoxy-Substituted Analogs

Cyclohex-2-enone (Aldrich) **25** was reacted with methanol under acid catalysis to afford an enantiomeric pair of methyl ethers **23b** (Scheme 2) where the 3-methoxy functional group was presumed to prefer an axial orientation (following the literature precedent of axial cyanide addition to a cyclohex-2-enone). The acetylide anion was prepared from TMS-acetylene, using n-BuLi in THF at 0 °C, and this reacted smoothly with ketone **23b** (with no detectable elimination of methanol to give cyclohex-2-enone in this case, *vide supra*). The diastereoisomeric mixture of propargyl alcohols **26** formed was not purified, but was efficiently deprotected under basic conditions using aqueous sodium hydroxide in methanol (95 %) to afford key acetylenic alcohol **24b** which incorporates the (C14/C16)-methoxy-substituted cyclohexane ring D of MLA **1**.

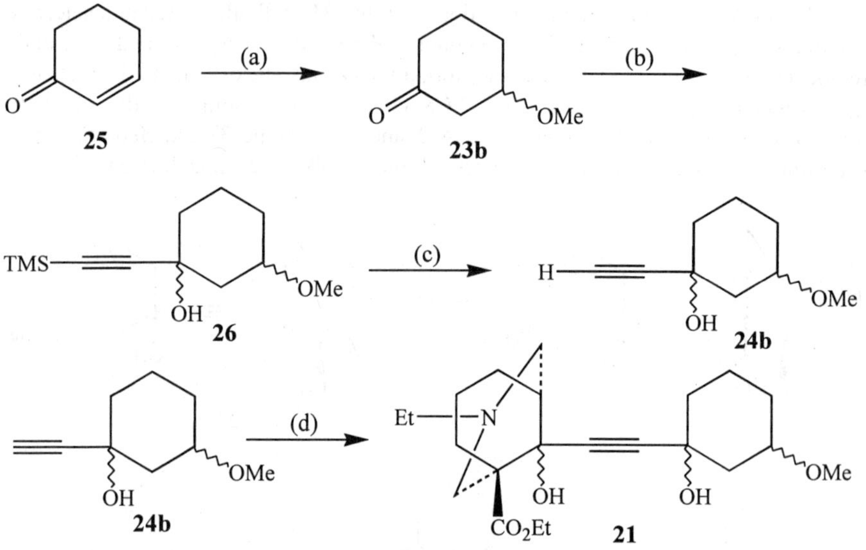

SCHEME 2

(a) MeOH, H^+ (b) TMS-acetylene, nBuLi, THF (c) NaOH, MeOH

(d) 2.4 nBuLi, bicyclic ketoester **15**, THF

Treatment of propargyl alcohol **24b** with 2.4 equivalents of n-BuLi gave the corresponding dianion which reacted (THF, 0 °C) with bicyclic β-ketoester **15** at the more electrophilic cyclic ketone functional group, in preference to the ethyl ester. After silica gel chromatography, bis-tertiary alcohol **21** was isolated as a colorless oil (~60 %). Bis-tertiary alcohol **21** is an advanced synthetic intermediate for the design and preparation of small molecules which will be useful as nAChR probes. This compound contains all of the carbon framework of MLA norditerpenoid rings A, D, and E, an appropriately located *N*-ethylated tertiary amine, tertiary alcohols at C5 (unrequired for the natural product series) and C8 (norditerpenoid numbering), and a C14/C16-methoxy substituent. Complete reduction (Scheme 3) of the C6-C7 alkyne functional group **21** (hydrogenation, 10 % Pd/C, EtOH, 15 °C, 60 %) gave alkane diol **20** as a colorless oil. Partial reduction of this alkyne **21**, in the presence of Lindlar's catalyst (*53*) or better with Pd/C poisoned with 10 % pyridine in ethanol (*52*), gave *Z*-alkene **27** as a colorless oil. These compounds now require functional group interconversion (ethyl ester into neopentyl-like alcohol) for final conversion into anthranilate esters. We were also able to take advantage of the propargylic alcohol functionality in **21** which was reduced with $LiAlH_4$ (*53*), with useful concomitant reduction of the ethyl ester, to afford *E*-alkene triol **28** (40 %). Thus, concise routes to synthetic AE-bicyclic analogs of MLA have been designed and developed.

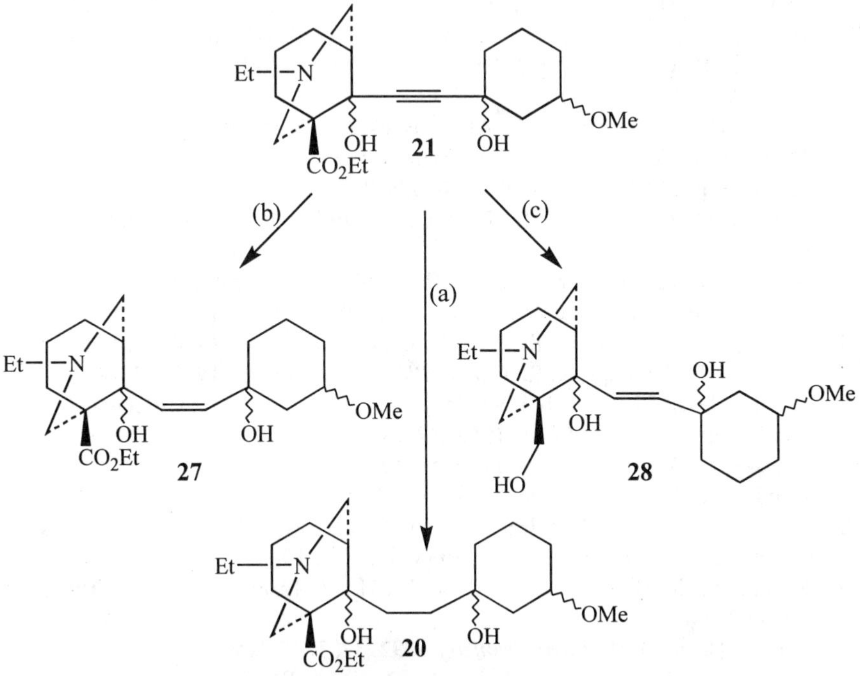

SCHEME 3

(a) H_2, Pd/C, EtOH (b) H_2, Lindlar's catalyst, EtOH (c) $LiAlH_4$, Et_2O

Acknowledgments

We wish to acknowledge financial support from Zeneca Agrochemicals and EPSRC through the CASE award scheme (WJT), Fondation pour la Recherche Médicale (GG), and The Wellcome Trust (Project Grants 036214 and 045023).

Literature Cited

1. Pelletier, S. W.; Mody, N. V.; Joshi, B. S.; Schramm, L. C. In *Alkaloids: Chemical and Biological Perspectives*; Pelletier, S. W., Ed.; Wiley-Interscience: New York, NY, 1984, Vol. 2; pp 205-462.
2. Pelletier, S. W.; Joshi, B. S. In *Alkaloids: Chemical and Biological Perspectives*; Pelletier, S. W., Ed.; Springer-Verlag: New York, NY, 1991, Vol. 7; pp 297-564.
3. Yunusov, M. S. *Nat. Prod. Reports* **1991**, *8*, 499-526.
4. Yunusov, M. S. *Nat. Prod. Reports* **1993**, *10*, 471-486.
5. Jennings, K. R.; Brown, D. G.; Wright, D. P. *Experientia* **1986**, *42*, 611-613.
6. Jennings, K. R.; Brown, D. G.; Wright, D. P.; Chalmers, A. E. *ACS Symposium Series* **1987**, *356*, 274-282.
7. Keeler, R. F. *Lloydia* **1975**, *38*, 56-61.
8. Manners, G. D.; Panter, K. P.; Pelletier, S. W. *J. Nat. Prod.* **1995**, *58*, 863-869.
9. Benn, M. H.; Jacyno, J. M. In *Alkaloids: Chemical and Biological Perspectives*; Pelletier, S. W., Ed.; John Wiley and Sons: New York, NY, 1983, Vol. 1; pp 153-210.
10. Plinius, G. *Naturalis Historia* AD77, Book 23, Chapter 13.
11. Macallan, D. R. E.; Lunt, G. G.; Wonnacott, S.; Swanson, K. L.; Rapoport, H.; Albuquerque, E. X. *FEBS Lett.* **1988**, *226*, 357-363.
12. Ward, J. M.; Cockcroft, V. B.; Lunt, G. G.; Smillie, F. S.; Wonnacott, S. *FEBS Lett.* **1990**, *270*, 45-48.
13. Wonnacott, S.; Albuquerque, E. X.; Bertrand, D. In *Methods in Neurosciences*; Conn, M., Ed.; Academic Press: New York, NY, 1993, Vol. 12; pp 263-275.
14. Kukel, C. F.; Jennings, K. R. *Can. J. Physiol. Pharmacol.* **1994**, *72*, 104-107; Kukel, C. F.; Jennings, K. R. *Can. J. Physiol. Pharmacol.* **1995**, *73*, 145.
15. Hardick, D. J.; Blagbrough, I. S.; Cooper, G.; Potter, B. V. L.; Critchley, T.; Wonnacott, S. *J. Med. Chem.* **1996**, *39*, 4860-4866.
16. Manske, R. H. *Can. J. Research* **1938**, *16B*, 57-60.
17. Goodson, J. A. *J. Chem. Soc.* **1943**, 139-141.
18. Coates, P. A.; Blagbrough, I. S.; Hardick, D. J.; Rowan, M. G.; Wonnacott, S.; Potter, B. V. L. *Tetrahedron Lett.* **1994**, *35*, 8701-8704.
19. Majak, W.; McDiarmid, R. E.; Benn, M. H. *J. Agric. Food Chem.* **1987**, *35*, 800-802.
20. Sun, F.; Benn, M. H. *Phytochemistry* **1992**, *31*, 3247-3250.
21. Pelletier, S. W.; Djarmati, Z.; Lajsic, S.; De Camp, W. H. *J. Am. Chem. Soc.* **1976**, *98*, 2617-2624.
22. Pelletier, S. W.; Mody, N. V.; Varughese, K. I.; Maddry, J. A.; Desai, H. K. *J. Am. Chem. Soc.* **1981**, *103*, 6536-6538.

23. Blagbrough, I. S.; Hardick, D. J.; Wonnacott, S.; Potter, B. V. L. *Tetrahedron Lett.* **1994**, *35*, 3367-3370.
24. Hardick, D. J.; Blagbrough, I. S.; Wonnacott, S.; Potter, B. V. L. *Tetrahedron Lett.* **1994**, *35*, 3371-3374.
25. Blagbrough, I. S.; Coates, P. A.; Hardick, D. J.; Lewis, T.; Rowan, M. G.; Wonnacott, S.; Potter, B. V. L. *Tetrahedron Lett.* **1994**, *35*, 8705-8708.
26. Coates, P. A.; Blagbrough, I. S.; Rowan, M. G.; Potter, B. V. L.; Pearson, D. P. J.; Lewis, T. *Tetrahedron Lett.* **1994**, *35*, 8709-8712.
27. Coates, P. A.; Blagbrough, I. S.; Lewis, T.; Potter, B. V. L.; Rowan, M. G. *J. Pharm. Biomed. Anal.* **1995**, *13*, 1541-1544.
28. Hardick, D. J.; Cooper, G.; Scott-Ward, T.; Blagbrough, I. S.; Potter, B. V. L.; Wonnacott, S. *FEBS Lett.* **1995**, *365*, 79-82.
29. Coates, P. A.; Blagbrough, I. S.; Rowan, M. G.; Pearson, D. P. J.; Lewis, T.; Potter, B. V. L. *J. Pharm. Pharmacol.* **1996**, *48*, 210-213.
30. Hardick, D. J.; Blagbrough, I. S.; Potter, B. V. L. *J. Am. Chem. Soc.* **1996**, *118*, 5897-5903.
31. Kraus, G. A.; Shi, J. *J. Org. Chem.* **1990**, *55*, 5423-5424.
32. Kraus, G. A.; Shi, J. *J. Org. Chem.* **1991**, *56*, 4147-4151.
33. Kraus, G. A.; Andersh, B.; Su, Q.; Shi, J. *Tetrahedron Lett.* **1993**, *34*, 1741-1744.
34. Baillie, L. C.; Bearder, J. R.; Whiting, D. A. *J. Chem. Soc. Chem. Commun.* **1994**, 2487-2488.
35. Jacyno, J. M.; Harwood, J. S.; Lin, N.-H.; Campbell, J. E.; Sullivan, J. P.; Holladay, M. W. *J. Nat. Prod.* **1996**, *59*, 707-709.
36. Wonnacott, S. *Trends in Neuroscience* **1997**, *20*, 92-98.
37. Eastham, H. M.; Reynolds, S. E.; Wolstenholme, A. J.; Wonnacott, S., University of Bath, unpublished data.
38. Stager, R. P.; Miller, E. B. *J. Org. Chem.* **1959**, *24*, 1214-1219.
39. Weiler, L. *J. Am. Chem. Soc.* **1970**, *92*, 6702-6704.
40. Huckin, S. N.; Weiler, L. *J. Am. Chem. Soc.* **1974**, *96*, 1082-1087.
41. Hauser, C. R.; Harris, T. M. *J. Am. Chem. Soc.* **1958**, *80*, 6360-6363.
42. Mukhopadhyay, T.; Seebach, D. *Helv. Chimica Acta* **1982**, *65*, 385-391.
43. Shimzu, B.; Ogiso, A.; Iwai, I. *Chem. Pharm. Bull.* **1963**, *11*, 333-336.
44. Kotani, R. *J. Org. Chem.* **1965**, *30*, 350-354.
45. Flannery, R. E.; Hampton, K. G. *J. Org. Chem.* **1972**, *37*, 2806-2890.
46. Buhler, J. D. *J. Org. Chem.* **1973**, *38*, 904-906.
47. Wittig, G.; Böll, W.; Krück, K.-H. *Chem. Ber.* **1962**, *95*, 2514-2525.
48. Shimzu, B.; Ogiso, A.; Iwai, I. *Chem. Pharm. Bull.* **1963**, *11*, 760-769.
49. Ogiso, A.; Shimzu, B.; Iwai, I. *Chem. Pharm. Bull.* **1963**, *11*, 770-774.
50. Ogiso, A.; Shimzu, B.; Iwai, I. *Chem. Pharm. Bull.* **1963**, *11*, 774-779.
51. Scharpwinkel, K.; Matull, S.; Schäfer, H. J. *Tetrahedron: Asymmetry* **1996**, *7*, 2497-2500.
52. Montalbetti, C.; Savignac, M.; Bonnefis, F.; Genêt, J. P. *Tetrahedron Lett.* **1995**, *36*, 5891-5894.
53. Walborsky, H. M.; Wüst, H. H. *J. Am. Chem. Soc.* **1982**, *104*, 5807-5808.

Chapter 21

Structure-Activity Relationshipsin 2,4-di-*tert*-Butyl-6-(4′-Substituted Benzyl)phenols as Chemosterilants for the House Fly (*Musca domestica* L.)

Jan Kochansky[1], Charles F. Cohen[2], and William Lusby[2]

[2]Insect Neurobiology and Hormone Laboratory, Western Regional Research Center, Agricultural Research Service, U.S. Department of Agriculture, Beltsville, MD 20705–2350

A series of 2,4-di-*tert*-butyl-6-(4′-X-benzyl)phenols was prepared, analogous to the compound Jurd 2644 (X = OCH_3) to investigate a possible Hammett relationship between the 4'-substituent and the chemosterilant activity of the compounds in adult *Musca domestica.* Fourteen compounds of this type were prepared, 12 of them new. Jurd 2419 (X = H) was found to be as active in our bioassay as J2644, as were four new compounds (X = $N(CH_3)_2$, CH_3, SCH_3, Cl). All prevented successful reproduction when administered at 30 mg/kg of diet. There was no apparent correlation between Hammett σ and activity. Active compounds all had small substituents and molecular weights below 350. The activity seemed to be determined by the bulk of X rather than by its electronic properties.

As part of a study of compounds responsible for the durability of wood from tropical trees of the genus *Dalbergia,* Jurd et al. (*1*) reported on the antimicrobial properties of obtusastyrene (4-cinnamylphenol). This study was soon extended to a series of synthetic analogs of obtusastyrene (*2*), and to benzylated phenols and benzodioxoles, summarized by Jurd and Manners (*3*). Specifically, certain benzyl-substituted di-*tert*-butylphenols (particularly Jurd 2644 (2,4-di-*tert*-butyl-6-(4′-methoxybenzyl)phenol) **1a)** were capable of sterilizing adult female house flies *(Musca domestica* L.) when fed a diet containing concentrations as low as 0.025% (*4*). This sterility was not permanent, since egg viability returned when treated flies were subsequently fed untreated diets. There was also a sterilizing effect on males. Jurd 2644 (**1a**) was the most active phenol, and it was used successfully for control of house flies in two screwworm rearing plants (*5*). The compound was non-mutagenic in the Ames tests

[1]Current address: Bee Research Laboratory, Agricultural Research Service, U.S. Department of Agriculture, Building 476, 10300m Baltimore Avenue, Beltsville, MD 20705-2350. The INHL has been disbanded.

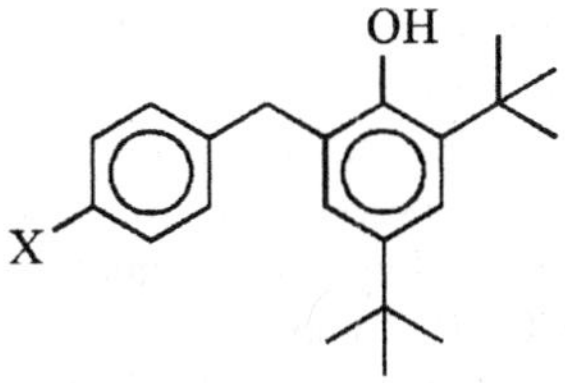

1

(*4*) and was therefore particularly attractive as a focus for further work. Chang et al. (*6*) confirmed the activity of J2644 in flies treated orally, topically, or by injection, and also noted the impermanence of its action. Matolczy et al. (*7*) replaced the methoxy group in **1a** with propargyloxy in an attempt to create a cytochrome P450 oxidase inhibitor, but the compound was inactive against the black blow fly, *Phormia regina,* at 10 g/l in the diet.

Chemosterilant activity of J2644 is not limited to house flies. It also sterilized face flies (*Musca autumnalis* DeGeer), (*8*), screwworm flies (*Cochliomyia hominivorax* (Coquerel)) treated orally (*9*) or topically (*10*), and the old-world screwworm fly (*Chrysomya bezziana* Villeneuve) (*11*). Activity, however, was not universal in diptera. J2644 and J2419 (2,4-di-*tert*-butyl-6-benzylphenol, **1b**) were tested against the tsetse fly (*Glossina morsitans morsitans* Westwood) and showed very low activity (*12*). Compounds J2644 and J2419 were also poorly active against pink bollworm (*Pectinophora gossypiella* (Saunders)) (*13*). While J2644 and J2419 interfered with mosquito larval development (at concentrations higher than those required by their methyl ethers) (*14*), treatment of mosquito eggs (*15*) or pupae (*16*) did not inhibit egg hatch to any useful extent.

It has been suggested (*3, 4*) that the mode of action of these compounds involves a microsomal oxidation to unstable *o*-quinone methides, followed by reaction of these intermediates with cell components to inhibit egg production or hatching. Among a series of papers published on chemical oxidation of this class of compounds are three on J2644 itself (*17–19*).

Since J2644 (X = OCH_3) was reported (*4*) to be four-fold more active than J2419 (X = H), we considered the possibility of a Hammett relationship (*20*) between activity and X. We realized, of course, that such a relationship neglects partition coefficients, absorption rates, activation, and other factors operative in biological systems. (What, after all, is the solvent polarity of a house fly?) In hopes that such complicating factors would not totally mask the effect of ring substituents on activity, and that we might find compounds more active than J2644, we prepared a series of analogs in which the substituent on the benzyl ring was varied from the strongly electron-donating $N(CH_3)_2$ group to strongly electron-withdrawing groups such as SO_2CH_3 and CO_2CH_3. Subsequent to the symposium presentation, the data on these compounds has been published (*21*).

Experimental

Synthesis and Characterization of the Chemicals Tested. Compounds with fairly strongly electron-donating substituents (OR, SCH_3, *tert*-butyl, $N(CH_3)_2$) were synthesized by suitable modifications of Jurd's method (*4*): reaction of a substituted benzyl alcohol with 2,4-di-*tert*-butylphenol in a mildly acidic medium (acetic acid or acetic/formic acid mixtures, sometimes containing a little oxalic acid) ("Method A"). Compounds having mildly electron-donating to mildly electron-withdrawing substituents (H, CH_3, Cl) were prepared by Green's method (*22*) using the phenol, the benzyl chloride, and $ZnCl_2$ in $CHCl_3$ ("Method B").

Attempts to prepare compounds having strongly electron-withdrawing substituents by either of these methods failed, since more strenuous reaction conditions led to loss of *tert*-butyl groups from the starting phenol. These compounds were therefore prepared by conversion of more easily accessible materials by oxidation (**1i, 1j**) or alkylation (**1n**) of more readily available materials.

The carbomethoxybenzylphenol was prepared by a roundabout route. Acylation of 2,4-di-*tert*-butylanisole with 4-trifluoromethylbenzoyl hexafluoroantimonate at 0° C in sulfolane (containing 10% dipropyl sulfone to depress the melting point) by the method of Olah et al. (*23*) gave the 4′-trifluoromethylbenzoyl 2,4-di-*tert*-butylanisole in 37% yield. Treatment of this material under Wolff-Kishner conditions with KOH and hydrazine in ethylene glycol gave a mixture of deoxygenated, more or less hydrolyzed products (unfortunately none with an intact CF_3 group). Fractional crystallization and selective remethylation gave compound **1m** in low yield.

Table I: Synthesis of **1** and Analogs

	Substituent X	Yield %	M. P. °C	Method (see text)
1a	OCH_3	58	82–84	A
1b	H	22	61–65	B
1c	OH	64	112–116	Demethylation of **1a** methyl ether (BBr_3)
1d	*n*-butoxy	47	54–59	A
1e	*n*-decyloxy	—	—	A
1f	$N(CH_3)_2$	7.6	114–116.5	A
1g	*tert*-butyl	65	55–69	A
1h	SCH_3	51	104–106.5	A
1i	$SOCH_3$	70	130–134	Oxidation of **1h** ($NaIO_4$)
1j	SO_2CH_3	45	128–132	Oxidation of **1h** (H_2O_2)
1k	CH_3	8	85–88	B
1l	Cl	6	90–92	B
1m	CO_2CH_3	—	106.5–108.5	See text.
1n	$N^+(CH_3)_3$	70	164–166	Methylation of **1f** (CH_3I)

Synthetic details have been published (*21*). Crude products were usually distilled, but even after distillation or silica gel chromatography, compounds with bulky or nonpolar substituents on the benzyl group crystallized very poorly, frequently only after weeks or months at 5° C, giving sticky solids of wider than normal melting ranges. These compounds, however, were adequately pure (>98% with the exception of **1c** >95%) by GC. Gas chromatography/mass spectrometry gave fragmentation patterns consistent with the expected structures. Preparation methods and melting points are summarized in Table I.

Bioassay. The house flies (NAIDM-l948 strain) were from our normal laboratory colony, reared by the CSMA procedure (*24*) on Purina 5060 Fly Larvae Media *[sic]*. The bioassay for chemosterilant activity was as described by Robbins *et al.* (*25*). Briefly, unfed house flies (100 flies, 50 each sex) <18 hours old were placed in a cage with water and treated diet (6grams; sucrose (42.5%), dry nonfat milk (42.5%), and dry whole egg (15%), coated with the experimental compound using acetone). Eggs were collected on days 6 and 12, and the flies were then provided with untreated diet for another week, after which time the final sample of eggs was collected. Aliquots of *ca.* 100 eggs were removed to determine hatchability. Duplicate egg samples were reared through on a sterile semi-defined medium for pupal and adult survival counts. Total egg production of treated or control flies was measured by using calibrated pipettes *(ca.* 10^4 eggs/mL). Separate controls were run for each batch of 4–5 compounds at each concentration. Experimental concentrations were decreased in a 1, 0.3, 0.1, ... series starting at 1%, and the minimum concentration at which no viable eggs were produced was recorded. We use the term "viable" to mean capable of hatching and developing to pupation *(vide infra)*.

Results

Bioassays. The results of the bioassays are summarized in Table II. The data for each concentration/week are in the form eggs per female (total eggs divided by number of surviving females)/percent hatch/number of pupae produced (average of two replicates). Dashes indicate inability to test since the previous number was zero.

Table II: Chemosterilant activity of compounds **1** against *Musca domestica* adults. Compounds administered in the diet at indicated concentrations

	Substituent X	Concentration (mg/kg diet)	Week 1	Week 2 (oviposition/hatch/pupation[a])	Week 3
1a	OCH_3	10^4	0/–/–	0/–/–	0/–/–[b]
		500	0/–/–	27/53/0	43/60/12
		300	67/40/0	70/64/0	61/94/23
		30	28/0/0	63/0/0	61/80/35
		10	14/69/49	79/94/32	77/85/49
1b	H	10^4	0/–/–	0/–/–	0/–/–
		10^3	3/0/–	43/0/–	32/58/0[b]
		300	6/68/0	51/23/0	37/83/8
		30	3/n.d./0	26/0/–	55/55/18[c]
		10	80/95/68	96/83/41	59/82/52

Table II, continued

	Substituent X	Concentration (mg/kg diet)	Week 1	Week 2	Week 3
			(oviposition/hatch/pupation[a])		
1c	OH	10^4	8/50/0	83/88/0	0/–/–[b]
		10^3	68/76/33	60/81/23	57/87/44
1d	*n*-butoxy	10^4	27/75/56	76/81/12	65/84/36
1e	*n*-decyloxy	10^4	6/94/60	88/89/44	103/86/58
1f	$N(CH_3)_2$	10^4	45/26/0	51/38/0	0/–/–[b]
		10^3	61/0/–	12/21/0	30/46/12
		300	76/40/0	60/67/0	55/98/24
		30	56/0/–	44/0/–	36/83/22
		10	6/100/0	73/87/27	82/84/56
1g	*tert*-butyl	10^4	21/84/46	65/89/48	66/85/35
1h	SCH_3	10^4	36/0/0	65/31/0	27/56/0
		10^3	14/6/0	53/55/0	22/63/0
		300	13/76/0	62/1/0	52/71/48
		30	20/0/–	57/0/–	77/82/49
		10	67/86/75	89/84/43	63/86/41
1i	$SOCH_3$	10^4	63/53/27	49/80/41	66/85/60
1j	SO_2CH_3	10^4	0/–/–	44/82/31	30/80/39
1k	CH_3	10^4	0/–/–	26/44/0	4/75/0
		10^3	47/10/0	0/–/–	31/65/31
		300	37/65/0	70/12/0	77/73/3
		30	70/0/–	81/0/–	89/0/–
		10	4/0/–	60/97/59	64/94/47
1l	Cl	10^4	0/–/–	0/–/–	0/–/–
		10^3	0/–/–	40/1/0	45/68/0
		300	28/56/0	55/16/0	68/82/25
		30	5/n.d./–	23/0/–	47/0/–
		10	75/97/75	104/93/53	79/92/50
1m	CO_2CH_3	10^4	16/28/8	33/44/2	6/0/–[b]
1n	$N^+(CH_3)_3$	10^4	20/20/0	43/72/20	55/89/50
Control[d]			61/92/50	78/88/50	66/87/50

[a]Eggs oviposited per female/% hatch/% pupation. n. d. not determined. Mean of two determinations.
[b]All males died.
[c]No adults emerged from 35 pupae in the two replicates.
[d]Means of all controls, standard deviations in text.
SOURCE: Adapted from ref. 21.

In contrast to Jurd's results (*4*), we found that the 4′-methoxybenzyl (**1a**) and benzyl (**1b**) compounds were equally active, both compounds completely inhibiting production of viable eggs at 30 mg/kg of diet (Figure 1). In Jurd's assay, the 4′-methoxybenzyl compound J2644 **(1a)** required 250 mg/kg of diet and the benzyl compound J2419 (**1b**) required 10^3 mg/kg to reduce pupation to zero. Compounds **1f** (X = $N(CH_3)_2$), **1h** (X = SCH_3), **1k** (X = CH_3), and **1l** (X = Cl) were as active as **1a** and **1b** in our assay, requiring only 30 mg/kg to inhibit successful reproduction.

We use the term "successful reproduction" to mean the production of "viable" eggs, which we define in turn as those that produce larvae which can develop to pupae (and usually adults). All of the compounds active at 30 mg/kg in our assay allowed, at this or higher concentrations, production of eggs which hatched. However, all larvae from these died, usually in the first instar. For the purposes of this assay, these eggs were regarded as "non-viable". In one case (compound **1b** at 30 mg/kg in the third week of the test) pupae were produced but no adults emerged. Jurd may have observed a similar effect, since his definition of hatch (*4*) is "proportion of progeny reaching the pupal stage from 100 eggs." This is reportedly common for chemosterilants in flies, with egg hatch alone not being considered to be a good criterion for sterilization (*26*, *27* and references therein). LaBrecque (*26*) also recommend the pupae/100 eggs standard.

With the exception of the 4'-hydroxy analog **1c**, all other compounds were inactive, i.e. viable eggs were produced at diet levels of 10^4 mg/kg. Analog **1c** prevented successful reproduction at 10^4 mg/kg but allowed it at 10^3 mg/kg. It was not tested at 3 x 10^3 mg/kg, but this concentration would only confirm the difference between "not very active" and "hardly active at all." Its activity is therefore shown as a range in Figure 1.

Under these bioassay conditions, control flies produced 61.0±17.4 eggs/female the first week, 78.5±14.8 the second, and 66.0±15.9 the third (N=13). Egg hatch for controls was 92.2±3.5% (N=12), 88.5±4.4%, and 87.9±5.3% respectively. Production of pupae and adults in the rear-through controls was 50±13 and 34±13 per replicate, respectively.

High concentrations (10^3-10^4 mg/kg) of **1a**, **1b**, **1c**, **1f**, **1h**, **1k**, and **1l** gave permanent inhibition of reproduction (no viable eggs produced the third week), although this was sometimes the result of complete male mortality. In all cases at lower concentrations, males survived, and viable eggs were usually produced in the third week of the test, during which the flies were offered untreated diet. No viable eggs were produced with compounds **1k** and **1l** at 30 mg/kg the third week, but both compounds had allowed formation of at least a few pupae at 300 mg/kg.

Several of these compounds, particularly **1k** and **1l** but to a lesser extent **1a**, **1b**, and **1f**, showed a "bimodal" activity curve. They were more active at high (>1000 mg/kg) and low (30 mg/kg) concentrations than at 300 mg/kg. Since the assays were not all run at the same time, it is unlikely to be a systematic error in the assay. A possible rationalization of these data is a slow spontaneous oxidation in the insect to an active species (quinone methide?) at high concentrations, and enzymatic activation only at lower concentrations. We have no data to support this speculation, but at least it seems plausible.

Figure 1 shows a plot of minimal effective concentration *vs.* the σ constants (*28*) for the various substituents. Data from Jurd et al. (*4*) are also plotted. None of the compounds with σ > 0.4 was active. Compounds having more electron-donating substituents seem to have activity based more on steric grounds, since compounds having large (*tert*-butyl and the two large alkoxy groups) substituents were inactive, whereas **11** (X = Cl, σ = 0.227) is as active as **1f** (X = $N(CH_3)_2$, σ = - 0.600). Compounds **1a, 1d,** and **1e** provide a steric probe. The 4′-methoxy compound **1a** was active, while the 4′-butoxy compound **1d** (σ = - 0.320) was inactive, as was **1e** with a probably similar σ for *n*-decyloxy. It seems unlikely that there would be a very large difference in polarity in such a large molecule with the addition of three CH_2 groups to **1a** to form **1d**, but the size does increase. The 4′-hydroxy compound **1c** (σ = - 0.357) is anomalous, its poor activity possibly due to hydrogen-bonding effects leading to poor absorption or inefficient receptor binding, or to metabolic conversion to an inactive conjugate.

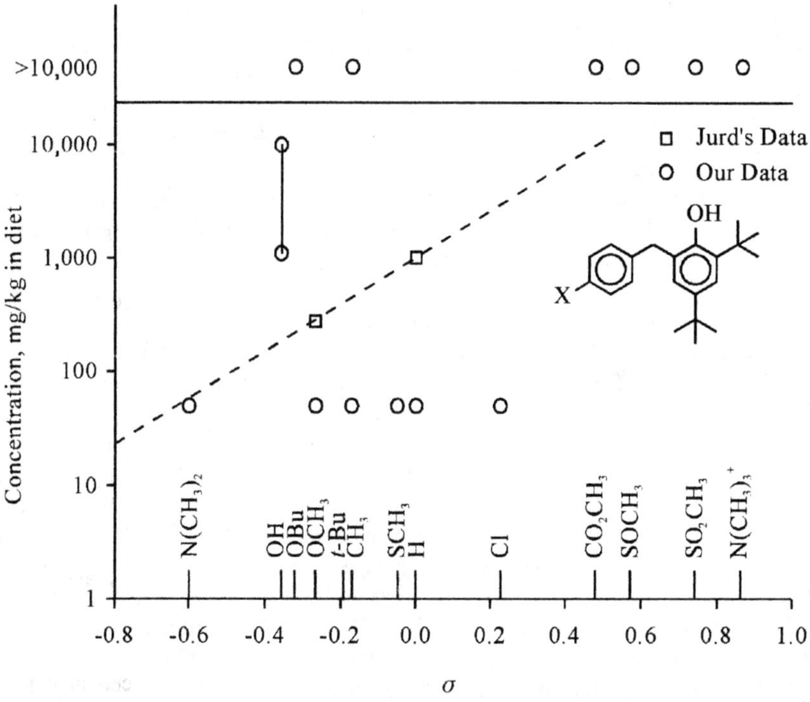

Figure 1. Plot of minimum sterilizing concentration (the lowest concentration at which eggs produced no "viable" larvae, see text) of compounds **1** vs. σ constants of substituents. The *n*-decyloxy compound **1c** is not plotted, since we could find no σ constant for it. It should be very close to that of the *n*-butoxy compound **1d,** and was also inactive at 1%.
(Reproduced from reference 21. Copyright 1995 American Chemical Society.)

Activity Correlations. The best correlation of activity appeared to be with molecular weight. The "receptor site" (*sensu lato*, this could be the channel by which the phenol is absorbed into the cell, the binding site of the activation enzyme system or the ultimate site of action of the activated phenol derivative) seems to be very sensitive to the size of the active species. Of six compounds active at 30 mg/kg, all have molecular weights between 296 and 342. Six compounds inactive at 10^4 mg/kg all had molecular weights from 352 to 452. Analog **1n** has a molecular weight of 354 for the cation, is inactive, but probably shouldn't be included in the correlation because of its charge. Compound **1c** again doesn't fit. The 4′-CF_3 compound (molecular weight 380.5, but similar in size to **1a** or **1k**) would have been an interesting compound to test in this regard, but we were unsuccessful in our attempts to synthesize it (as well as the corresponding 4′-nitro analog) in the time available.

Casual inspection of the relationship between structure and calculated (HyperChem, release 4 with ChemPlus extensions, release 1, Hypercube, Inc., Waterloo, Ontario) values for various QSAR parameters generally supported the size/activity correlation. The dividing line used was activity at 30 mg/kg in the diet. Presentation is in the form parameter (active compound values, inactive compound values, value for *p*-OH compound **1c**): Solvent accessible surface area, Å^2 (530-626, 603-845, 540), Van der Waals surface area, Å^3 (343-403, 387-552, 352), solvent accessible volume, Å^3 (307-352, 344-482, 314), log P (octanol/water) (6.8-7.6, 8.05-10.43, 6.81), and refractivity, Å^3 (95-109, 108-143, 97). There was no obvious correlation with polarizability. These data, in spite of some overlap between active/inactive value ranges, support the smaller, not-too-lipophilic nature of active materials. **1c** is anomalous in essentially all of these tests.

Conclusions

It is clear that no good correlation of activity with Hammett σ value exists for this series of compounds in this bioassay. All active compounds had small substituents and correspondingly small molecular weights (under 350). Among the new compounds prepared, four (**1f**, X = $N(CH_3)_2$, **1h**, X = SCH_3, **1k**, X = CH_3, and **1l**, X = Cl) were as active as J2644 and J2419, but none exceeded their activity. Because certain compounds in this work allowed female flies to produce eggs which hatched but failed to produce surviving larvae, care should be taken in interpreting previous data on these or similar compounds.

Acknowledgments

Mention of a commercial product does not constitute an endorsement by the U.S. Department of Agriculture.

Literature Cited

1. Jurd, L., King, Jr., A. D., Mihara, K. and Stanley, W. L. *Appl. Microbiol.* **1971**, *21*, 507–510.
2. Jurd, L., Stevens, K. L., King, Jr., A. D., and Mihara, K. *J. Pharm. Sci.*, **1971**, *60*, 1753–1755.
3. Jurd, L. and Manners, G. D. *J. Agric. Food Chem.*, **1980**, *28*, 183–188.

4. Jurd, L., Fye, R. L and Morgan, Jr., J. *J. Agric. Food Chem.*, **1979**, *27*, 1007–1016.
5. Rawlins, S. C., Woodard, D. B., Coppedge, J. R. and Jurd, L. *J. Econ. Entomol.*, **1982**, *75*, 728–732.
6. Chang, S. C., Borkovec, A. B. and DeMilo, A. B. *J. Econ. Entomol.*, **1980**, *73*, 745-747.
7. Matolczy, G., Feyereisen, R., Van Mellaert, H., Pál, A., Varjas, L., Bélai, I. and Kulcsár, P. *Pestic. Sci.*, **1986**, *17*, 13–24.
8. Broce, A. B. and Gonzaga, V. G. *J. Econ. Entomol.*, **1987**, *80*, 37–43.
9. Rawlins, S. C., Jurd, L. and Snow, J. W. *J. Econ. Entomol.*, **1979**, *72*, 674–677.
10. Rawlins, S. C. and Jurd, L. *J. Econ. Entomol.*, **1981**, *74*, 215–217.
11. Pound, A. A. and Spradbery, J. P.. *J. Aust. Ent. Soc.* **1984**, *23*, 99–103.
12. Langley, P. A., Trewern M. A. and Jurd, L. *Bull. Ent. Res.*, **1982**, *72*, 473–481.
13. Flint, H. M., Jurd, L. and Merkle, J. R. *J Econ. Entomol.*. **1980**, *73*, 710–714
14. Dame, D. A. and Jurd, L. *Mosq. News*, **1983**, *43*, 50–54.
15. Nelson, F. R. S., Mohamed, A. K. and Vattikutti, P. *J. Am. Mosq. Control Assoc.*, **1985**, *1*, 240–242.
16. Nelson, F. R. S. and Hoosseintehrani, B. *J. Econ. Entomol.*, **1982**, *75*, 877–878.
17. Jurd, L. and Wong, R. Y. *Aust. J. Chem.*, **1981**, *34*, 1633–1644.
18. Jurd, L. and Wong, R. Y. *Aust. J. Chem.*, **1981**, *34*, 1645–1654.
19. Wong, R. Y. and Jurd, L. *Aust. J. Chem.*, **1984**, *37*, 2593–2597.
20. Taft, Jr., R. W. In: *Steric Effects in Organic Chemistry,* Newman, M. S., Ed., John Wiley & Sons, Inc., 2nd printing, New York, NY., 1963, chapter 13, pp. 556–675.
21. Kochansky, J., Cohen, C. F., and Lusby, W. R. *J. Agric. Food Chem.* **1995**, *43*, 2974–2980.
22. Green, H. A., U.S. Patent 3,193,526, **1965**.
23. Olah, G. A., Lukas, J. and Lukas, E.. *J. Am. Chem. Soc.*, **1969**, *91*, 5319–5323.
24. Chemical Specialties Manufacturers Association. *Soap and Sanitary Chem. Blue Book,* **1955**, 249–250, 267.
25. Robbins, W. E., Kaplanis, J. N., Thompson, M. J., Shortino, T. J. and Joyner, S. C. *Steroids,* **1970**, *16*, 105–125.
26. LaBrecque, G. C. In: *Principles of Insect Chemosterilization.* LaBrecque, G. C. and Smith, C. N., Eds. Appleton-Century Crofts, New York, 1968, Chapter 3, pp. 41–98.
27. Fye, R. L., LaBrecque, G. C. and Gouck, H. K. *J. Econ. Entomol.* **1966**, *59*, 485–487.
28. Jaffe, H. H. *Chem. Revs.*, **1953**, *53*, 191–261.

Control of Plant Fungi

Chapter 22

Oxazolidinones: A New Class of Agricultural Fungicides

Jeffrey A. Sternberg[1], Detlef Geffken[2], John B. Adams, Jr.[3], Douglas B. Jordan[3], Reiner Pöstages[2], Charlene G. Sternberg[3], Carlton L. Campbell[3], William K. Moberg[3], and Robert S. Livingston[3]

[1]Agricultural Products, Experimental Station, duPont and Company, Building 402, Wilmington, DE 19880

[2]Institut für Pharmazeutische Chemie, Universität Hamburg, Bundesstrasse 45, 20146 Hamburg, Germany

[3]Agricultural Products, Stine-Haskell Research Center, duPont and Company, Building 300, Newark, DE 19714

5-Methyl-5-(4-phenoxyphenyl)-3-phenylamino-2,4-oxazolidinedione, DPX-JE874, is a new agricultural fungicide under development by DuPont. DPX-JE874 is a member of a new class of oxazolidinone fungicides which demonstrate excellent control of plant pathogens in the Ascomycete, Basidiomycete, and Oomycete classes which infect grapes, cereals, tomatoes, potatoes, and other crops. The synthesis, mode of action, and structure-activity relationships of two types of oxazolidinones, 2-thioxo-4-oxazolidinones and 2,4-oxazolidinediones, will be discussed.

The discovery of new safe and environmentally benign agricultural products is becoming increasingly more challenging with rising business and regulatory pressures. As a result, DuPont, like many other large agricultural chemical companies, is pursuing several discovery strategies in its research efforts. One strategy that has proven particularly successful in recent years has been the procurement and screening of novel compounds from collaborators at universities and research institutes around the world. This strategy has provided a large and diverse array of novel compounds at a relatively low cost.

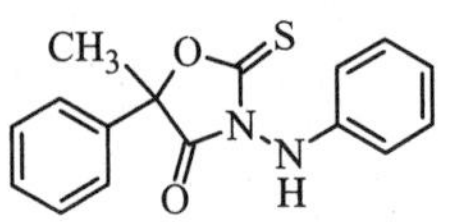

1

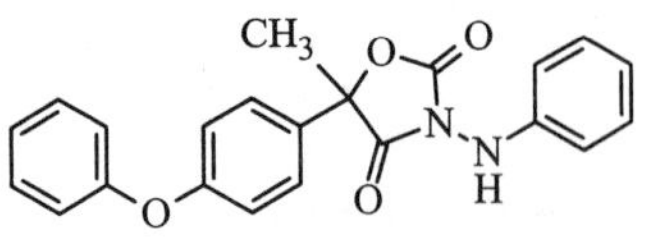

DPX-JE874

DuPont's entry into the oxazolidinone area resulted from the procurement of 5-methyl-5-phenyl-3-phenylamino-2-thioxo-4-oxazolidinone, **1**, from Professor Detlef Geffken then at the University of Bonn (*1*). An extensive analog program was initiated shortly after the fungicidal activity of this material was discovered through routine greenhouse testing. The program eventually culminated in the advancement of 5-methyl-5-(4-phenoxyphenyl)-3-phenylamino-2,4-oxazolidinedione, DPX-JE874 (famoxadone), to commercial development (*2,3*). This paper describes the synthesis of various oxazolidinone ring systems and the development of the structure-activity relationships which led to the discovery of DPX-JE874.

Mode of Action

DPX-JE874 is a potent inhibitor of mitochondrial electron transport blocking the function of ubiquinol:cytochrome c oxidoreductase (bc_1, complex III, EC 1.10.2.2). According to the criteria used for classifying natural product inhibitors of bc_1, oxazolidinones belong to Group I whose other members include myxothiazol, strobilurins, and stigmatellins (*4*). Whereas all of these materials bind in the Q_o domain of the enzyme complex, oxazolidinones are chemically distinct and represent a new class within Group I.

Biological effects observed for oxazolidinones were consistent with inhibition of mitochondrial function. When zoospores of the plant pathogen *Phytophthora infestans* were treated with DPX-JE874 or an analog, oxygen consumption ceased and zoospores lost motility within seconds. Within minutes the zoospores disintegrated. When mycelia of *P. infestans* were given a DPX-JE874 analog, protein and nucleic acid biosyntheses were curtailed. DPX-JE874 was also a potent inhibitor of *Saccharomyces cerevisiae* growth when the yeast was grown on nonfermentable carbon sources, but was 1000-fold weaker as an inhibitor when the yeast was grown on fermentable carbon sources. Thus, the classical whole-organism test confirms that the primary mode of action for DPX-JE874 is inhibition of mitochondrial respiration.

Synthesis

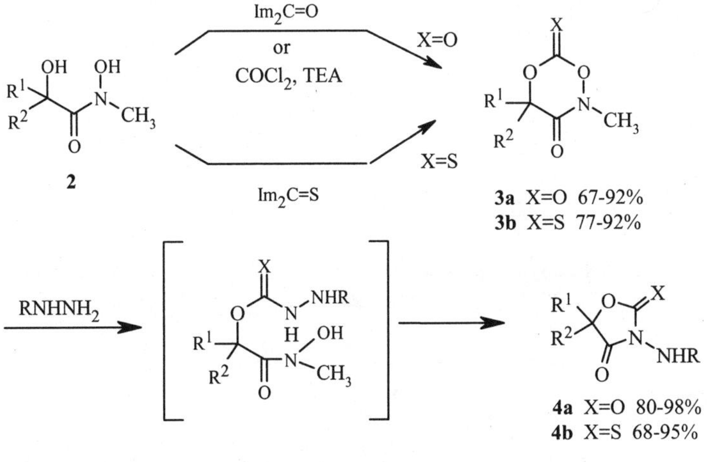

Figure 1. Initial syntheses of 3-amino-oxazolidinones

The syntheses of various types of oxazolidinones as novel heterocyclic ring systems had been investigated for several years (*1,5-11*). The syntheses employed to prepare the 3-amino derivatives are illustrated in Figure 1 (*7,8*).

Treatment of *N*-methyl-2-hydroxy-hydroxamic acid **2** with 1,1'-carbonyldiimidazole (CDI) afforded the dioxazinedione **3a**. The cyclization could also be accomplished using phosgene and triethylamine albeit in lower yield. Reaction of the hydroxamic acid **2** with 1,1'-thiocarbonyldiimidazole (TCDI) gave the corresponding thioxodioxazinone **3b**. The dioxazinedione **3a** or **3b** was then treated with a monosubstituted hydrazine to give the 3-amino-oxazolidinone **4a** or **4b**, respectively. The five-membered ring structure was confirmed by ^{1}H NMR and IR analyses. The isomeric six-membered ring oxadiazinones were not observed (*12,13*).

The initial syntheses afforded relatively high yields of oxazolidinones, but the multi-step preparation of the starting hydroxamic acids hindered the analog effort. A simpler route was envisaged involving treatment of 2-hydroxy-hydrazides with a carbonylating agent as illustrated in Figure 2.

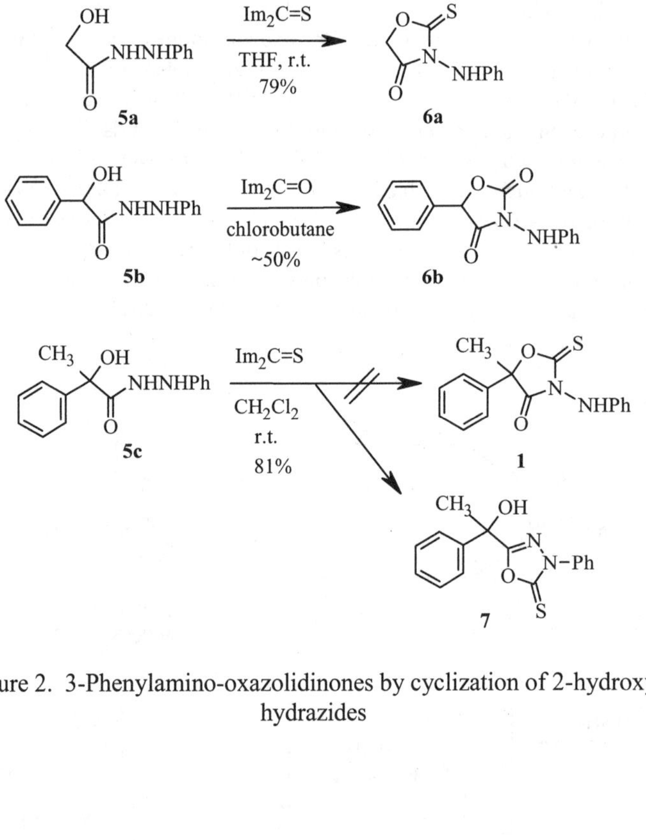

Figure 2. 3-Phenylamino-oxazolidinones by cyclization of 2-hydroxy-hydrazides

The glycolic acid hydrazide **5a** reacted smoothly with TCDI to form the five-membered ring 2-thioxo-4-oxazolidinone **6a** in good yield. The reaction between TCDI and the mandelic acid hydrazide **5b** did not proceed as cleanly and a number of products were formed. Reaction with CDI, however, produced mainly the desired 2,4-oxazolidinedione **6b**, although other products were observed. Interestingly, hydrazides bearing a tertiary alcohol such as **5c** reacted in an entirely different manner. Treatment with TCDI afforded the oxadiazol-2(3H)-one **7** in high yield without any of the desired 2-thioxo-4-oxazolidinone **1**. Apparently in this case, acylation of the aniline nitrogen occurred faster than reaction with the alcohol and subsequent cyclization formed the five-membered ring oxadiazol-2(3H)-one (*14*).

Attempts to methylate the 2-thioxo-analog of **6b** to prepare the 5-methyl-5-phenyl derivative **1** using lithium diisopropylamide and iodomethane were unsuccessful (*15*). The 5,5-disubstituted compounds were of interest because of their significantly higher fungicidal activity, and therefore alternate syntheses were pursued.

An efficient "one-pot" synthesis of 5,5-disubstituted 2-thioxo-4-oxazolidinones was developed starting with readily available 2-hydroxy-esters (Figure 3). Sequential treatment of the ester **8** with potassium *t*-butoxide, carbon disulfide, ethyl chloroformate, and finally phenylhydrazine afforded the 2-thioxo-4-oxazolidinone in good to excellent yield. The procedure was fairly general and the product could often be purified without chromatography. The only disubstituted 2-hydroxy-esters which seemed to work poorly in this reaction were those bearing strongly electron-withdrawing groups (e.g., R^1 = CF_3), powerful electron-donating groups (e.g., R^2 = 4-$(CH_3)_2$N-Ph), or those with sterically hindered groups (e.g., R^2 = 2,5-diCH_3-Ph).

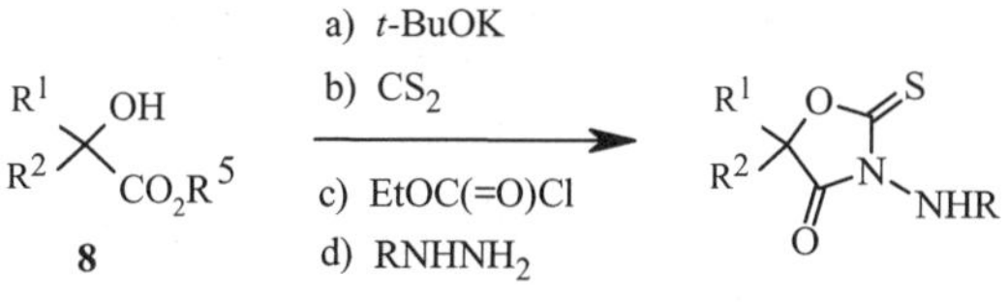

Figure 3. "One-pot" synthesis of 5,5-disubstituted 2-thioxo-4-oxazolidinones

Attempts to prepare the corresponding 2,4-oxazolidinediones using the same basic methodology were unsuccessful. Treatment of the 2-hydroxy ester **8** with potassium *t*-butoxide, then carbonyl sulfide instead of carbon disulfide, and finally an *S*-alkylating agent, failed to afford the desired thiolcarbamate. Carbonyl sulfide was apparently insufficiently electrophilic to react with the alkoxide.

2,4-Oxazolidinediones could however be obtained in high yield by hydrolysis of the corresponding 2-thioxo-4-oxazolidinone. Aqueous potassium peroxymonosulfate (Oxone) and aqueous silver nitrate (*1,16*) both accomplished the transformation cleanly at room temperature (Figure 4). Not only did this procedure provide a route to 2,4-oxazolidinediones in high yield, it also provided both the oxo- and thioxo-compounds for biological evaluation from a single synthetic sequence.

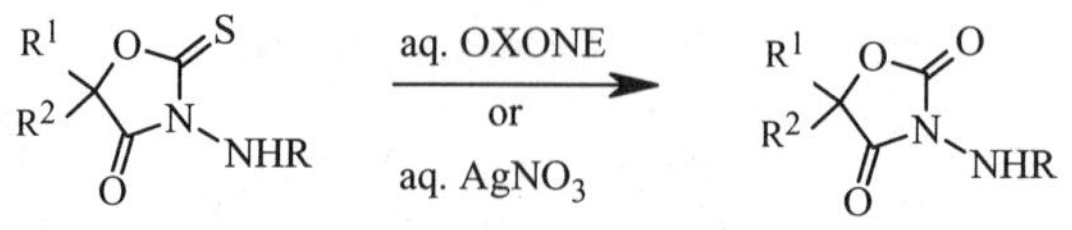

Figure 4. Conversion of 2-thioxo-4-oxazolidinones to 2,4-oxazolidinediones

Based on considerations of biological activity, cost, and waste, interest shifted from the 2-thioxo-4-oxazolidinones to the 2,4-oxazolidinediones. It therefore became desirable to develop a synthesis that provided these compounds directly and avoided the inefficiency of introducing a sulfur atom and then subsequently removing it. Biological results also suggested that 5-alkyl, 5-aryl, and 3-phenylamino substituents were desirable for maximum fungicidal activity, and therefore process efforts were directed towards 2,4-oxazolidinediones with this substitution pattern.

A potentially efficient route was envisaged starting with 1,3-dioxolan-2,4-diones **9** which are readily available from α-hydroxy carboxylic acids (Figure 5). We hoped treatment with a hydrazine would lead to attack at the carbonate carbonyl to afford the carbazate intermediate **10**, which in turn could be cyclized to the oxazolidinedione. However, it is known that reaction of 5,5-dimethyl-1,3-dioxolan-2,4-dione with alcohols and amines occurs exclusively at the other carbonyl to give esters and amides (*17*). The same regioselectivity was observed in our labs using 5-methyl-5-phenyl-dioxolan-2,4-dione and phenylhydrazine. Only the phenyl hydrazide **11** (R = Ph) was obtained.

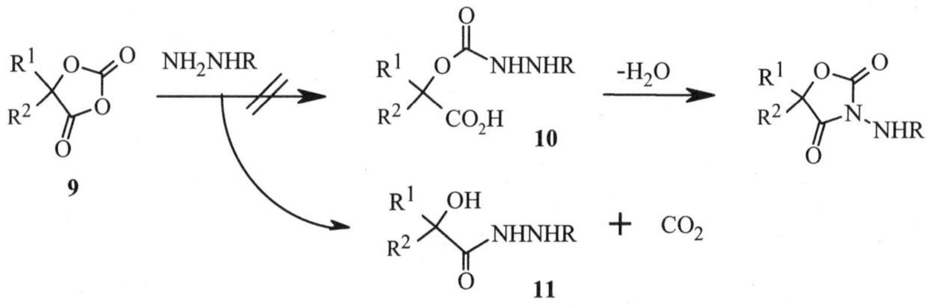

Figure 5. Reaction of 1,3-dioxolan-2,4-diones with hydrazines

2-Hydroxy esters were attractive as starting materials due to their potentially low cost and accessibility by a number of synthetic routes (*18*). The first attempt to convert the esters directly to 2,4-oxazolidinediones involved treatment with phosgene and *N,N*-diethylaniline (DEA) with a catalytic amount of pyridine (Figure 6). The resulting chloroformate **12** was not purified but treated directly with phenylhydrazine and *N,N*-diisopropylethylamine. Cyclization of the resulting carbazate was slow at room temperature, and therefore the reaction was heated to reflux in the presence of triethylamine (TEA). Yields were generally low, several by-products were observed (e.g., the acrylate **13**), and what appeared to be unreacted starting material was generally recovered despite efforts to drive the phosgenation to completion. We now believe starting material was in fact consumed and the intermediate chloroformate **12** decomposed to the tertiary chloride under the reaction conditions. The tertiary chloride then hydrolyzed back to the starting 2-hydroxy ester during the aqueous work-up.

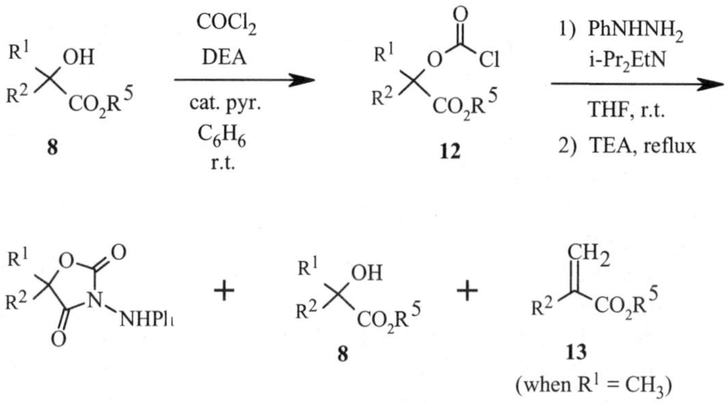

Figure 6. Preparation of 2,4-oxazolidinediones with phosgene

A more successful approach to 2,4-oxazolidinediones involved treatment of the 2-hydroxy-ester **8** first with CDI to form the acylimidazole **14** (Figure 7). The acylimidazole was not isolated but treated directly with phenylhydrazine and acetic acid to give the 2,4-oxazolidinedione. The intermediate anilinocarbamate **15** was not observed with 2,2-disubstituted 2-hydroxy-esters.

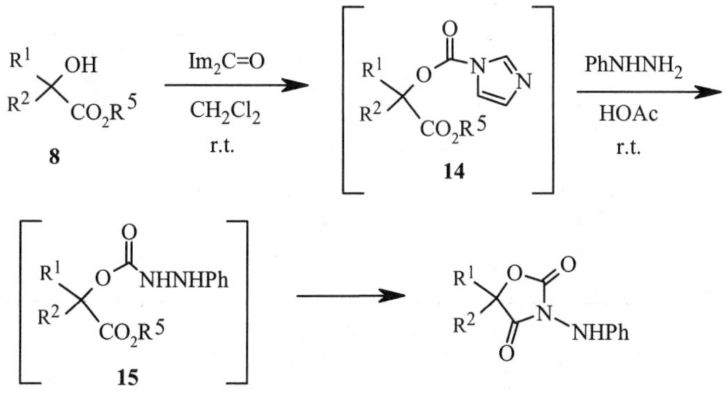

Figure 7. Preparation of 2,4-oxazolidinediones with CDI

In general, yields were good and the products could be purified without chromatography. Acetic acid appeared to be critical to the success of the reaction as no product was observed without it. It was postulated that acetic acid protonated the imidazolyl ring and thereby activated the carbonyl to nucleophilic attack. Stronger acids such as HCl gave no reaction, presumably due to irreversible protonation of the phenylhydrazine which rendered the hydrazine non-nucleophilic.

Structure-Activity Relationships

Over 700 oxazolidinone analogs were prepared and tested for fungicidal activity. Greenhouse inoculations were made by using aqueous spore suspensions of the plant pathogen, and incubation was conducted under conditions favorable for disease development. Visual assessments were made in percent leaf area affected.

Structure-activity relationships were developed using 5-methyl-5-phenyl-3-phenylamino-2-thioxo-4-oxazolidinone **1** as the base structure and varying the substituents on the heterocyclic ring.

The first substituent investigated was R^1, which is methyl in **1**. The data in Table I illustrate the significant effect the size of this group has on biological activity. For example, changing R^1 from methyl to ethyl reduced preventive control of *P. infestans* and *Plasmopara viticola* 8-fold and 4-fold, respectively (**1** vs. **17**). The R^1 = *n*-butyl compound **19** showed no activity at the rates shown. The R^1 = hydrogen compound **16** was also significantly less active. These data and others suggested that methyl was the optimal substituent at this position.

Table I. Effect of Various R^1 Groups on Preventive Control of *P. infestans* on Tomatoes and *P. viticola* on Grapes

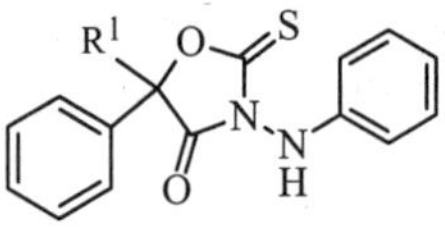

Cmpnd	R^1	*P. infestans* % Control at 200 ppm	*P. viticola* % Control at 10 ppm
1	CH_3	100	97
16	H	86	8
17	CH_2CH_3	33	23
18	$CH_2CH{=}CH_2$	0	41
19	$CH_2(CH_2)_2CH_3$	0	0
20	Ph	0	24

The effect of varying R[3], the group attached to the oxazolidinone ring nitrogen atom, was investigated next (Table II). The data for Compounds **21-25** suggest an anilino-type moiety at this position is critical for good activity. Two compounds with R^3 = OPh were also prepared and these too were significantly less active than the corresponding R^3 = NHPh compounds, especially against *P. infestans*.

The data for compounds **26-35** indicate that fungicidal activity generally decreased when substituents of any type were introduced on the anilino ring, often dramatically. The only substituted compounds which initially appeared as active were the 2-CH_3, 3-F, and 4-F compounds **26**, **30**, and **33**. However advanced tests indicated that they too were less active than the unsubstituted anilino-compound. As with R^1, little structural variability appeared to be tolerated at R^3 and the optimal substituent was that present in the lead compound **1**.

Table II. Effect of Various R^3 Groups on Preventive Control of *P. infestans* on Tomatoes and *P. viticola* on Grapes

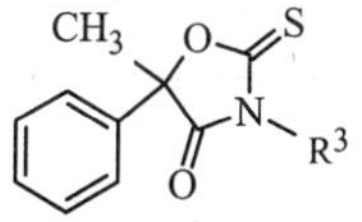

Cmpnd	R^3	*P. infestans* % Control, 200 ppm	*P. viticola* % Control, 40 ppm
1	NHPh	100	100
21	Ph	0	15
22	CH_2Ph	0	43
23	$NHCH_2Ph$	15	5
24	NH(c-hexyl)	0	65
25	NH(2-pyridyl)	25	68
26	NH(2-CH_3-Ph)	92	100
27	NH(2-F-Ph)	74	95
28	NH(2-Cl-Ph)	38	99
29	NH(3-CH_3-Ph)	90	91
30	NH(3-F-Ph)	100	100
31	NH(3-Cl-Ph)	94	97
32	NH(3-CH_3O-Ph)	94	97
33	NH(4-F-Ph)	97	100
34	NH(4-CH_3O-Ph)	0	73
35	NH(4-CF_3-Ph)	0	89

Somewhat surprisingly, the hydrogen on the exocyclic nitrogen atom is preferred but not essential for fungicidal activity. As indicated in Table III, the *N*-methyl compound **36** was also active although less so. The *N*-acetyl and *N*-ethyl compounds **37** and **38** were significantly less active as were oxazolidinones with even larger *N*-alkyl groups. Once again, the substituent in the lead compound **1** was determined to be best at this position.

Table III. Effect of R^4 on Preventive Control of *P. infestans* on Tomatoes and *P. viticola* on Grapes

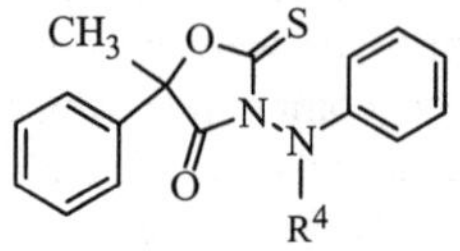

Cmpnd	R^4	*P. infestans* % Control, 200 ppm	*P. viticola* % Control, 40 ppm
1	H	100	100
36	CH_3	65	82
37	CH_2CH_3	24	NT[1]
38	$C(=O)CH_3$	23	45

[1] NT = Not tested

Whereas R^1, R^3 and R^4 were somewhat limited in their structural variability, a diverse group of substituents could be tolerated at the R^2 position as indicated in Table IV. Aryl groups were found to be preferred but the alkyl-substituted compounds **39** and **40** were also quite active. In general, ortho-substitution on the R^2 phenyl ring was found to be detrimental to activity (e.g., see **43**). The R^2 = 2-PhO-Ph 2-thioxo-4-oxazolidinone (i.e., the 2-phenoxy, sulfur analog of DPX-JE874) was inactive on *P. infestans* and showed minimal activity on *P. viticola* at 200 ppm. Only a fluorine atom at the ortho-position improved fungicidal control (see **42**).

Compounds with a fluorine or chlorine at the 3-position of the R^2 phenyl ring were more active than the unsubstituted compound (**44**, **45** vs. **41**), however other meta-substituted compounds were generally less active. The R^2 = 3-PhO-Ph 2-thioxo-4-oxazolidinone analog of DPX-JE874 was found to be less active than the corresponding 4-PhO-Ph 2-thioxo-compound on both *P. infestans* and *P. viticola*.

Substitution at the para-position was preferred and compounds with a wide variety of substituents were found to be highly active. The most active compounds prepared early on were the 2,4-difluoro- and 4-phenoxy-compounds **52** and DPX-JE874, respectively. DPX-JE874 was preferred because of its lower cost of manufacture.

Substitution on the phenoxy-ring was also studied and a wide variety of substituents appeared to be tolerated (Table V). High activity was maintained even with large groups at the 3-position as illustrated by **58** and **59**. The 4-position appeared less variable. The 4-Ph and 4-*t*-Bu compounds **63** and **64** were inactive on *P. infestans* at 200 ppm, although the 4-Br compound **61** was quite active. In summary, some of the substituted compounds were nearly as active as DPX-JE874, but none were significantly more active or possessed a desirable attribute lacking in DPX-JE874. Furthermore all of the substituted materials would have been more expensive to prepare commercially.

Table IV. Effect of R^2 on Preventive Control of *P. infestans* on Tomatoes and *P. viticola* on Grapes

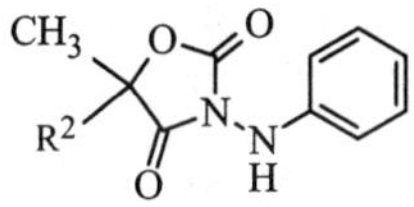

Cmpnd	R^2	*P. infestans* % Control, 200 ppm	*P. viticola* % Control, 10 ppm
39	$CH_2(CH_2)_4CH_3$	100	94
40	cyclohexyl	48[1]	29
41	Ph	88	48
42	2-F-Ph	100	100
43	2-CH_3-Ph	17	44[2]
44	3-F-Ph[3]	92	91
45	3-Cl-Ph	100	86
46	3-CH_3-Ph	42	8
47	3-CF_3O-Ph	60	6
48	4-F-Ph	93	56
49	4-Cl-Ph	76	17
50	4-CH_3O-Ph	15	25
51	4-CF_3O-Ph	97	100
DPX-JE874	4-PhO-Ph	95	100
52	2,4-diF-Ph	100	100

[1] Data for 40 ppm
[2] Data for 200 ppm
[3] Data for the 2-thioxo-4-oxazolidinone

Table V. Effect of R^5 on Preventive Control of *P. infestans* on Tomatoes and *P. viticola* on Grapes

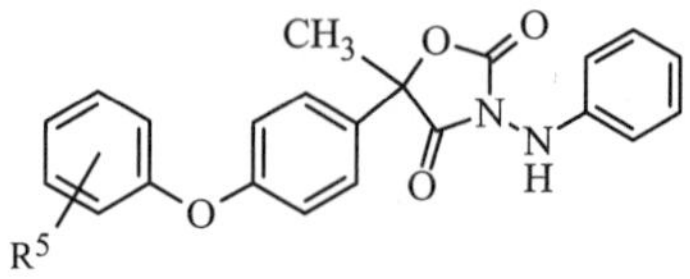

Cmpnd	R^5	*P. infestans* % Control, 200 ppm	*P. viticola* % Control, 2 ppm
53	2-*i*-Pr	0	84
54	2-Cl	98	68
55	2-F	100	82
56	3-F	98	97
57	3-Cl	92	100
58	3-PhO	64	100[1]
59	3-*t*-Bu	74	100[2]
60	4-Cl	97	74
61	4-Br	97	77
62	4-CH_3O	68	99
63	4-Ph	0	96[2]
64	4-*t*-Bu	0	21

[1] Data at 5 ppm
[2] Data at 40 ppm

Summary

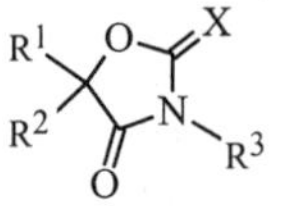

$R^1 = CH_3$, R^2 = Ph, R^3 = NHPh, X = S **1**

$R^1 = CH_3$, R^2 = 4-PhO-Ph, R^3 = NHPh, X = O **DPX-JE874**

An extensive analog program began at DuPont after testing the novel 2-thioxooxazolidinone **1** acquired through our Technology Transfer effort. Several syntheses were developed during the course of the program. Structure-activity relationships indicated that R^1 and R^3 were sensitive to structural variation, whereas compounds with a wide variety of R^2 groups were fungicidally active. In the end, a 4-phenoxy-phenyl group was found to be optimal at this position. In addition, the 2,4-oxazolidinedione ring system (X = O) was found to be preferable to the 2-thioxo-4-oxazolidinone structure (X = S). This work led to the advancement of DPX-JE874 to commercial development for the control of fungal infections of grapes, cereals, tomatoes, potatoes, and other crops. Additional work in the area, e.g., ring analogs, heterocyclic substituents, and process development, will be the subject of future publications (*19-22*).

Acknowledgments

The authors would like to thank the many members of Chemical Discovery who provided useful insights and suggestions to the project, most notably Robert J. Pasteris and Kevin Kranis, and the associate scientists for their technical assistance. We also thank John Groce and Gina Blankenship for technical assistance in providing rapid, quality NMR analyses, and the entire Plant Disease Control group for biological evaluations.

Literature Cited

1. Geffken, D. *Z. Naturforsch., B: Anorg. Chem., Org. Chem.* **1983**, *38B*, 1008.
2. Joshi, M. M.; Sternberg. J. A. *Brighton Crop Protection Conf. - Pests and Diseases*, **1996**, in press.
3. Adams, J. B., Jr.; Geffken, D.; Rayner, D. R. PCT Int. Patent Appln. WO 90/12791, 1990.
4. von Jagow, G.; Link, T. A. In *Methods in Enzymology*; Fleischer, S. and Fleischer, B., Eds.; Academic Press: New York, 1986, Vol. 126; pp 253.
5. Geffken, D. *Chem. Ztg* **1979**, *103*, 299.
6. Geffken, D. *Arch. Pharm. (Weinheim, Ger.)* **1980**, *313*, 817.
7. Geffken, D. *Synthesis* **1981**, 38.
8. Geffken, D. *Arch. Pharm. (Weinheim, Ger.)* **1982**, *315*, 802.
9. Geffken, D. *Liebigs Ann. Chem.* **1982**, 211.
10. Geffken, D.; Strohauer, K. *Arch. Pharm. (Weinheim, Ger.),* **1986**, *319*, 577.
11. Geffken, D. *Z. Naturforsch., B: Chem. Sci.* **1987**, *42*, 1202.
12. Hauser M. *J. Org. Chem.* **1966**, *31*, 968.
13. Kametani, T. *Heterocycles* **1978**, *9*, 1031.
14. Hetzheim, A.; Mockel, K. *Adv. Heterocyclic Chem.* **1966**, *7*, 183.
15. For alkylation of heterocyclic mandelic acid derivatives, see Frater, G.; Muller, U.; Gunther, W. *Tetrahedron Lett.* **1981**, 4221.
16. Geffken, D. *Arch. Pharm. (Weinheim, Ger.),* **1979**, *312*, 363.
17. Tighe, B. J. *Chemistry and Industry* **1969**, 1837.
18. Sternberg, J. A.; Sun, K.-M.; Toji, M.; Witterholt, V. United States Patent 5,552,554, 1996.
19. Campbell, C. L.; Gross, C. M.; Sternberg, J. A.; Sun, K.-M. PCT Int. Patent Appln. WO 93/18016, 1993.
20. Sun, K.-M. PCT Int. Patent Appln. WO 93/24467, 1993.
21. Campbell, C. L. PCT Int. Patent Appln. WO 93/22299, 1993.
22. Adams, J. B., Jr.; Sternberg, J. A. PCT Int. Patent Appln. WO 92/16515, 1992.

Chapter 23

Pyrimidine Fungicides: A New Class of Broad Spectrum Fungicides

Zen-Yu Chang, Pamela J. Delaney, John O'Donoghue, and RuPing Wang

Agricultural Products, Stine-Haskell Research Center, duPont and Company, Building 300, Newark, DE 19714

Broad spectrum fungicidal activities were shown by a class of pyrimidines substituted with nitrogen heterocycles. Biological data, synthesis, and structure-activity relationships are discussed

The discovery of fungicidal activity of compounds **1**, **2** (*1*) and **3** (Ferimzone®) (*2*) prompted us to undertake a major research program to explore the structure-activity of this class of chemistry. Based on the common structural features of these three compounds, we designed compounds represented by the general structure **4.** We have focused our optimization effort in the A-ring, B-ring and the aromatic or benzyl moieties shown in structure **4**. Many of the compounds represented by **4** provided 100% control of economically important diseases such as wheat foot rot, wheat glum blotch, and rice blast in our discovery screens at a 200 PPM rate. Several compounds proved sufficiently active that they were evaluated in field tests to assess the potential for commercial development.

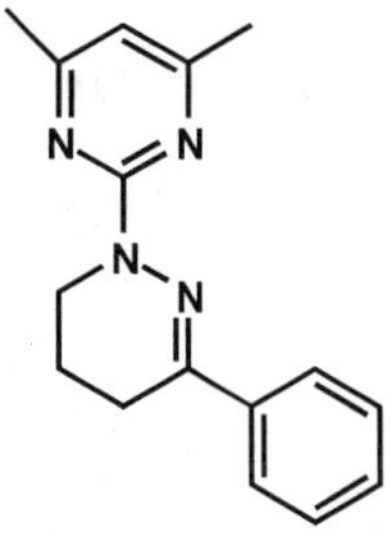

1

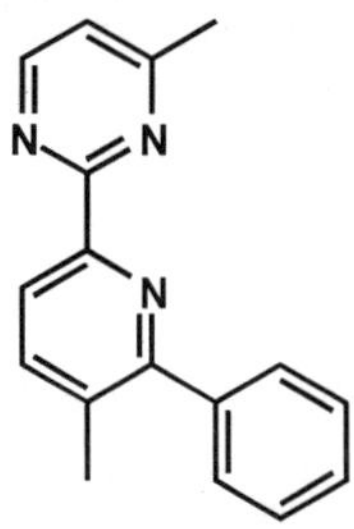

2

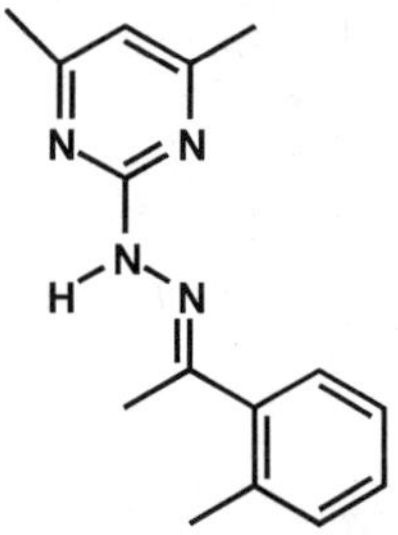

3

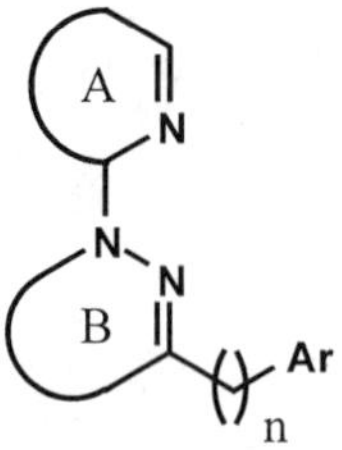

n = 0, 1

4

Synthesis

Synthesis of different sub-classes of compounds of generic structure **4** will be discussed. The A-ring and the aromatic moiety come primarily from commercial sources and do not warrant much discussion. Rather, we will concentrate our discussion in the variation of B-ring. Pyrazolines **5** with a 5-membered B-ring were prepared by condensation of the corresponding heterocyclic hydrazines with substituted 3-chloropropiophenones or 3-*N,N*-dimethylpropiophenones (Scheme 1). 3-*N,N*-Dimethylpropiophenones were prepared via the Mannich reaction (*3*).

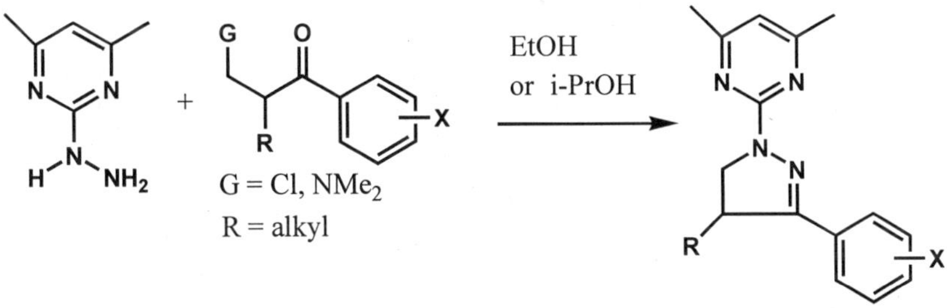

Scheme 1

Tetrahydropyridazines **6** with a 6-membered B-ring were prepared by condensation of the hydrazines with substituted 4-chlorobutyrophenones. Friedel-Crafts acylation of substituted benzenes with 4-chlorobutyrl chloride provided the requisite butyrophenones (Scheme 2).

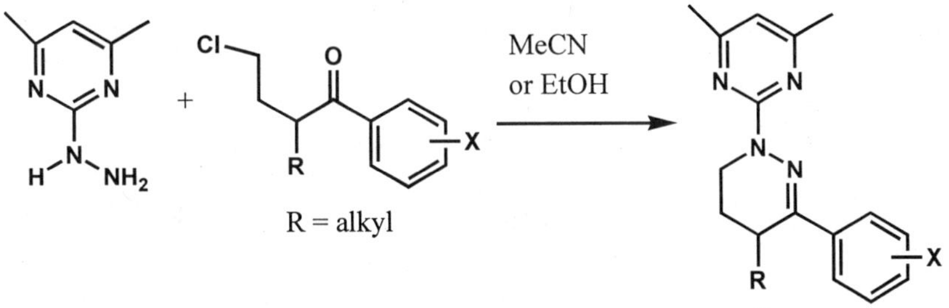

Scheme 2

1,4-Dihydropyridazines **7** were prepared by condensation of the pyrimidinyl-hydrazines with benzoylpropionic acids to give pyridazinones. Reduction with one equivalent of lithium aluminum hydride (*4*) gave the intermediate aminal. Subsequent treatment with methanesulfonyl chloride in pyridine afforded the desired 1,4-dihydropyridazines **7** (Scheme 3).

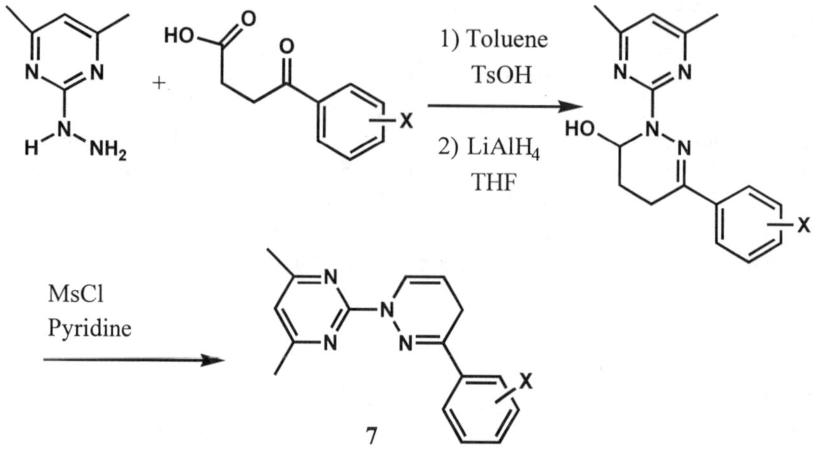

Scheme 3

The synthesis of 1,6-dihydropyridazines is shown in Scheme 4. Condensation of the pyrimidinylhydrazines with α,β-diketones gave the corresponding hydrazones as mixtures of E/Z isomers. Treatment of the mixtures with sodium hydride followed by addition of vinyltriphenylphosphonium bromide (*5*) provided the 1,6-dihydropyridazines **8**. The latter cyclization reaction did not work when the R group was 2-pyridyl or 2-furanyl.

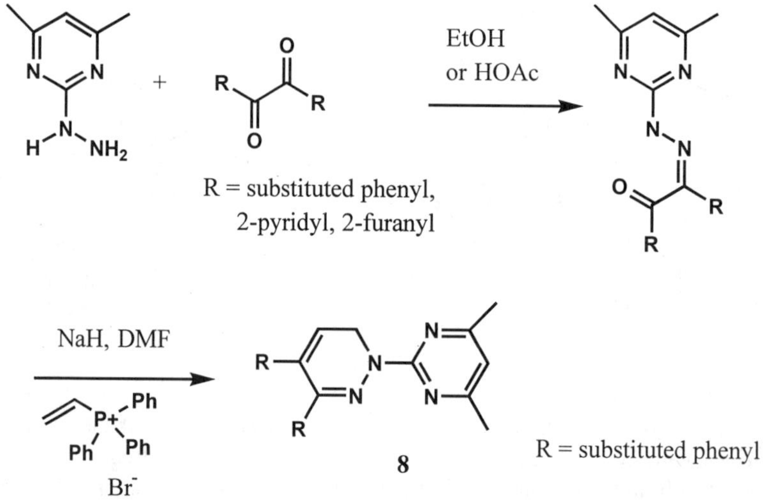

Scheme 4

For the synthesis of 7-membered B-rings tetrahydrodiazepines **9,** condensation of the pyrimidinylhydrazines with substituted 5-chlorovalerophenones was followed by treatment with sodium hydride (Scheme 5).

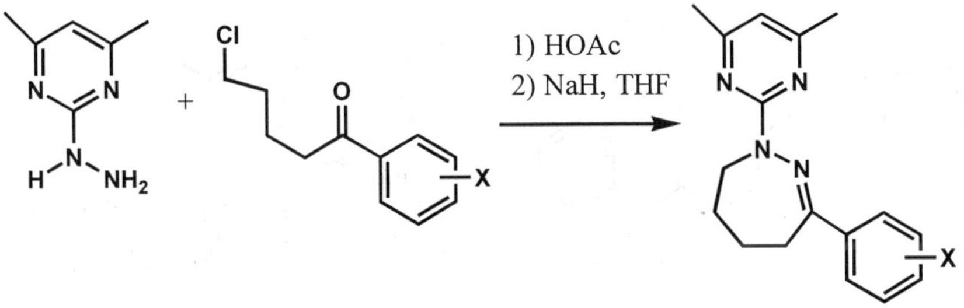

9

Scheme 5

The preparation of pyrazolines **5** in which the phenyl group is replaced by a benzyl group is shown in Scheme 6. The requisite chloroketones were prepared by coupling reactions between benzyl bromide and 3-chloropropionyl chloride catalyzed by a palladium reagent and zinc powder (*6*). Subsequent condensation with the pyrimidinylhydrazine provided the desired benzylpyrazolines **10**.

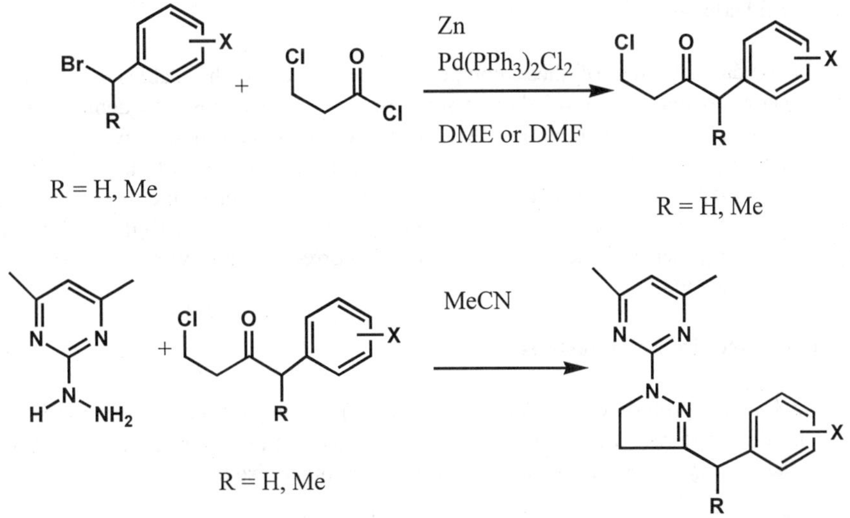

10

Scheme 6

When condensations carried out using 5-chloro-1-phenyl-2-pentanone prepared analogously according to the butanones of Scheme 6 and 2,6-dimethylpyrimidinyl hydrazine, the desired benzylpyridazine **11** was not isolated. Instead, cyclic hydrazine **12** with an exocyclic double bond was isolated (Scheme 7).

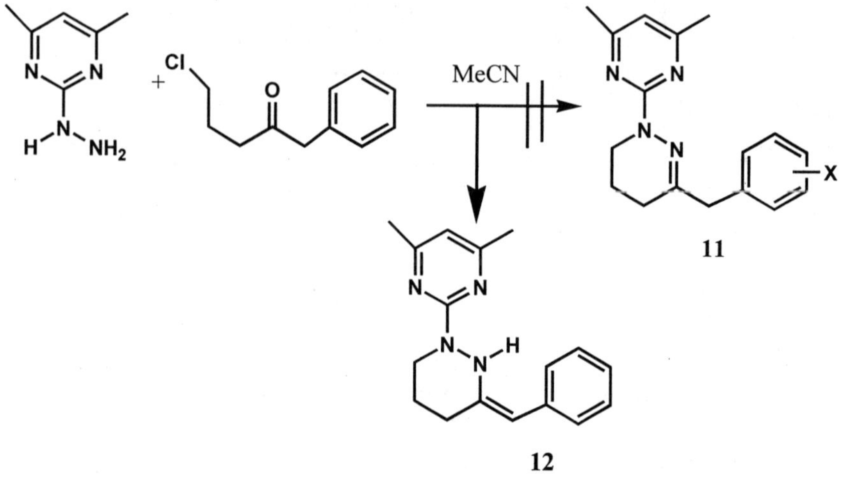

Scheme 7

Biological Data

The fungicidal activities of these compounds are shown in the Tables I-IV. The Tables show preventive control though some of these compounds also demonstrated curative efficacy (data not shown). The greenhouse disease control data were used to develop structure-activity relationships. Many heterocycles in place of the pyrimidine were also prepared via Schemes 1-6. These heterocycles include pyridine, quinazoline, benzothiazole, benzoxazole, and pyridazine. The activities of these compounds were generally less than those of the corresponding pyrimidines and the data is not presented.

Structure-Activity Relationships

For the A-ring, pyrimidine gave the best activity. For substitution on the pyrimidine ring, 4,6-dimethyl- was better than 4-methyl and unsubstituted pyrimidine in overall fungicidal activity when the rest of the molecules are otherwise the same.

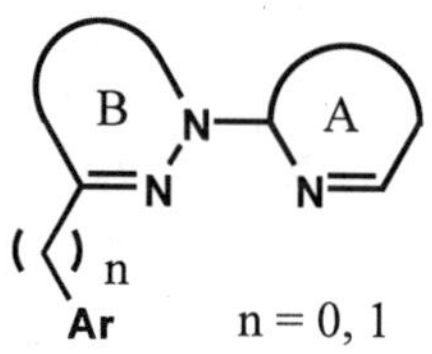

For the B-ring, tetrahydropyridazines (Table II) gave better activity than pyrazolines (Table I) and tetrahydrodiazepines (Table III). Among the 6-member rings, the 1,4,5,6-tetrahydropyridazine ring was better than the two types of dihydropyridazines (Table IV)

In the pyrazoline series, the methylene group between the aromatic ring and B-ring afforded compounds at comparable fungicidal activities to those without the methylene. However, since the synthesis was more difficult, only a few compounds were prepared (Table I).

Table I. Biological data

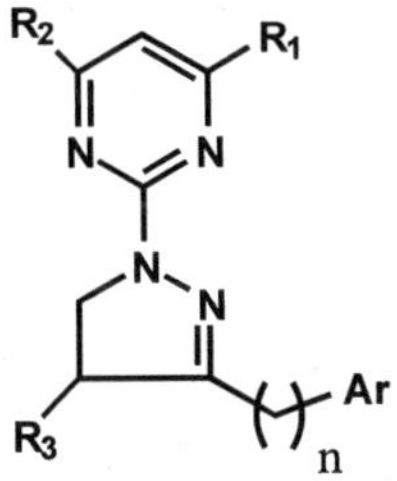

#	R_1	R_2	R_3	n	Ar	Percent Disease Control at 200 PPM WPM	WLR	WFR	WGB	RCB
1	Me	H	H	0	1-naphthalenyl	NT	NT	100	100	86
2	Me	Me	H	0	1-naphthalenyl	85	25	NT	100	96
3	Me	Me	H	0	2-Me-Ph	92	62	100	NT	87
4	Me	Me	H	0	2-Cl-Ph	90	84	100	NT	62
5	Me	Me	H	0	3-Cl-Ph	63	95	75	NT	99
6	Me	Me	H	0	3-Me-Ph	20	53	100	100	93
7	Me	Me	H	0	4-Me-Ph	39	45	100	96	NT
8	Me	Me	H	0	2-OMe-Ph	28	36	84	NT	100
9	Me	H	Me	0	Ph	91	99	NT	NT	NT
10	H	H	H	1	3-Me-Ph	96	89	0	93	0
11	Me	H	H	1	3-Me-Ph	98	89	49	97	74
12	Me	Me	H	1	3-Me-Ph	98	98	61	100	91

NT -- Not Tested.

WPM - Wheat Powdery Mildew (*Erysiphe graminis*)
WLR - Wheat Leaf Rust (*Puccinia recondita*)
WFR - Wheat Foot Rot (*Pseudocercosporella herpotrichoides*)
WGB - Wheat Glum Blotch (*Septoria nodorum*)
RCB - Rice Blast (*Pyricularia oryzae*)

For the aromatic group, phenyl appeared to be superior to heterocyclic aromatic groups. In the tetrahydropyridazine series (Table II), alkyl and halo substitutions on the phenyl ring were better than alkoxy and trifluoromethyl groups in overall activities. A methyl or ethyl group as R^3 group in the structure shown in Table II also improved the activity. Compound **14** from Table II was the most active compound and was therefore tested in the field. Unfortunately, the promising greenhouse activity of this class of chemistry was not maintained in the field.

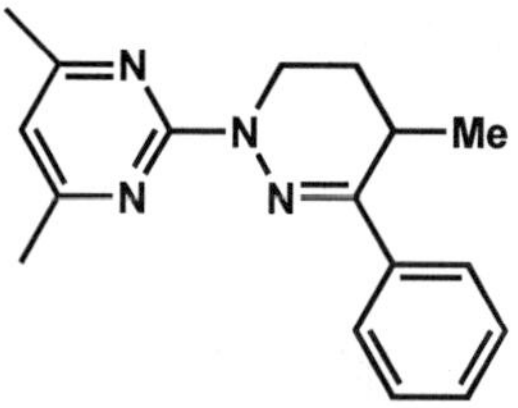

Table II. Biological data

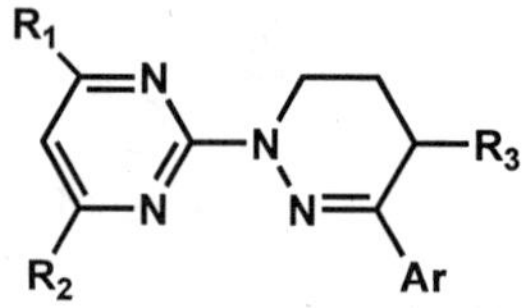

#	R_1	R_2	R_3	Ar	Percent Disease Control at 200 PPM WPM	WLR	WFR	WGB	RCB
1	Me	H	H	Ph	88	100	97	NT	NT
2	Me	Me	H	Ph	96	99	62	NT	60
3	Me	Me	H	4-F-Ph	NT	89	38	NT	61
4	Me	Me	H	3-Cl-Ph	97	21	92	100	98
5	Me	Me	H	4-Cl-Ph	91	100	100	100	100
6	Me	Me	H	4-OH-Ph	57	79	NT	NT	91
7	Me	Me	H	4-OMe-Ph	66	98	85	100	99
8	Me	Me	H	4-OPh-Ph	72	86	79	100	90
9	Me	Me	H	2-Me-Ph	80	41	87	100	90
10	Me	Me	H	3-Me-Ph	94	99	98	100	85
11	Me	Me	H	4-Me-Ph	89	100	74	97	100
12	Me	Me	H	4-*n*-Pr-Ph	99	100	92	98	99
13	Me	Me	H	4-*tert*-Bu-Ph	83	97	100	NT	99
14	Me	Me	Me	Ph	98	67	59	100	99
15	Me	Me	Me	Ph	98	67	59	100	99
16	Me	Me	Et	Ph	95	93	62	100	93
17	Me	Me	Me	3-Me-Ph	99	99	77	100	96
18	Me	Me	H	3-CF_3-Ph	71*	53*	0*	100*	35*
19	Me	Me	H	2-thienyl	46	67	100	100	57

* indicate disease control at 40 PPM.

Table III. Biological data

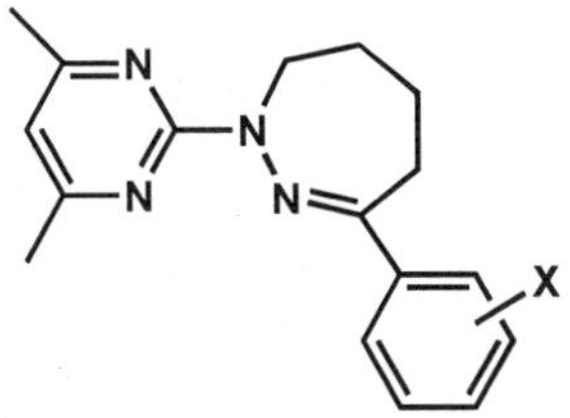

#	X	Percent Disease Control at 200 PPM WPM	WLR	WFR	WGB	RCB
1	4-Cl	89	83	100	*95	71
2	4-OMe	60	66	39	99	73
3	4-Me	79	70	41	100	82
4	H	96	83	33	100	83

* indicate disease control at 40 PPM.

Table IV. Biological data

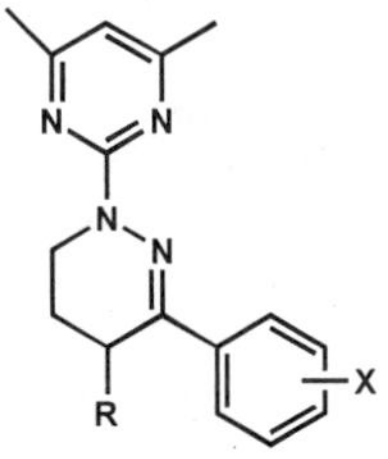

For compounds # 1-4

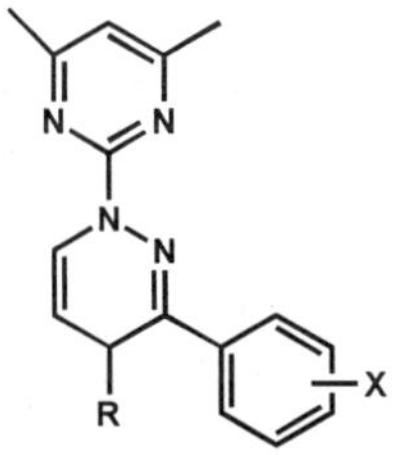

For compounds # 5-7

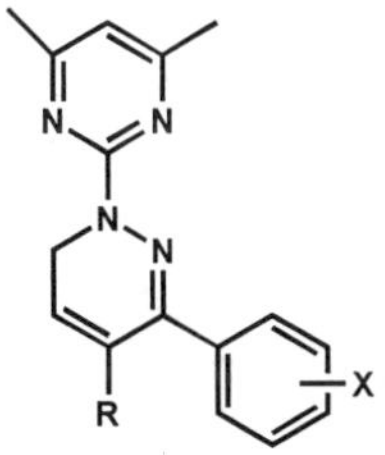

For compounds # 8-10

#	R	X	Percent Disease Control at 200 PPM WPM	WLR	WFR	WGB	RCB
1	H	H	96	99	62	NT	60
2	H	4-Me	89	100	74	97	100
3	H	4-OMe	66	98	85	100	99
4	H	3,4-diMe	91	98	NT	100	90
5	H	H	0	97	0	3	13
6	H	3,4-diMe	60	83	0	95	67
7	4-Me-Ph	4-Me	57	0	0	0	0
8	Ph	H	24	77	0	78	24
9	4-Me-Ph	4-Me	NT	80	48	58	32
10	4-OMe-Ph	4-OMe	84	88	61	35	0

Conclusions

The pyrimidinyltetrahydropyridazines discussed demonstrated broad spectrum of fungicidal activity, especially against wheat foot rot, wheat glum blotch and rice blast, in the greenhouse. Dimethyl substitution on the pyrimidinyl group boosted the fungicidal activity. The greenhouse activity was not maintained in the field.

Acknowledgments

The authors wish to thank those biologists who contributed to this work. We would also like to thank John Daub, Ed Adams, Debbie Frasier and Simon Xu for sharing ideas in this area of chemistry.

Literature Cited

1. JP03044304, (Takeda, 1989)
2. EP259139 (Sumitomo, 1987)
3. Maxwell, C. E. *Org. Synth.* **III**, **1955**, 305.
4. Aubagnac, J. L.; Elguero, J.; Jacquier, R.; Robert, R. *Bull. Soc. Chim. France*, **1972**, 2859.
5. Schweizer, E. E.; Kopay, C. M. *J. Org. Chem.* **1972**, *37*, 1561.
6. Sato, T.; Naruse, K.; Enokiya, M; Fujisawa, T. *Chem. Lett.* **1981**, 1135.

Chapter 24

Cyprodinil: A New Fungicide with Broad-Spectrum Activity

Urs Müller, Adolf Hubele, Helmut Zondler, Jürg Herzog

Research and Development BU Fungicides,Novartis Crop Protection AG, Rosental, CH–4002 Basel, Switzerland

Cyprodinil is a new fungicide for the control of a broad spectrum of plant diseases. Control of major cereal diseases and of *Botrytis cinerea ssp.* on grapes and vegetables are the major strengths of cyprodinil. The careful analysis of physico-chemical properties of certain sulfonylureas and the consistent testing of all compounds as fungicides, herbicides and insecticides led to the discovery of the class of compounds and led to cyprodinil. Aspects of chemistry, structure activity considerations, ecological and safety assessments are presented to demonstrate the unique properties of cyprodinil which make it a versatile new tool for the management of plant diseases.

Cyprodinil - CGA 219 417 is a novel fungicide recently introduced into the market. It is applied against a broad spectrum of plant diseases in many crops, such as in cereals: *Pseudocercosporella herpotrichoides, Erysiphe graminis, Pyrenophora teres, Rynchosporium secalis and Leptosphaeria nodorum*; in grapes, vegetables, field corps and strawberries: *Botrytis spp., Alternaria spp.*; and in deciduous fruit: *Venturia inaequalis, Alternaria spp.* and *Monilinia spp.* (*1*).

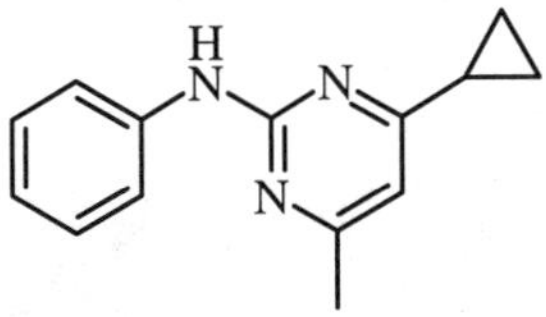

Figure 1: Cyprodinil; CGA 219 417 (*2*)

Cyprodinil belongs to the new class of anilinopyrimidine fungicides (*2*). Two other compounds - pyrimethanil (2-anilino-4,6-dimethyl-pyrimidine) (*3,4*) and mepanipyrim (2-anilino-4-methyl-6-(1-propynyl)-pyrimidine) (*5,6*) - have been announced or have already reached the market (Figure 2).

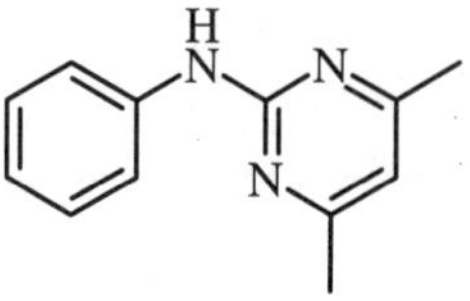

6

pyrimethanil

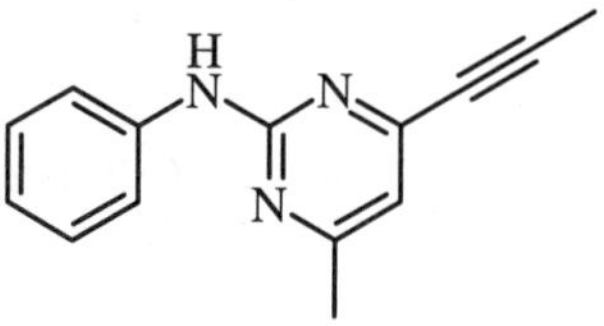

7

mepanipyrim

Figure 2: Structures of pyrimethanil (*3,4*) and mepanipyrim (*5,6*)

At former CIBA Crop Protection, this class of compounds was found in a typically serendipitous way. In the early eighties we were intensively searching for new sulfonylurea herbicides. The compound CGA 143 686 **1** (Figure 3) proved to be an interesting herbicide, especially in greenhouse trials. Selected for field trials, it was decided to study the hydrolytic behavior of this sulfonylurea herbicide. The expected 2-nitro-benzosulfonamide **2** and 2-amino-4 methyl-6-difluoromethoxy-pyrimidine **3**

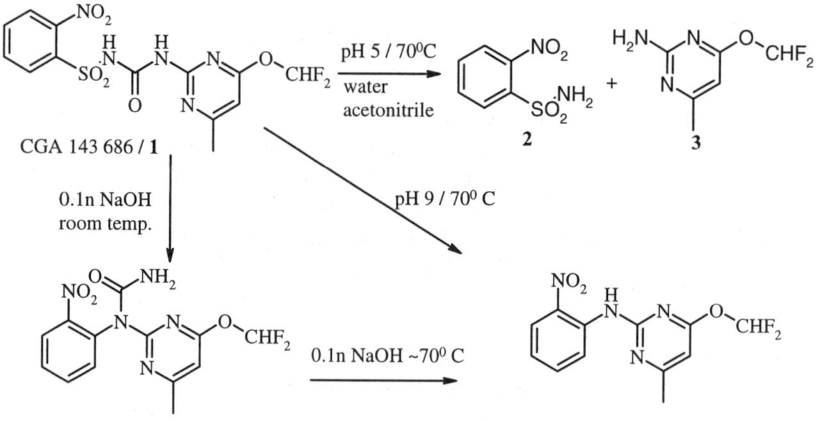

Figure 3 : Hydrolysis of CGA 143 686 (*7*)

were found in experiments typically carried out at pH 5 in water/acetonitrile at 70° C (*7*) (Figure 3). However under strongly basic conditions,0.1 n NaOH, the urea derivative **4** was isolated, while at pH 9, 70° C 2-(2-nitro-anilino)-4-methyl-6-difluoromethoxy-pyrimidine **5** was formed. As shown in a separate experiment, the urea derivative **4** can hydrolyze further to the pyrimidine derivative **5**. Only sulfonylureas bearing a 2-nitro substituent showed this hydrolytic behavior, which prompts the following hypothesis: nucleophilic attack by the 2-amino group of the pyrimidine displaces the sulfonylgroup, an intermediate which spontaneously looses sulfur dioxide leading to the urea derivative **4** (Figure 4).

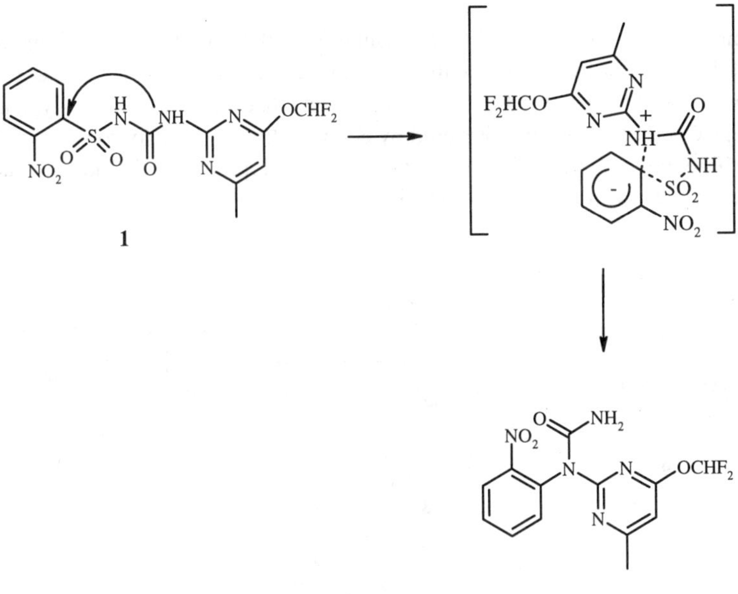

Figure 4 : Proposed mechanism of the hydrolysis of CGA 143 686.

Depending on conditions, the urea derivative **4** can hydrolyze further to the 2-anilinopyrimidine derivative **5**. Similar results where later found by Schneiders and Brown studying the environmental properties of sulfonylurea herbicides (*8, 9*).

Compounds **4** and **5** were tested as herbicides and as fungicides. The urea derivative **4** showed an interesting activity against broadleaf weeds with safety to rape. The fungicidal activity was weak for both compounds and was not pursued further at first. Attempts to optimize the herbicidal activity followed two directions: First following the scheme in Figure 3 a number of *o*-nitro analogues of type **4** and **5** were prepared in reasonable yields. None of these compounds showed interesting fungicidal activity. Second starting with the readily available anilinopyrimidines for which almost any substituent was accessible, the bridge amino-group was derivatized. One of the first compounds prepared was CGA 176440 **6** (Figure 2). This compound showed a good activity against *Botrytis cinerea* on grapes in our screens. Both its structure and fungicidal activity were known from the literature (*4,10*). Nevertheless, we saw opportunities to find new and potentially more active compounds. Shortly after the optimization project was started, similar compounds were published in a patent application by Kumiai Chemical Industry (*6*). 2-Anilino-4-methyl-6-(1-propynyl)-pyrimidine **7**, common name mepanipyrim, was later announced as a development compound (*5*).

Chemistry

2-Anilino-pyrimidines are readily accessible by either condensing aryl-guanidines with β-diketones or by reacting anilines with pyrimidines bearing a leaving group in the 2-position. Many methods to synthesize β-diketones are known. In a typical procedure, CGA 219417 **11** was prepared by condensation of 1-cyclopropyl-butane-1,3-dione **10**, obtained by a Claisen condensation of cyclopropyl-methyl-ketone and ethylacetate **8** (*11*), with phenylguanidine **9** (Figure 5).

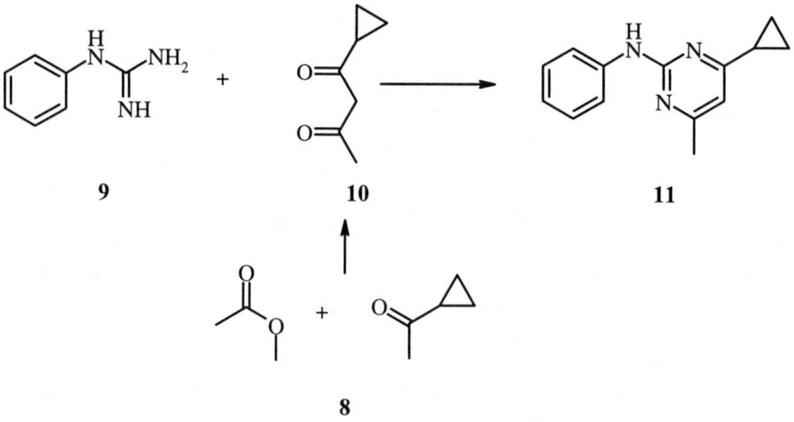

Figure 5: Synthesis of CGA 219 417, cyprodinil (*2*)

On reacting phenylhydrazines **12** with 2-halo-pyrimidines **13** in the presence of strong bases like potassium t-butoxide, N-phenyl-N-pyrimidin-2-yl-hydrazines **14** are formed in good yield (*12*) (Figure 6). The N,N-phenyl-pyrimidinyl-hydrazines **14** could be further derivatized, e.g., via formation of hydrazones, reductive alkylation with aldehydes or acylation. The 1,2-di-substituted hydrazine derivatives **15** are formed under acidic reaction conditions (*13*). N-Hydroxy derivatives **17** are prepared in a similar fashion by reacting N-aryl-hydroxylamine compounds **16** under acidic conditions (*14*). Compounds **17** were again further derivatized at the hydroxyl group.

The new hydrazine and hydroxylamine derivatives **14** and **17** are stable in crystalline form and can be stored at room temperature. When exposed to light, the N-amino and N-hydroxy compounds decompose to the corresponding anilino compounds in good yield (*15*). The N-acylated compounds tend to hydrolyze in aqueous solutions (Table 1).

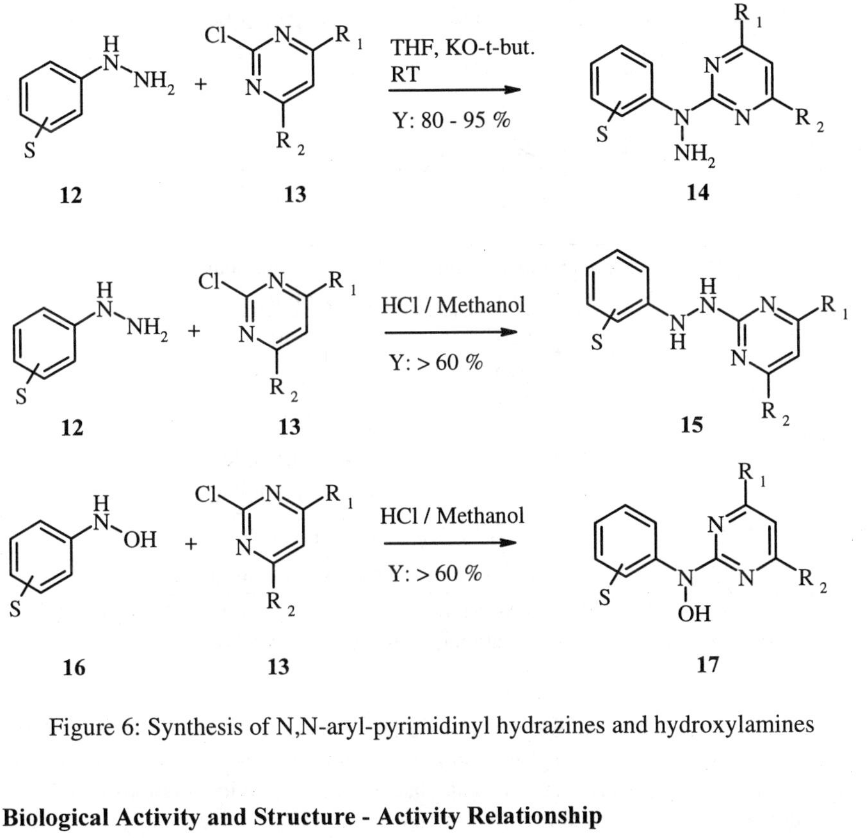

Figure 6: Synthesis of N,N-aryl-pyrimidinyl hydrazines and hydroxylamines

Biological Activity and Structure - Activity Relationship

The compounds synthesized were tested in the greenhouse against a variety of pathogens relevant to agriculture. Structure-activity relationships were assessed with three main pathogens: *B. cinerea* on apple fruit, *E.graminis* on barley and *V. inaequalis* on apple seedlings. In a first study of derivatives with different substituents at the bridging nitrogen, it became clear that activity is linked to stability (Table 1). The structure-activity relationship of compounds **14** and **17** followed the same rules as the corresponding unsubstituted compounds (see below). Labile derivatives can therefore be considered as prodrugs. Compounds bearing stable substituents like methyl are inactive. 1-Aryl-2-pyriminidin-2-yl hydrazines **15** are inactive. The optimization of structure and activity concentrated on the variation of the aryl and the pyrimidine part of the lead structure CGA 176440 **6**. A large number of compounds bearing different substituents on the phenyl and the pyrimidine rings were synthesized. It soon became clear that only a narrow selection of substituents at either of the two rings showed interesting activity compared to the lead structure. The results from a representative selection of compounds are summarized in Table 2. Besides hydrogen, only fluorine is tolerated on the aniline. Derivatives with the pyrimidine ring

Table 1: Activity and Stability of N-substituted derivatives

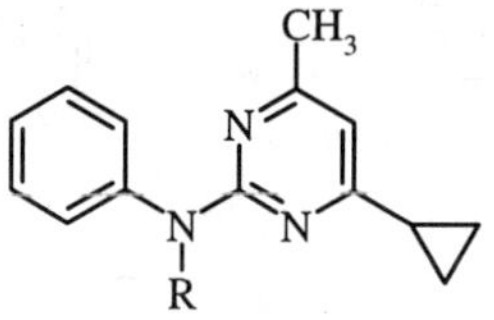

R	EC_{80} (*B.cinerea*/apple fruit)	Physico-chemical stability	t 1/2 (*15*)
H	2.3	stable	-
CH_3	> 200	stable	-
$COCH_3$	~ 10 a)	rapid hydrolysis	n.a. b)
OH	11	photodegradation	n.a.
NH_2	4	photodegradation	3 h
$NHCH_3$		photodegradation	3h
$N(CH_3)_2$	35	photodegradation	0.75 h

a) estimated value derived from the 4,6-dimethylpyrimidine derivative
b) no measured values available, qualitative result analyzed by t.l.c.

were generally inactive. Activity is also quite sensitive to substituents R_2 and R_4. Small changes, like cyclopropyl to i-propyl or methyl to ethyl decrease the activity drastically. Attempts were made to calculate structure-activity relationships; however, due to the limited variability in structure and activity, no reasonably well fitting equation could be found to predict the activity of the various substituent patterns reliably. Data published by Krause *et al.* on *Phytophtera infestans* would only partly fit for our activity profiles (*16*). Since no data on enzyme inhibition are available yet, factors like metabolic degradation, penetration and transport behavior might blur a clear structure-activity relationship.

Mode of Action

In the search for novel fungicides, interest is not only on novel structures, but also on compounds with a new mode-of-action, in particular inhibitors of fungal biochemical pathways with no analogy in mammals. When tested on *B.cinerea,* the class of the

Table 2: Structure and activity of selected anilino-pyrimidines

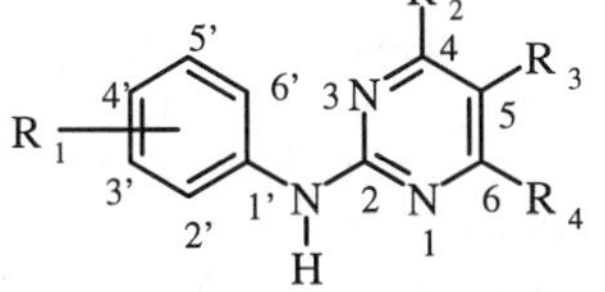

R_1	R_2	R_3	R_4	Sum EC_{80} (ppm)[a)]	
3' - F	CH_3	H	Cyclopropyl	26.6	
4' - F	CH_3	H	Cyclopropyl	28.9	
3',5' - F	CH_3	H	Cyclopropyl	78.6	
3',4' - F	CH_3	H	Cyclopropyl	82.2	
3' - Cl	CH_3	H	Cyclopropyl	524.0	
4' - Cl	CH_3	H	Cyclopropyl	> 600	
H	CH_3	H	Cyclopropyl	47.0	cyprodinil
H	CH_3	H	2-CH_3-cyclopropyl	39.5	
H	Cl	H	Cyclopropyl	41.2	
H	CH_3	H	Propargyl	65.1	mepanipyrim
H	CH_3	H	CN	65.2	
H	CH_3	H	$CH_2OC_2H_5$	121.9	
H	CH_3	H	CH_3	205.2	pyrimethanil
H	Cyclopropyl	H	Cyclopropyl	307.7	
H	C_2H_5	H	C_2H_5	390.7	
H	i-C_3H_7	H	CH_3	575.0	
H	CH_3	CH_3	CH_3	400[b)]	
H	CH_3	Cl	CH_3	400[b)]	
H	CH_3	S-C_2H_5	CH_3	400[b)]	

a) *Botrytis cinerea*/apple fruit; *Erysiphe graminis*/barley; *Venturia inaequalis*/apple. 200 ppm highest concentration tested b) *Botrytis cinerea* /apple fruit; *Erysiphe graminis* / barley

anilinopyrimidines did not show any effect on respiration, lipid peroxidation, osmotic stability or on the biosynthesis of protein, RNA, DNA, chitin, melanin or ergosterol (*17, 3, 5*). Both pyrimethanil and mepanipyrim have been reported to inhibit the secretion of cell wall degrading enzymes such as cutinase, pectinase, cellulase, lipases and proteases, all of which are required for infection (*18, 19, 20*). The same effects could also be measured with cyprodinil (*21*). Masner *et al.* reported the inhibition of methionine biosynthesis as a possible new mode of action (*17*). It could be shown that methionine reverses the toxicity of CGA 219 417 to *B. cinerea.* Pyrimethanil and mepanipyrim showed similar effects. These findings are supported by Leroux *et al.* (*22*). On incubating $Na_2[^{35}S]O_4$ together with pyrimethanil, a decrease of $[^{35}S]$-

methionine and, simultaneously, an increase of [^{35}S]-cystathionine was observed, suggesting that the primary target of pyrimidineamine fungicides could be the cystathione β-lyase. However, conclusive results with isolated cystathione β-lyase were not given. These findings indicate the value of cyprodinil and the other anilino-pyrimidine fungicides in two ways: a) a biochemical pathway specific for plants and micro-organisms seems to be affected, and b) no other known fungicide seems to have the same mode of action. Both factors are desirable for safety in mammals and for anti-resistance strategies, especially in the control of *B. cinerea* tolerant to dicarboximides, benzimidazoles or diethofencarb. Indeed, anilinopyrimidine fungicides fully control these resistant strains of *B. cinerea* (*1, 3*). Cross resistance was neither observed against prochloraz or DMI resistant strains of *P. herpotrichoides* (wheat-type and rye-type strains) nor against DMI or morpholine resistant strains of *E. graminis* (*23, 24*). These data impressively indicate the lack of a cross-resistance potential with current market products.

Cyprodinil, CGA 219417, is a systemic product that is taken up well into leaves after foliar application. It is transported within the plant throughout the leaf tissue and acropetally in the xylem. In key target pathogens such as *P. herpotrichoides*, *E. graminis* and *B. cinerea*, site of action studies have shown cyprodinil to be effective by inhibiting both the penetration and mycelial growth on the surface of and inside the leaves (*23*).

The broad spectrum of activity against major pathogens in major crops, favorable environmental profile and safety properties for users and consumers (*1, 23*) and a new mode of action controlling resistant strains of pathogens make cyprodinil a valuable product in the fight for disease management in agricultural production.

Literature Cited

1. Heye, U.J., Speich, J., Siegle, H., Wohlhauser, R., Hubele, A., CGA 219417 - A Novel Broad-Spectrum Fungicide. *Proc. Brighton Crop Protection Conference - Pests and Diseases;* **1994,** Vol. 2, 501-508.
2. Hubele, A., EPA 310 550, (28.09.87/11.04.88); Publ. 05.04. 89.
3. Neumann, G.L., Winter, E.H., Pittis, J.E., Pyrimethanil: A New Fungicide. *Proc. Brigthon Crop Protection Conference - Pests and Diseases,* **1992**, Vol. 1, 395-402.
4. Franke, F., Kiepel, M., Krause, G., Lehmann, H., Brämer, B., DDR 151 404 (21.10.81).
5. Maeno, S., Miura, I., Mepanipyrim (KIF-3535), A New Pyrimidine Fungicide. *Brighton Crop Protection Conference - Pests and Diseases;* Proc. Vol. 3, 415-422, 1990.
6. Ito,S., Masuda, K., Kusano, S., Nagata, T., Kojima, Y. Sawai, N., Maeno, S., EPA 224 339 (Prior. 30.10.85, 29.07.86) Publ. 03.06.87.
7. Stamm, E., Internal Communication CIBA- GEIGY, Plant Protection.
8. Schneiders, G. E., Koeppe, M. E., Naidu, M.V., Brown, A. M., Mucha, C. F., *J. Agric. Food Chem.*, **1993**, *41*, 2404-2410.

9. Brown, H. M., Kearny, P. C., in *Synthesis and Chemistry of Agrochemicals II*, Baker, D. R., Fenyes, J. G., Moberg, W.K. Eds. ACS Symposium Series 443; American Chemical Society: Washington DC, 1991; pp 32-49.
10. St. Angerstein, *Chem. Ber.*, **1901**, *34*, 3961.
11. Cannon, G.W., Ellis, R.C., Leal, J.R., *Org. Synthesis*, *Coll.Vol. IV*, 597.
12. Zondler, H. , Hubele, A., EPA 358 609 (09.09.88) Publ. 14.03.90.
13. Zondler, H.,Internal Report, CIBA-GEIGY Plant Protection, **1988**.
14. Zondler, H., EPA 441747 (07.02.90) Publ. 14.08.91.
15. B. Frei, Internal Communication CIBA-GEIGY, **1989**.
16. Krause,M., Klepel, M., Jumar, A., Franke, R., *Tag.-Ber.,Akad. Landwirtsch.-Wiss. DDR*, **1984**, *222*, 229-232.
17. Masner, P., Muster, P., Schmid, J., *Pestic. Sci.*, **1994**, *42*, 163-166.
18. Miura, I. , Kamakura T., Maeno Sh., Hayashi Sh., Yamaguchi I.,*Pestic.Bioch.&Physiol.*, **1994**, *48*, 222-228.
19. Daniels, A., Lucas, J.A., *Pestic. Sci.*, **1995**, *45*, 33-41.
20. Milling R.J., Richardson C.J., *Pestic. Sci.*, **1995**, *45*, 43-48.
21. Pillonel, Ch., Internal communication CIBA-GEIGY, **1995**.
22. Fritz, R., Lanen, C., Colas, V., Leroux P., *Pestic. Sci.*, **1997**, *49*, 40-46.
23. Heye, U.J., Speich,J., Siegle, H., Steinemann, A., Forster, B. , Knauf-Beiter, G., Herzog, J., Hubele, A., *Crop Protection*, **1994,** *13 (7)*, 541-549.
24. Hilber,U., Schüepp, H., *Pestic. Sci.*, **1996**, *47*, 241-247.

Chapter 25

Pyridinylpyrimidine Fungicides: Synthesis, Biological Activity, and Photostability of Conformationally Constrained Derivatives

John P. Daub and Donna L. Piotrowski

Agricultural Products, Stine-Haskell Research Center, duPont and Company, P.O. Box 30, Newark, DE 19714

A class of pyridinylpyrimidine fungicides represented by compound **1** is presented. Synthetic methods for the preparation of these novel compounds and structure-activity relationships are discussed. These compounds show broad spectrum control of plant pathogens with excellent activity on wheat eyespot (*Pseudocercosporella herpotrichoides*), wheat leaf blotch (*Septoria nodorum*) and rice blast (*Pyricularia oryzae*). The impact of photolability on the commercial viability of this class of fungicides is highlighted.

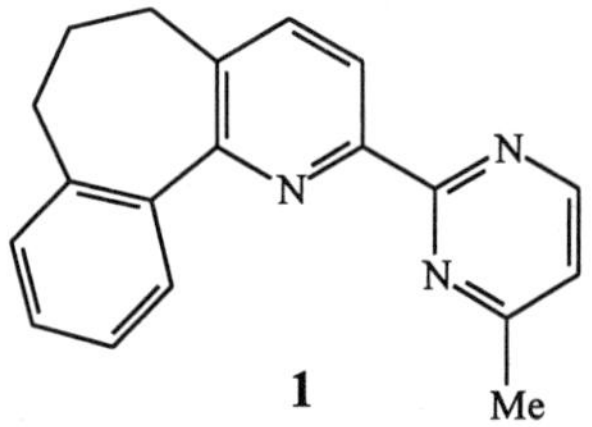

1

Our investigation of the class of fungicides represented by compound **1** was directed by the hypothesis that the mode-of-action involves chelation to a metalloenzyme. Molecular modeling suggested that the phenyl ring needs to be twisted out of plane relative to the pyridine for chelation to occur (Kleier, D. A., DuPont Company, unpublished data). Therefore, our strategy for the preparation of novel fungicides was to incorporate a bridge between the phenyl and pyridine rings to control the dihedral angle between these two rings in order to maximize biological activity. For example, the pyrimidine fungicide **2,** which was discovered by Katoh *et al.* (*1*), is transformed into the fused-tricycle **3** by incorporation of a trimethylene bridge as illustrated below. Analysis of Dreiding models indicated that the dihedral angles with a trimethylene and tetramethylene bridge are about 50° and 100°, respectively. Calculations on bridged bipyridines by Thummel suggest these angles to be about 45° and 58°, respectively, due to conjugation effects (*2*). Field efficacy of compound **1** was significantly lower than predicted by greenhouse testing which we attributed to photolability since the half-life observed in laboratory tests was 16 hours under artificial sunlight. Therefore, we also investigated the photostability of this class of fungicides.

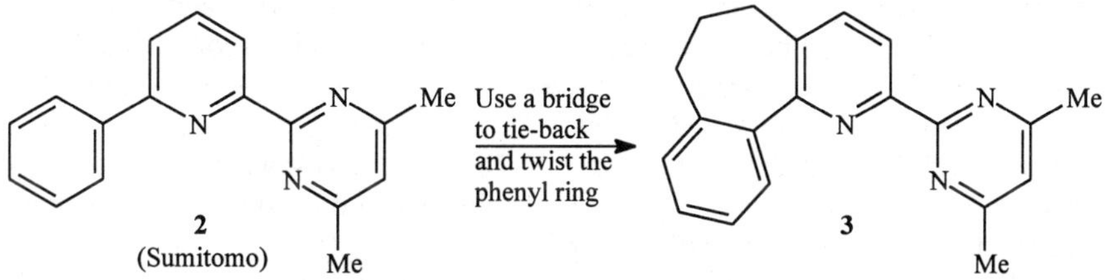

Results and Discussion

Synthetic Routes. Some methods for the preparation of these bridged compounds have been previously described by Daub *et al.* (*3*). Below, we provide details for the preparation of several bridged fungicides to illustrate these methods. The seven-membered ring fluoride **8** was prepared from 8-fluoro-1-benzosuberone (**4**) as shown below. The pyridoannulation method of Chelucci *et al.* (*4*) was used for the first three steps with the exception that we substituted 1,3-dimethyl-3,4,5,6-tetrahydro-2(1*H*)-pyrimidinone (DMPU) for hexamethylphosphoramide. The novel benzoazetine **6**

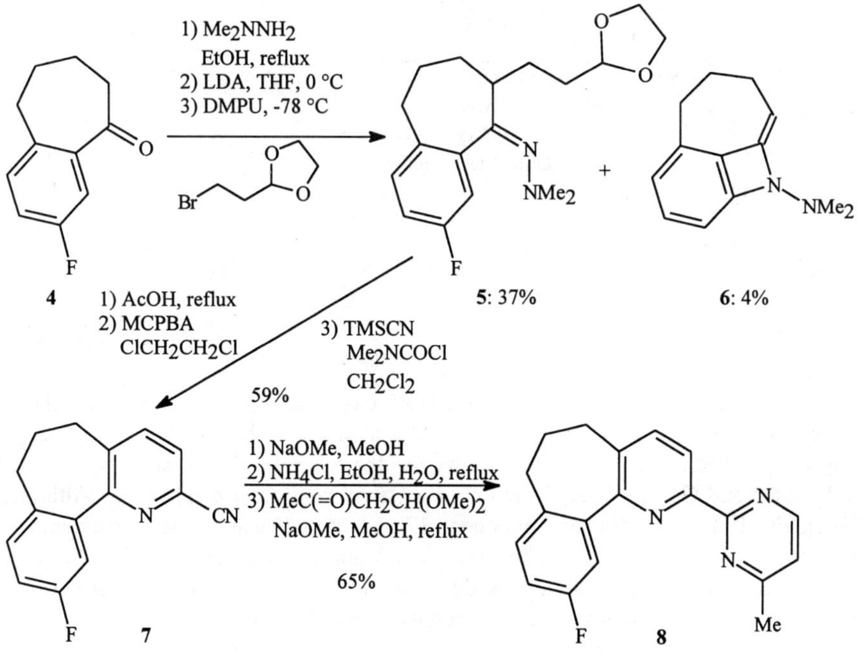

was obtained in the alkylation step and its structure was confirmed by hydrolysis to a hydrazine-substituted benzosuberone. Presumably, the mechanism for the formation of this benzoazetine proceeds via a benzyne intermediate. The cyanopyridine **7** was prepared from the intermediate *N*-oxide using the highly regioselective method of Fife (*5*). Finally, the pyrimidine ring was built using standard methods (*1,3*). The trimethylene- and ethylene-bridged compounds **1** and **51** (see Table I under Structure-Activity Relationships) were prepared from 1-benzosuberone and 1-tetralone, respectively, using this linear sequence. The propenylene-bridged compound **52** (see Table I) was prepared from compound **1** by radical bromination with *N*-bromosuccinimide followed by hydrobromide elimination using 1,8-diazabicyclo[5.4.0]undec-7-ene (DBU). The ethenylene-bridged compound **54** (see Table I) was prepared from 7,8-benzoquinoline using thc lattcr part of the above sequence starting at the oxidation/Fife reaction steps.

The halogenated compounds **12-15** were prepared as shown below. The known (*3*) tricyclic pyridine **9** was chlorinated to provide the 10-chloro and 8-chloro tricyclic pyridines **10a** and **10b** in 32% and 15% yields, respectively. Similarly, bromination of pyridine **9** afforded the 10-bromo and 8-bromo tricyclic pyridines **11a** and **11b** in 47% and 21% yields, respectively. Individually, these chlorides **10a,b** and bromides **11a,b** were converted to the bridged fungicides **12-15** by the same methods described above for the preparation of the fluoride **8**. We found that chlorides **12** and **13** could also be prepared directly from the unsubstituted compound **1** in 32% and 24% yields, respectively, along with a trace (4%) of the 9-chloro isomer **17a** using the same chlorination method.

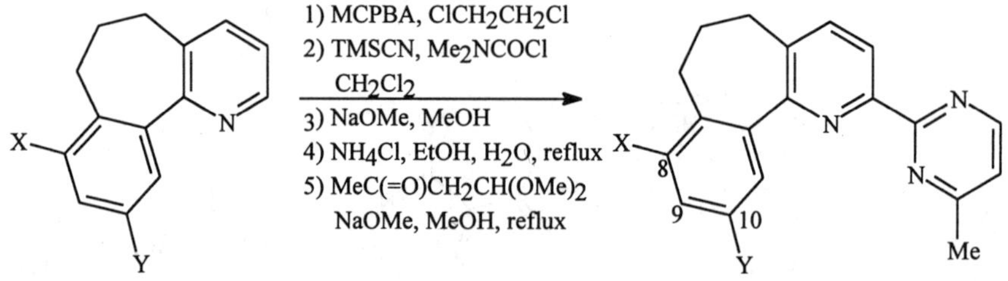

5% aq. NaOCl, TFA → **9**: X = Y = H; **10a**: X = H, Y = Cl, 32%; **10b**: X = Cl, Y = H, 15%

NBS, TFA, RT → **11a**: X = H, Y = Br, 47%; **11b**: X = Br, Y = H, 21%

12: X = H, Y = Cl, 28%
13: X = Cl, Y = H, 11%
14: X = H, Y = Br, 58%
15: X = Br, Y = H, 27%

In order to prepare 9-substituted derivatives, we used the procedure shown below. The previous route required eight steps to convert a benzosuberone to a target pyridinylpyrimidine. In this sequence, we use the effective pyridoannulation method of Jameson and Guise (*6*) requiring only two steps from a benzosuberone. Although the yields of the 9-substituted compounds **17a-c** were rather low, the convergency of the method saved much effort. This pyridoannulation method was also used to prepare the tetramethylene- and methylene-bridged compounds **50** and **53** (see Table I) from 7,8,9,10-tetrahydro-5(6*H*)-benzocyclooctenone and 1-indanone, respectively.

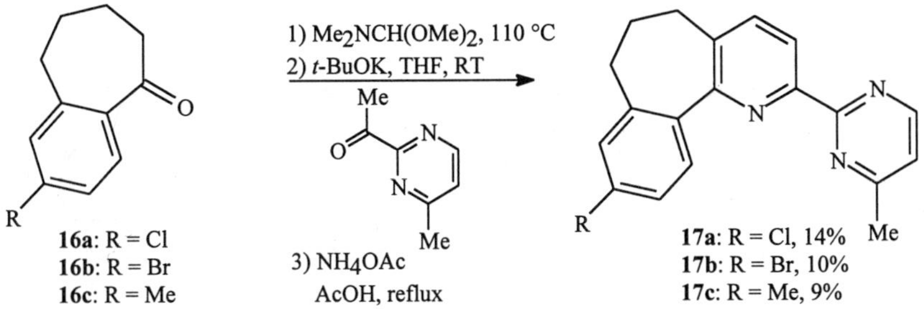

Palladium couplings using the 10- and 8-bromides **14** and **15** provided very efficient access to several other substituents. As shown below, we converted the 10-bromide **14** to the 10-phenyl and 10-naphthalenyl compounds **18** and **19** via Suzuki coupling. Similarly, the 8-bromide **15** was coupled with phenylboronic acid to give the 8-phenyl compound **20**. The 10-cyanide **21** and the silylacetylene **22** were also prepared via palladium couplings. The silylacetylene **22** was deprotected to give the desired 10-acetylene **23** by treatment with potassium hydroxide in methanol.

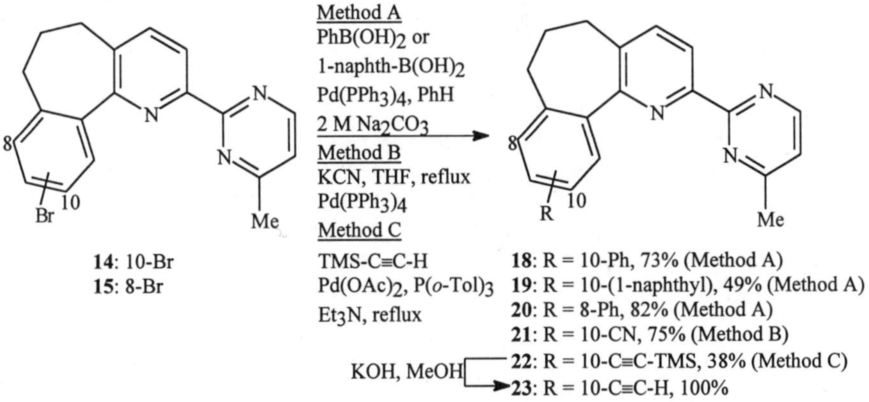

Our strategy for the preparation of 11-substituted compounds used the pyridine nitrogen to direct chemistry to the *ortho* position. Compound **1** was cyclopalladated with palladium(II) acetate to give a crude palladium complex. This crude complex was treated with 3-chloroperoxybenzoic acid followed by reduction with sodium borohydride as described by Grigor *et al.* (*7*) to afford the phenol **24** in 28% yield. This phenol was of high interest since it was designed to increase photostability through a photoenolization process. Bromination of the crude palladium complex under the conditions of Horino *et al.* (*8*) afforded the 11-bromide **25** in 18% yield. This bromide showed slow interconversion of conformations by NMR due to the steric hindrance between the bromine and the nitrogen lone-pair. In contrast, the phenol **24** showed rapid interconversion of conformations presumably due to hydrogen bonding between the phenolic hydrogen and the pyridine nitrogen.

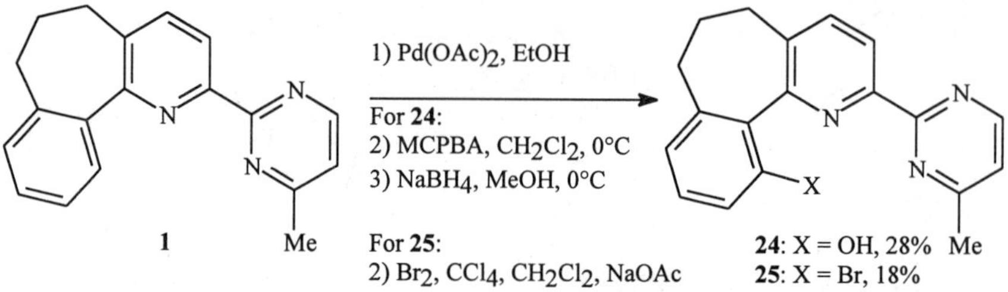

The following scheme illustrates the preparation of several heterocycle combinations which also possess a 1,4-orientation of two nitrogens for chelation. The bipyridines **27a-c** were prepared using the pyridoannulation method of Jamcson and Guise (*6*). The pyridinylpyrimidines **28a-c**, wherein the location of the two heterocycles have been reversed relative to compound **1**, and the bipyrimidine **29** were

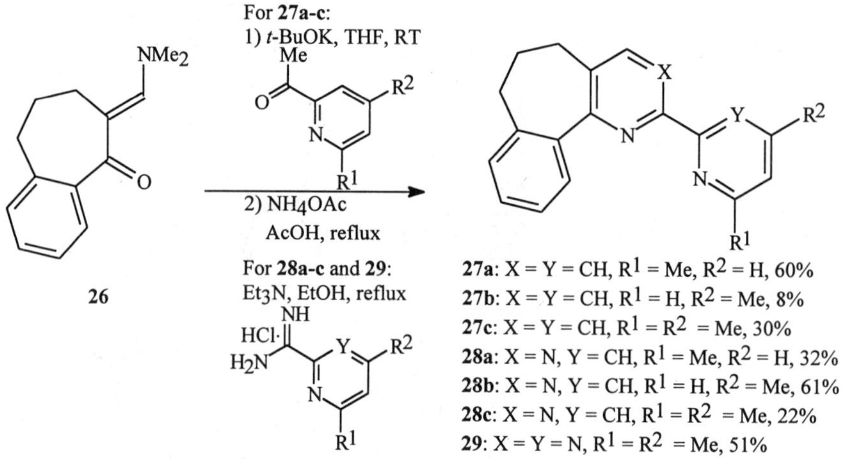

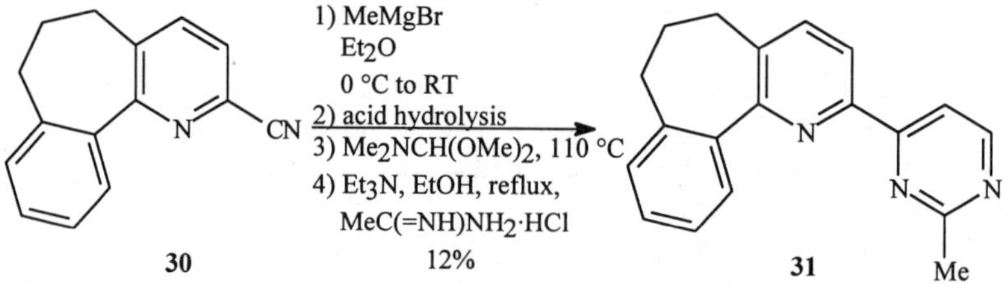

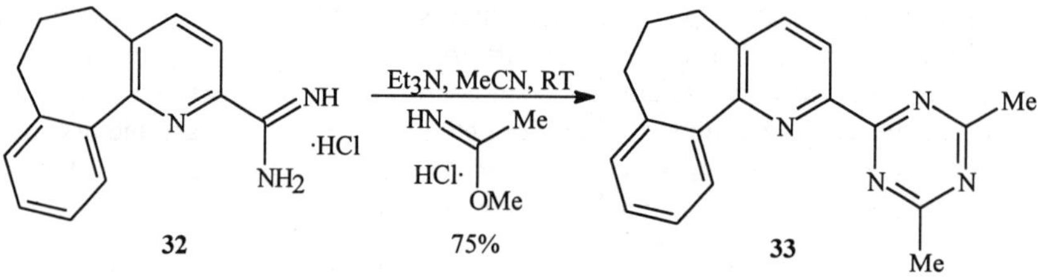

prepared by condensation of the vinylogous amide **26** with amidines. The pyridinylpyrimidine **31** where the pyrimidine is attached through the 4-position was prepared from cyanopyridine **30**. The pyridinyltriazine **33** was prepared by the condensation of the amidine **32** with methylacetimidate.

Next, we turned our attention to optimizing the substitution of the pyrimidine ring of compound **1**. Notably, we incorporated alkoxy, alkylthio and alkylamino groups onto the pyrimidine ring to increase the basicity of the pyrimidine electron lone-pairs and presumably increase the chelation ability. As illustrated below, most of these compounds were prepared by standard methods of pyrimidine synthesis. The dimethoxypyrimidine **49** was prepared using the pyridoannulation method of Jameson

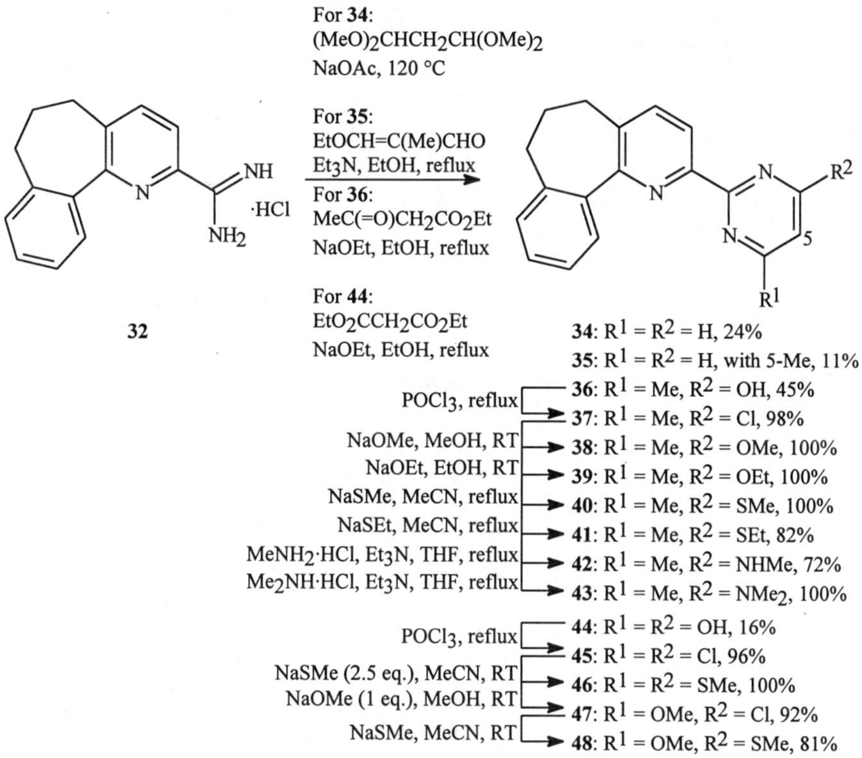

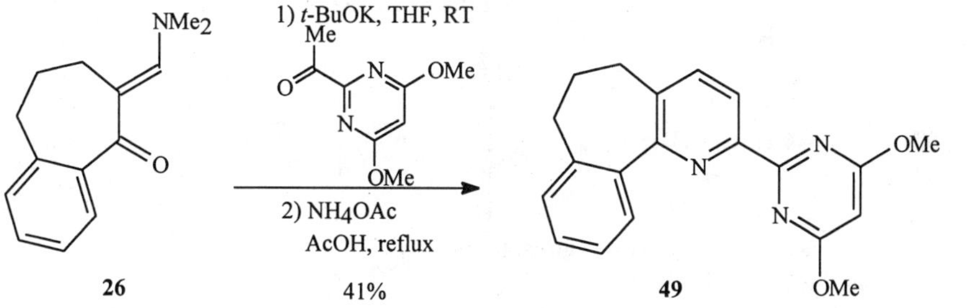

and Guise (*6*) in good yield. The di(thio)alkoxypyrimidine compounds **46**, **48** and **49** were of interest for the increased basicity of the pyrimidine nitrogens as well as for the possibility of increased photostability since they lack pyrimidine methyl groups which could be involved in radical and other photodegradative processes.

Structure-Activity Relationships. Tables I-V below show the preventive control of wheat eyespot, wheat leaf blotch and rice blast observed for the compounds described above. These compounds show broad-spectrum control of plant pathogens (*3*) but these three pathogens are the strength of this class. In all five tables, the trimethylene bridged compound **1** is shown in the first line as a benchmark since it was overall the most active compound prepared. In addition, these tables show the photolysis half-life ratio relative to compound **1** since we believed that photolysis impacted the field efficacy of this class of fungicides. Table I shows the biological control of compounds with various bridges. The compounds are arranged in descending overall activity. Note that an asterisk in the tables indicates that the compound was tested at 25 g/ha. The two most active compounds are clearly the trimethylene bridged compound **1** and the tetramethylene bridged compound **50**. These two are the only ones with the phenyl ring significantly twisted out of planarity relative to the pyridine ring supporting our working hypothesis. Compounds **51-54** have much smaller dihedral angles between the phenyl and pyridine rings which, we believe, lowers activity. Most notably, the ethenylene bridged compound **54** wherein the dihedral angle is constrained to 0° is not active at 100 g/ha on any of the pathogens.

Table I. Effect of Ring Size and Conformation

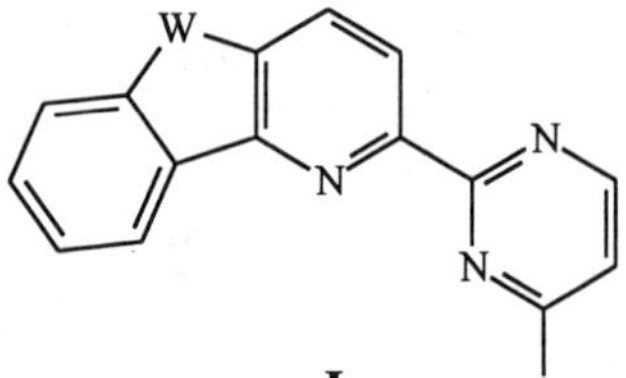

I

Cmpd No.	W	% Control (I = less than 50%) Wheat Eyespot 100 g/ha	Wheat Leaf Blotch 100 g/ha	Rice Blast 100 g/ha	Photolysis $t_{1/2}$ Ratio
1	$CH_2CH_2CH_2$	94	100*	92	1.0
50	$CH_2CH_2CH_2CH_2$	88	100*	67	0.79
51	CH_2CH_2	--	99	I	0.39
52[1]	$CH{=}CH{-}CH_2$	I	97	I	--
53	CH_2	I	90	I	0.09
54	$CH{=}CH$	I	I	I	0.54

[1] Compound is a mixture of two regioisomers.
* Compound tested at 25 g/ha.

Table II compares the activity of compounds with the optimal trimethylene bridge versus the Sumitomo fungicides **2** (*1*), **55** (*9*) and **56** (*10*). The bridged compounds generally show higher activity. Most notable is the comparison of the bridged compound **28a** to the unbridged compound **55** wherein the central ring is a pyrimidine and the terminal ring is a pyridine.

Table II. Comparison with Sumitomo Fungicides

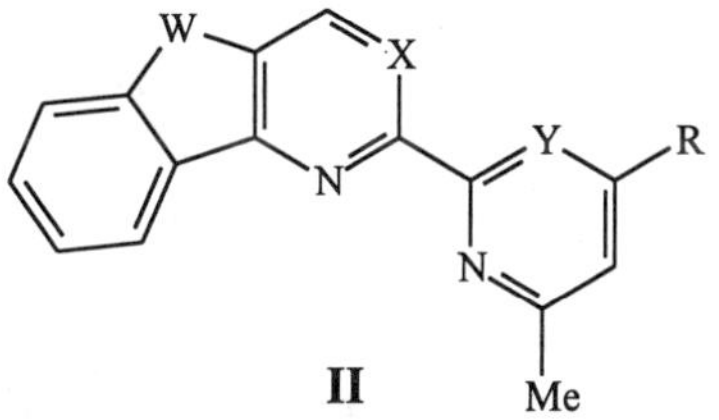

II

Cmpd No.	W	% Control (I = less than 50%) Wheat Eyespot 100 g/ha	Wheat Leaf Blotch 100 g/ha	Rice Blast 100 g/ha
	X = CH, Y = N, R = Me (Pyr-Pyrim)			
2	H, H	83	--	79
3	$CH_2CH_2CH_2$	I	100*	93
	X = N, Y = CH, R = H (Pyrim-Pyr)			
55	H, H	I	97	I
28a	$CH_2CH_2CH_2$	75	95*	80
	X = Y = CH, R = H (Pyr-Pyr)			
56	H, H	53	--	92
27a	$CH_2CH_2CH_2$	89	97*	94

* Compound tested at 25 g/ha.

Table III shows various heterocycle combinations with the optimal trimethylene bridge. The comparison of bipyridines **27a-c** shows the importance of a methyl group at the 6-position of the terminal pyridine ring for activity. Similarly, the pyridinylpyrimidine **28a** with a methyl group at the 6-position of the terminal pyridine ring is clearly more active than the 4-methyl (**28b**) and 4,6-dimethyl (**28c**) analogs. The heterocycle combinations are arranged in descending activity based on the most active compounds prepared within each set. Pyridinyltriazine **33** has dimethyl substitution so a more appropriate comparison may be with compounds **3** (Table V), **27c** and **28c** which would move this heterocycle combination above the bipyridines in relative activity. Compound **31,** where the pyrimidine ring is connected through the 4-position, is much less active than its isomer, compound **1**. Bipyrimidine **29** was not active at 100 g/ha on any of the three pathogens. All of the heterocycle combinations tested for photostability showed lower stability than pyridinylpyrimidine **1**.

Table III. Effect of Heterocycle Combinations

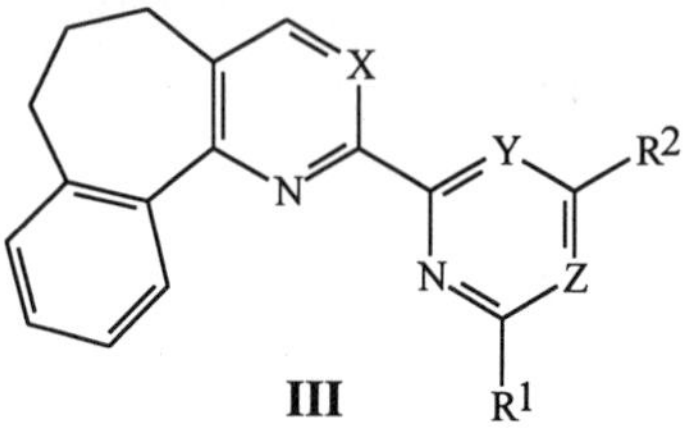

III

Cmpd No.	R^1	R^2	% Control (I = less than 50%) Wheat Eyespot 100 g/ha	Wheat Leaf Blotch 100 g/ha	Rice Blast 100 g/ha	Photolysis $t_{1/2}$ Ratio
	X = CH, Y = N, Z = CH (Pyr-Pyrim)					
1	Me	H	94	100*	92	1.0
	X = Y = Z = CH (Pyr-Pyr)					
27a	Me	H	89	97*	94	0.53
27b	H	Me	64	I	I	--
27c	Me	Me	93	99	71	--
	X = N, Y = CH, Z = CH (Pyrim-Pyr)					
28a	Me	H	75	95*	80	0.87
28b	H	Me	I	I	I	--
28c	Me	Me	I	99	57	--
	X = CH, Y = Z = N (Pyr-Triaz)					
33	Me	Me	I	98*	60	0.81
	X = Y = CH, Z = N (Pyr-4-Pyrim)					
31	Me	H	I	88	71	--
	X = Y = N, Z = CH (Pyrim-Pyrim)					
29	Me	Me	I	I	I	--

* Compound tested at 25 g/ha.

Table IV shows the activity for compounds with the optimal bridge and optimal heterocycle combination wherein the phenyl ring is substituted. Compounds with small substituents (halogen, methyl and ethynyl) show comparable activity to compound **1** but were slightly less photostable. Surprisingly, the 10-cyano compound

21 is an exception with much lower activity. The larger phenyl group is acceptable at the 10-position (compound **18**) though it loses activity on wheat eyespot. A phenyl group in the 8-position (compound **20**) greatly reduces activity. Phenol **24**, which was designed to be more photostable, showed *exceptional* photostability but, unfortunately, showed much lower activity than compound **1**. The hydrochloride salt **57** and the copper(II) chloride complex **58** derived from compound **1** have equal activity to compound **1** but much lower photostability.

Table IV. Effect of Substitution on Phenyl Ring

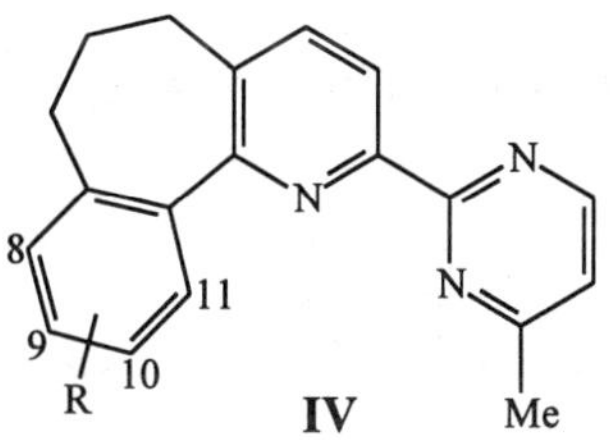

IV

Cmpd No.	R	% Control (I = less than 50%) Wheat Eyespot 100 g/ha	Wheat Leaf Blotch 100 g/ha	Rice Blast 100 g/ha	Photolysis $t_{1/2}$ Ratio
1	H	94	100*	92	1.0
13	8-Cl	98	98*	89	0.52
15	8-Br	I	100*	75	0.75
17a	9-Cl	93	99*	94	0.65
17b	9-Br	I	98*	98	0.50
17c	9-Me	I	97*	98	0.44
8	10-F	76	100*	94	0.49
12	10-Cl	98	98*	93	0.58
14	10-Br	92	100*	96	0.65
25	11-Br	77	99*	53	0.82
57	H (HCl Salt)	84	100*	91	0.15
58	H ($CuCl_2$ Complex)	100	99*	93	<0.1
23	10-C≡CH	50	99*	93*	--
22	10-C≡C-TMS	I	89	I	--
18	10-Ph	I	96*	95	0.78
19	10-(1-Naphth)	I	85	I	--
20	8-Ph	I	96	I	--
21	10-CN	I	99	I	--
24	11-OH	I	95	I	44

* Compound tested at 25 g/ha.

Table V shows the activity for compounds with the optimal bridge and optimal heterocycle combination with various substituents on the pyrimidine ring. The first section of this table compares hydrogen and methyl substitution. The second section compares heteroatom substitution where the other substituent is methyl. The third section compares di-heteroatom substitution. Within each section, the compounds are arranged by descending activity. The first section shows that at least one methyl group is necessary for high activity. Clearly, it is preferable for a methyl to be at the 4-position (compounds **1** and **3**) over the 5-position (compound **35**). The chloromethyl compound **59** was inactive at the high rate. The second and third sections show that methoxy and thiomethoxy substitution result in very similar

Table V. Effect of Substitution on Pyrimidine Ring

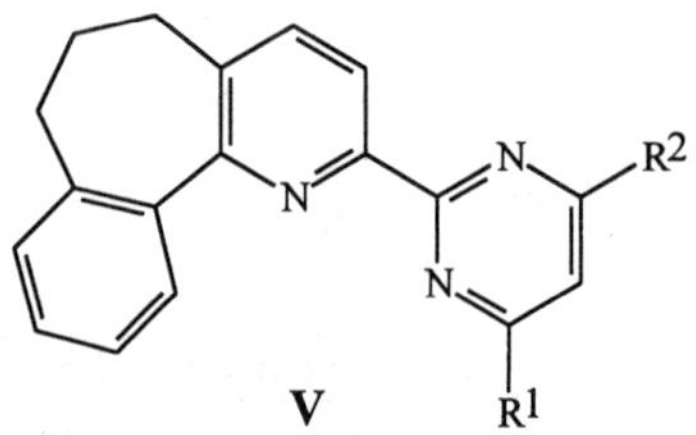

V

Cmpd No.	R^1	R^2	% Control (I = less than 50%) Wheat Eyespot 100 g/ha	Wheat Leaf Blotch 100 g/ha	Rice Blast 100 g/ha	Photolysis $t_{1/2}$ Ratio
1	Me	H	94	100*	92	1.0
3	Me	Me	I	100*	93	0.62
35	H (5-Me)	H	77	90*	98	--
34	H	H	54	98	94	0.38
59	CH_2Cl	H	I	I	I	--
40	Me	SMe	75	100*	92	0.42
38	Me	OMe	--	98*	ca. 80	1.3
41	Me	SEt	I	82*	I	--
39	Me	OEt	I	99	I	--
42	Me	NHMe	I	90	I	--
43	Me	NMe_2	I	69	I	--
37	Me	Cl	I	58	I	--
36	Me	OH	I	I	I	--
49	OMe	OMe	I	94*	86	1.8
48	OMe	SMe	I	96*	53	--
46	SMe	SMe	I	97*	I	0.44
47	OMe	Cl	I	99	I	--
44	OH	OH	I	I	I	--
45	Cl	Cl	I	I	I	--

* Compound tested at 25 g/ha.

activity relative to compound **1** but with a reduced spectrum. It should be noted that methoxy substitution in compounds **38** and **49** resulted in slightly improved photostability as compared to compound **1**. Ethoxy (compound **39**) and thioethoxy (compound **41**) substituents significantly lower activity, presumably due to their size. Methylamino (compound **42**) and dimethylamino (compound **43**) substituents result in even lower activity. With the exception of compound **47**, chloro and hydroxy substitution results in little to no activity at 100 g/ha.

Conclusion

We suggest that the activity in the pyridinylpyrimidine fungicide area is sensitive to the dihedral angle between the phenyl and pyridine rings with a trimethylene as the optimal bridge between these rings. The heterocycle combination of a central pyridine and a terminal pyrimidine is optimal for biological activity level and spectrum. Activity was not particularly affected by substitution with small groups on the phenyl ring, and 4-methyl substitution on the terminal pyrimidine ring was optimal. A solution to the probable photostability problem in field tests for compound **1** was found in phenol **24**, but the level of biological activity was not acceptable.

Acknowledgments

We wish to express our gratitude to all of the biologists who conducted fungicidal evaluations on these compounds. We also wish to express our gratitude to the following individuals: Dr. Michael C. Klapproth for coordinating the biological aspects of this program; Dr. Daniel A. Kleier for computational contributions and discussions; Anton S. Burr and Denis L. Perkins for the photostability studies; and John C. Groce, Jr. for variable temperature NMR studies.

Literature Cited

1. Katoh, T.; Maeda, K.; Shiroshita, M.; Yamashita, N.; Sanemitsu, Y.; Inoue, S. (Sumitomo Chemical Co.) European Patent Publication EP-A-259,139.
2. Thummel, R. P. *Tetrahedron* **1991**, *47*, 6851.
3. Daub, J. P.; Finkelstein, B. L.; Kleier, D. A. (DuPont Company) World Patent Publication WO 93/14080.
4. Chelucci, G.; Gladiali, S.; Marchetti, M. *J. Heterocycl. Chem.* **1988**, *25*, 1761.
5. Fife, W. K. *Heterocycles* **1984**, *22*, 93.
6. Jameson, D. J.; Guise, L. E. *Tetrahedron Lett.* **1991**, *32*, 1999.
7. Grigor, B. A.; Nielson, A. J. *J. Organomet. Chem.* **1977**, *129*, C17.
8. Horino, H.; Inoue, N. *J. Org. Chem.* **1981**, *46*, 4416.
9. Katoh, T.; Maeda, K.; Shiroshita, M.; Yamashita, N.; Sanemitsu, Y.; Inoue, S. (Sumitomo Chemical Co.) European Patent Publication EP-A-270,362.
10. Katoh, T.; Shiroshita, M.; Takano, J.; Maeda, K.; Yamashita, N. (Sumitomo Chemical Co.) Japanese Kokai Patent Publication JP 1,261,305.

Chapter 26

4-Arylalkoxyquinazolines with Antifungal Activity

M. C. H. Yap, B. A. Dreikorn, L. N. Davis, R. G. Suhr, S. V. Kaster, N. V. Kirby, G. Paterson, P. R. Graupner, and W. R. Erickson

Discovery Research Center, DowElanco, 9330 Zionsville Road, Indianapolis, IN 46268-1503

4-Arylalkoxyquinazolines are a class of highly active wide spectrum fungicides against cereal and broad leaf pathogens. Activity optimisation through an extensive rationally designed, SAR program yielded a 4-(4-trifluoroethoxyphenethoxy)-8-fluoroquinazoline which, unfortunately, proved too phytotoxic in the field. Safening was achieved through introduction of a methyl group at position 6 of the quinazoline but at the expense of reduced biological spectrum. Molecular modeling predictions suggesting that biological spectrum could be restored by replacement of the phenyl group by "aza-benzenes" were vindicated.

Evaluation of biological activity in rationally-designed and 'randomly' selected molecules represent two important areas of the Fungicide Discovery program in DowElanco. The lead molecule described in this paper was discovered through the random screening process. 4-Phenethylaminoquinoline (Compound 5, Table I) displayed moderate control of downy mildew in squash and grape. The rational design, structure-activity program which followed uncovered compounds with activity across a much broader spectrum of disease including glume and leaf blotch in wheat, powdery mildew, leaf rust, rice blast and apple scab (*1*). In addition, many of the same compounds possessed potent insecticidal activity against leafhoppers, Southern armyworm, codling moth and two spotted spider mite.

Wheat powdery mildew is a major disease. Consequently, a considerable research effort concentrated on elucidating structure-activity relationships against this pathogen.

The initial focus of the structure-activity relationship study was on the importance of the tether nitrogen atom of the phenethylamino "tail" of the lead molecule (2). The wheat powdery mildew activity of four pairs of 4-phenethoxy and 4-phenethylaminoquinolines bearing a variety of substituents on the benzene ring were

Table I - Differences in PMW Activity Between Oxygen and Nitrogen Tethered Quinolines

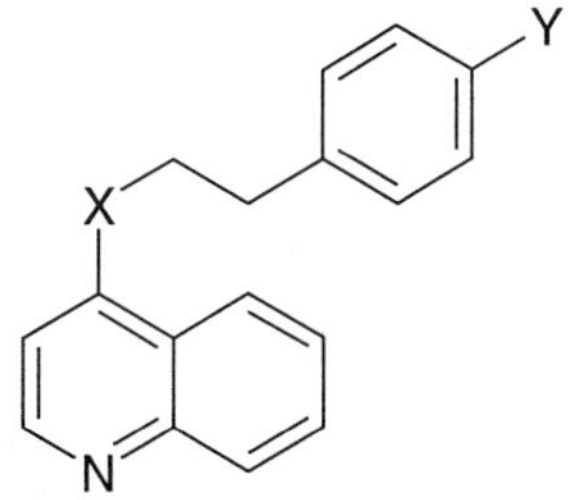

#	X	Y	LC_{90} / ppm
1	NH	Me	>500
2	O	Me	105
3	NH	Cl	>500
4	O	Cl	411
5	NH	H	>500
6	O	H	382
7	NH	OEt	>500
8	O	OEt	72

compared. The data in Table I shows that the ethers were clearly superior to the amines, irrespective of the *para* substituent on the benzene ring of the "tail". The differences are especially marked in Compounds 7 and 8 where the oxygen-tethered analog was seven times more active than the nitrogen-tethered compound. Other tether atoms including sulphur and carbon have also been investigated but none proved superior to oxygen.

Another area that was investigated was replacement of quinoline with other nitrogen-bearing aromatic heterocycles. With the exception of quinazoline, other heterocycles did not have comparable activity or biological spectrum. These include pteridine, the naphthyridines, purine, cinnoline, pyrimidine and pyridine (*3*).

Table II compares the differences between quinolines and quinazolines. With the exception of the highly active Compounds 15 and 16, quinazolines appear to have a clear advantage over quinolines. Like the data shown in Table I, these results also indicate the importance of benzene ring substituents on overall activity.

Substitution on the Quinazoline Ring

Having established the ideal heterocycle and tether atom, we proceeded to investigate the influence of substituents on the quinazoline. Given that activity has already been shown to be highly dependent on functionalities on the benzene ring "tail", we were careful to ensure that a wide range of "tail" substituents were studied.

Holding the substitution on the "tail" constant with a tertiary butyl group at the *para* position on the benzene ring, introduction of any substitution at position 2 of the quinazoline completely destroyed activity (Table III). Replacement of tertiary butyl with either alkoxy or aryloxy ethers (Table III) did not alter the detrimental effects of substitution at position 2. The intolerance of position 2 towards substitution remains unchanged with a simple unsubstituted phenyl "tail" (Table III).

Table II - Differences in PMW Activity Between Quinolines and Quinazolines

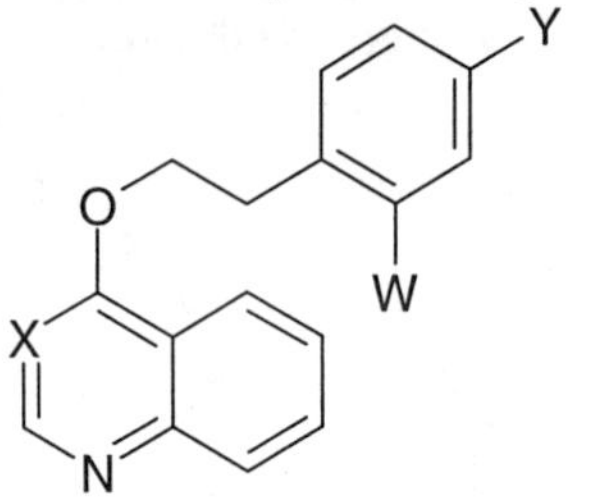

#	W	X	Y	LC_{90} / ppm
9	F	N	F	52
10	F	CH	F	212
11	H	N	Cl	45
12	H	CH	Cl	411
13	H	N	H	49
14	H	CH	H	381
15	H	N	OCH_2CF_3	4
16	H	CH	OCH_2CF_3	6

Table III - Effect of 2-Substitution on Activity of 4-Phenethoxyquinazolines

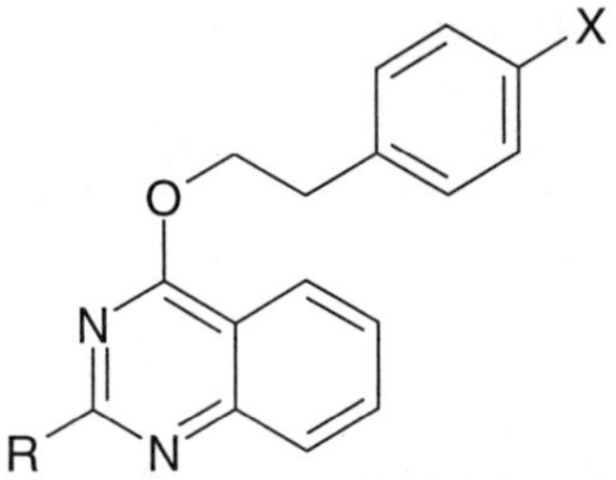

#	R	X	LC_{90} / ppm
17	H	*tert*-Butyl	87
18	F	*tert*-Butyl	>500
19	Cl	*tert*-Butyl	>500
20	CF_3	*tert*-Butyl	>500
21	CCl_3	*tert*-Butyl	>500
22	SCH_2Ph	*tert*-Butyl	>500
23	H	4-CF_3PhO	5
24	F	4-CF_3PhO	>500
25	H	4-Cl PhO	83
26	Cl	4-Cl PhO	>500
27	H	OCH_2CF_3	4
28	CF_3	OCH_2CF_3	>500
29	H	4-Cl PhO	83
30	CCl_3	4-Cl PhO	>500
31	H	H	21
32	Cl	H	>500
33	Ph	H	>500

Substitution at position 5 of quinazoline was also not beneficial to activity. Synthetic difficulties precluded the study of a wider range of substituents. Given that the limited data shown in Table IV did not reveal any easily discernible trends, our attentions were quickly switched to other portions of the molecule which appeared to offer more promising results.

Table IV - Effect of 5-Substitution on PMW Activity of Various 4-Phenethoxyquinazolines

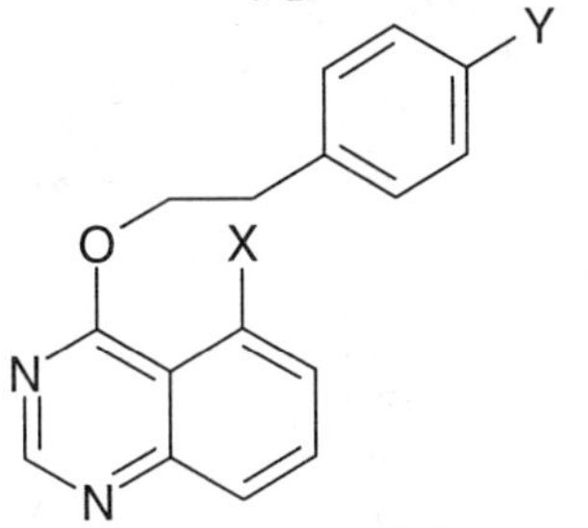

#	X	Y	LC_{90} / ppm
34	H	OCF_3	82
35	F	OCF_3	91
36	Cl	OCF_3	411
37	H	OCF_2CF_2H	17
38	F	OCF_2CF_2H	82
39	Me	OCF_2CF_2H	23
40	H	CF_3	11
41	F	CF_3	90
42	Cl	CF_3	137
43	Me	CF_3	72
44	H	*tert*-Bu	359
45	F	*tert*-Bu	>500
46	Cl	*tert*-Bu	>500
47	Me	*tert*-Bu	>500

Quinazolines functionalised at the 6 position were studied in great detail. The methyl analog (Table V, Compound 49) was highly active. The ethyl analog (Compound 50) also had almost identical PMW activity but there was a sharp loss for higher alkyls. Given that both 6-methyl and 6-ethylquinazolines were clearly superior to the unsubstituted parent (Compound 48), we initially speculated that the lower alkyls may be metabolised into the actual active moiety. Unfortunately, neither the carboxaldehyde (Compound 61) nor the carboxylic acid (Compound 62) showed any activity. This does completely rule out the hypothesis as metabolism may occur close to the site of action.

Table V - PMW Activity of 6-Substituted Quinazolines

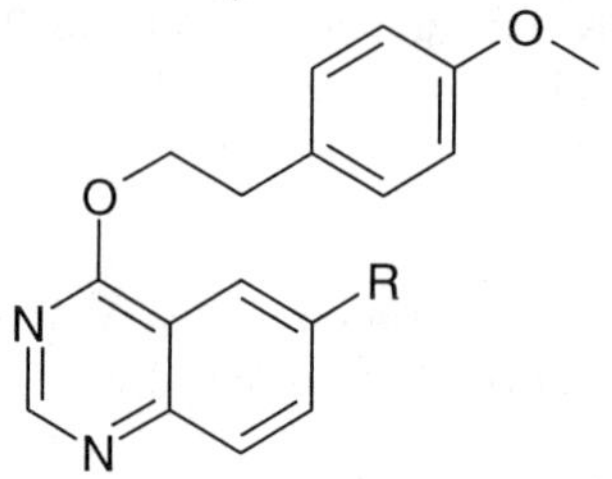

#	R	LC_{90} / ppm
48	H	90
49	Me	9
50	Et	11
51	*n*-Pr	25
52	*n*-Bu	>500
53	F	251
54	Br	>500
55	I	>500
56	NO_2	382
57	NH_2	>500
58	$NHSO_2Me$	>500
59	NHCOMe	>500
60	NHCOPh	>500
61	CHO	>500
62	COOH	>500

Other than methyl, ethyl and *n*-propyl, other substituents at position 6 of quinazoline were inactive. Even the introduction of a small fluorine moiety diminished fungicidal activity.

Substitution at position 7 on the quinazoline did not follow a consistent trend (Table VI). With the *tert*-butyl substituents on the "tail", the introduction of fluorine here was beneficial (Compounds 63 and 65). However, the identical substitution in the trifluoroethoxy series (Compounds 69 and 71) resulted in a 20-fold loss of activity.

Introduction of a methyl at position 7 caused a 5-fold loss of activity in the trifluoroethoxy series (Compounds 69 and 72) compared to the unsubstituted parent Only a small decrease was seen in the *tert*-butyl series (Compounds 63 and 66).

With the exception of quinazolines with methyl substitutents at position 6, analogs with fluorine at position 8 were the most active of the series (Table VII). Unfortunately, like modifications at position 6, changes otherwise at this position diminished activity. Even replacement of fluorine with chlorine (Compound 88) was not tolerated. Neither was replacement by ethers (Compounds 80-82, 91,95). In the trifluoroethoxy series, the trifluoromethyl analog (Compound 78) was equally as active as fluorine but the loss of activity in the *tert*-butyl series (Compound 92) was slightly greater. There was a greater loss in the trifluoromethoxy series (Compound 96) with trifluoromethyl replacing fluorine at the 8 position of the quinazoline.

Table VI - PMW Activity of 7-Substituted Quinazolines

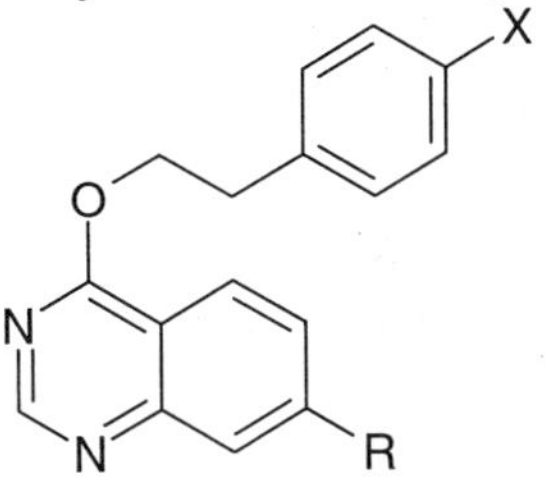

#	R	X	LC_{90} / ppm
63	H	*tert*-Bu	79
64	Cl	*tert*-Bu	>500
65	F	*tert*-Bu	32
66	Me	*tert*-Bu	111
67	NO_2	*tert*-Bu	>500
68	CF_3	*tert*-Bu	>500
69	H	OCH_2CF_3	4
70	Cl	OCH_2CF_3	>500
71	F	OCH_2CF_3	81
72	Me	OCH_2CF_3	19
73	NO_2	OCH_2CF_3	>500
74	NH_2	OCH_2CF_3	>500
75	$NHSO_2Me$	OCH_2CF_3	>500

The effect of the length of the polymethylene "bridge" on activity was also investigated. As Table VIII indicates, the best analogs (Compounds 99 and 105) had a two-carbon "bridge". Table VIII also shows a sharp loss of activity with the propyl chain and an improvement with the butyl chain. The reason for this is not known.

Substitution on the Aromatic 'Tail'

The structure-activity relationship study thus far has concentrated on compounds bearing substitutents at the para position of the "tail" benzene ring. The data clearly indicate that these substitutents had tremendous impact on activity. In order to determine ideal positions for these "tail" substitutents, the PMW activities of the *ortho* and *meta* analogs were compared with the *para* compounds (Table IX).

With a methoxy substituent (Compounds 109 and 110), there was a small preference for *para* over *ortho* substitution. However, with the highly active trifluoromethoxy group trifluoromethoxy group, the advantage of *para* substitution was clearly demonstrated (Compounds 111 and 112).

Activity differences between *meta* and *para* substituted analogs were more clear cut. There was at least a three-fold decrease in PMW activity as a result of having a phenyl substituent at the *meta* position (Compounds 118 and 119, Table IX Other substituents imparted even greater differences.

Table VII - PMW Activity of 8-Substituted Quinazolines

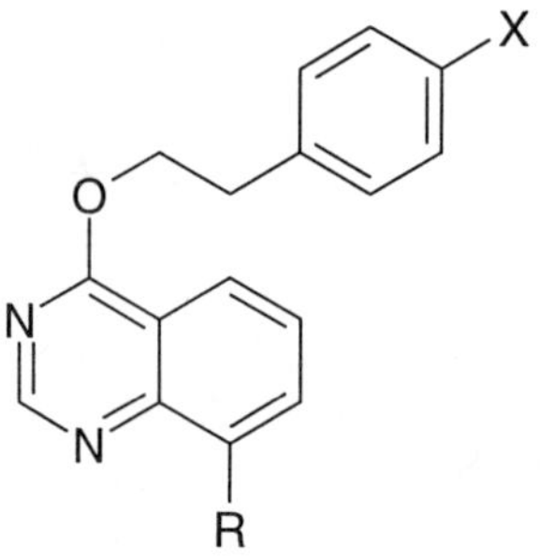

#	R	X	LC_{90} / ppm
76	H	OCH_2CF_3	4
77	F	OCH_2CF_3	2
78	CF_3	OCH_2CF_3	3
79	Me	OCH_2CF_3	11
80	OEt	OCH_2CF_3	>500
81	OMe	OCH_2CF_3	21
82	OCF_3	OCH_2CF_3	17
83	CH_2Cl	OCH_2CF_3	411
84	$CHCl_2$	OCH_2CF_3	>500
85	CHO	OCH_2CF_3	>500
86	CH_2CN	OCH_2CF_3	>500
87	H	*tert*-Bu	57
88	Cl	*tert*-Bu	413
89	F	*tert*-Bu	16
90	Me	*tert*-Bu	>500
91	MeO	*tert*-Bu	111
92	CF_3	*tert*-Bu	29
93	H	OCF_3	73
94	F	OCF_3	3
95	MeO	OCF_3	20
96	CF_3	OCF_3	72
92	OCF_3	OCF_3	276

The work on identifying the ideal heterocycle, tether heteroatom and length of the “bridge” has revealed that substituents on the benzene "tail" has a highly important impact on activity against powdery mildew in wheat. Having identified the ideal position for introducing the substituent, a great deal of synthetic effort was directed at the preparation of more *para* analogs to investigate the importance of steric size, lipophilicity and electronegativity. Work in this area was also driven by the broad spectrum fungicidal and insecticidal activity in these compounds.

Table VIII - Effect of "Bridge" Length on PMW Activity

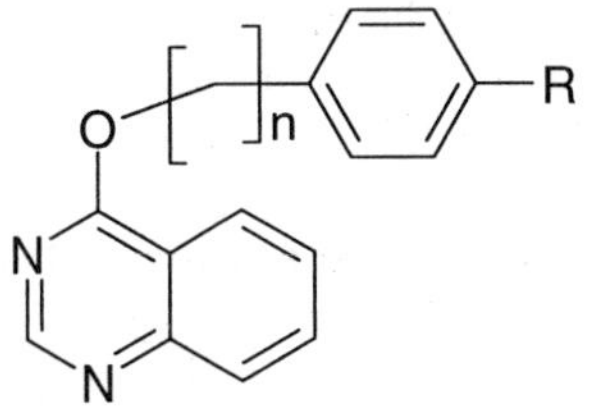

#	R	n	LC_{90} / ppm
98	H	1	411
99	H	2	31
100	H	3	>500
101	H	4	89
102	H	5	87
103	*tert*-Bu	0	>500
104	*tert*-Bu	1	>500
105	*tert*-Bu	2	58
106	*tert*-Bu	3	111
107	*tert*-Bu	4	61

Table IX - Differences in PMW Activity Between *Ortho*, *Meta* and *Para* Substituted "Tails"

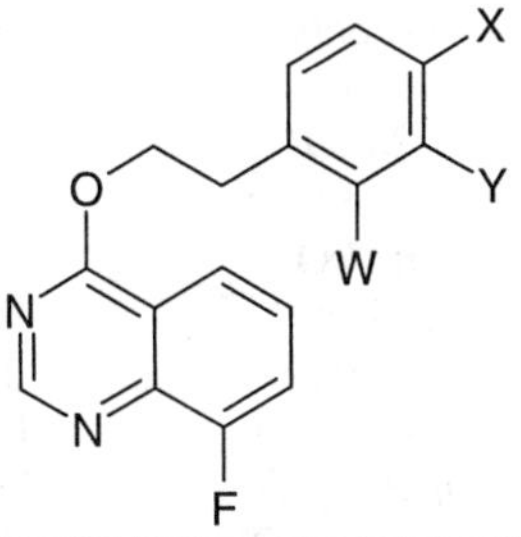

#	W	X	Y	LC_{90} / ppm
108	H	H	H	114
109	H	OMe	H	19
110	MeO	H	H	32
111	H	OCH_2CF_3	H	2
112	OCH_2CF_3	H	H	>500
113	H	H	H	115
114	H	OEt	H	4
115	H	H	OEt	109
116	H	OCH_2CF_3	H	2
117	H	H	OCH_2CF_3	119
118	H	Ph	H	28
119	H	H	Ph	107
120	H	4-Cl-PhO	H	5
121	H	H	4-Cl-PhO	>500

Table X shows the structure-activity relationships of alkyl and alkoxy groups. It appears on primary inspection that the *iso*-Propyl analog (Compound 124) deserves further attention. However, results from other substituted quinazolines do not reveal unexpectedly enhanced activity with this functionality.

Table X - Effect of Alkyl and Alkoxy Substituents on PMW Activity

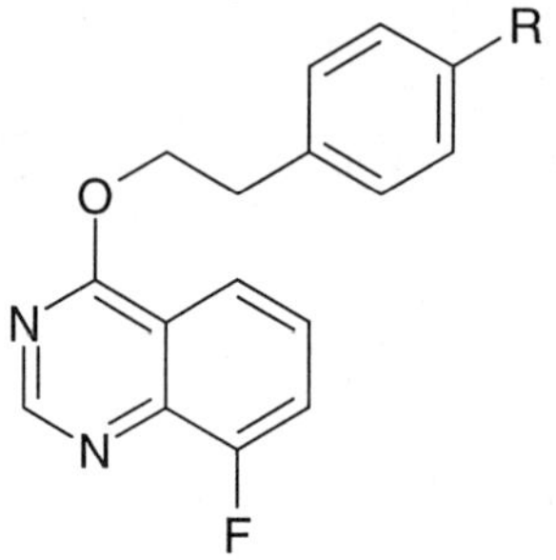

#	R	LC_{90} / ppm
122	H	33
123	Me	17
124	*iso*-Pr	4
125	*iso*-Bu	409
126	*tert*-Bu	16
127	*n*-Pentyl	28
128	Ph	26
129	MeO	5
130	EtO	4
131	*n*-PrO	8
132	*n*-BuO	9
133	*tert*-BuO	5
134	*n*-PentylO	17
135	*n*-HexylO	51
136	*n*-HeptylO	27
137	*n*-OctylO	28
138	*n*-DecylO	27
139	BnO	8
140	PhO	7

The relatively small influence of size and lipophilicity of the *para* substituent is reinforced with the alkyl ethers (Table X). There was little difference between groups as diverse as methoxy (Compound 129) and the butoxys (Compounds 132 and 133). There was a loss in activity with pentoxy (Compound 134) but thereafter there was little difference with even the *n*-decyloxy (Compound 138). Even the bulky phenoxy (Compound 140) and the benzyloxy (Compound 139) maintained good activity.

When the para substituents are carboxylic esters, there were no discernible trends (Table XI). The compounds were clearly less active than the ethers. The fluoroalkyl esters were especially poor.

The good activity and the apparent steric tolerance of the alkoxy ethers led us to examine fluoroalkoxy and fluoroalkylthio ethers. The resultant analogs were some of the most highly active molecules of this class of compounds (Table XI). The lack of dependence on steric size was also seen. In spite of extensive testing, we were unable to discern any relationships between the extent of fluorination and PMW activity. The most active fluorination pattern is represented by Compound 152.

The good activity of simple ethers also extends into the poly-oxy ether series (Table XI). The reasons for the poorer activity of the ethoxy and propoxy analogs (Compounds 157 and 158) are not known.

Table XI- Effect of Ester Substituents on the "Tail" on PMW Activity

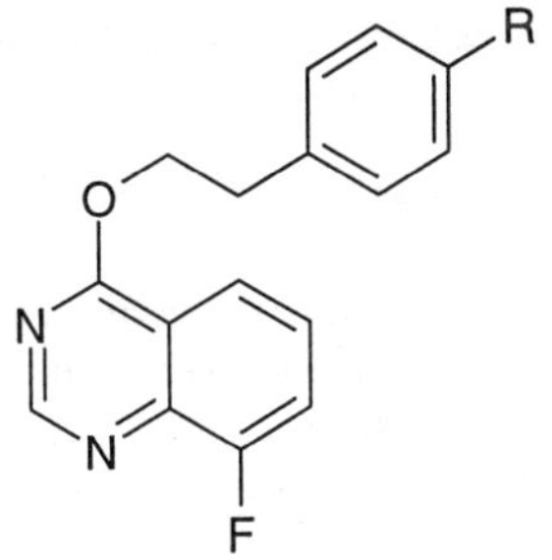

#	R	LC_{90} / ppm
141	C(O)OMe	>500
142	C(O)O-*iso*-Pr	29
143	C(O)O-*n*-Bu	>500
144	C(O)O-*tert*-Bu	89
145	C(O)O-*n*-Heptyl	261
146	$C(O)OCH_2CF_3$	>500
147	$C(O)OCH(CF_3)_2$	>500
148	$C(O)OC_6F_6$	>500
149	$OCHF_2$	5
150	OCF_3	5
151	OC_2F_5	16
152	OCH_2CF_3	2
153	OCH_2CF_2H	4
154	SCF_2CF_2H	26
155	$OCH_2C_2F_4H$	6
156	OCH_2CH_2OMe	7
157	OCH_2CH_2OEt	26
158	OCH_2CH_2OPr	27
159	OCH_2CH_2OBu	7
160	$OCH_2CH_2CH(OEt)_2$	1
161	F	28
162	Cl	8
163	Br	22
164	OH	>500
165	TMS	4
166	CF_3	5

Table XI also shows the activity of some other *para* substituents. Although the chloro (Compound 162), trimethylsilyl (Compound 165) and trifluoromethyl (Compound 166) analogs all appeared to be highly active, performance did not match the fluoroalkoxy compounds when in advanced level tests. Consequently, the analog that was eventually selected for further testing was Compound 152.

The data presented in Tables I - XI represents the activity of quinazolines when tested against a single disease (wheat powdery mildew) in the greenhouse. This structure-activity relationship study was able to demonstrate that the best quinazoline was Compound 152, the 8-fluoro-4-(4-trifluoroethoxyphenethoxy)quinazoline.

However, when this quinazoline was tested in the field, it occasionally showed phytotoxicity. Attempts to safen the compound or to discover the reasons responsible for the phytotoxicity were unsuccessful. Consequently, we were forced to redesign an inherently crop-safe compound.

Disubstitution on the Quinazoline Ring

The compounds discussed thus far in this structure-activity relationship study had a single substituent on the benzo ring of the heterocycle. Di-substituted analogs had received less attention as a result of studies in the early days of the project in which these compounds demonstrated poor to moderate activity (Table XII).

Table XII - PMW Activity of "early" 6,8-Disubstituted Quinazolines

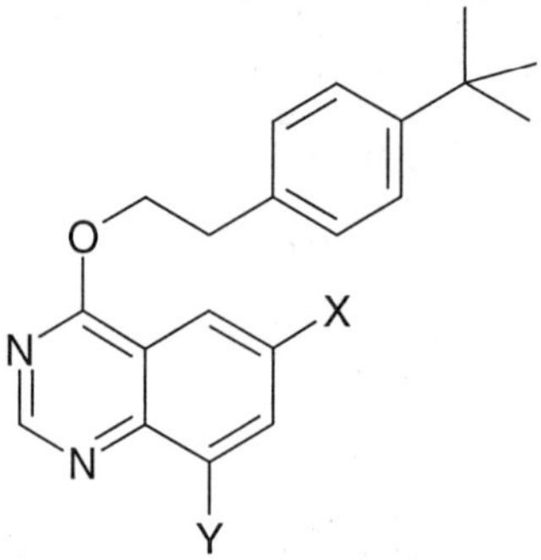

#	X	Y	LC_{90} / ppm
167	H	H	77
168	Cl	Cl	>500
169	Br	Br	>500
170	Me	Br	>500
171	Me	Me	>500

Re-examination of crop-safety data revealed that methyl substitution at the 6 position of the quinazoline ring did not impart any phytotoxicity. Given that structure-activity relationship studies have already shown that the most active quinazolines possess a 8-fluoro moiety, we speculated that di-substitution might yield a highly active yet non-phytotoxic compound. Initially, a small number of analogs bearing the 8-fluoro-6-methylquinazoline motif were prepared. When these were found to be extremely active and crop-safe in greenhouse tests, a more extensive variety of 6,8-disubstituted quinazolines were designed by utilising the activity and crop-safety data of the simple 6- and 8-substituted compounds that had previously been studied (*See Dreikorn B. A. et al., this publication).*

The data in Table XIII clearly illustrate that the activity of 6,8-disubstituted quinazolines is not always additive relative to their mono-substituted "parents". Compounds 177 and 184, like their respective 8-fluoro and 6-methyl or 6-methoxy "parents", both had good PMW activity. However, 6-fluoro-8-methylquinazoline (Compound 179) is considerably less active than either the 6-fluoro or 8-methylquinazoline "parent". As a result of this variation, we were forced to be highly selective in preparing new analogs. Only functional groups which had proven activity with a wide variety of "tails" or had been predicted by molecular modeling to be extremely interesting were prepared.

Table XIII - PMW Activity of Rationally-Designed 6,8-disubstituted Quinazolines

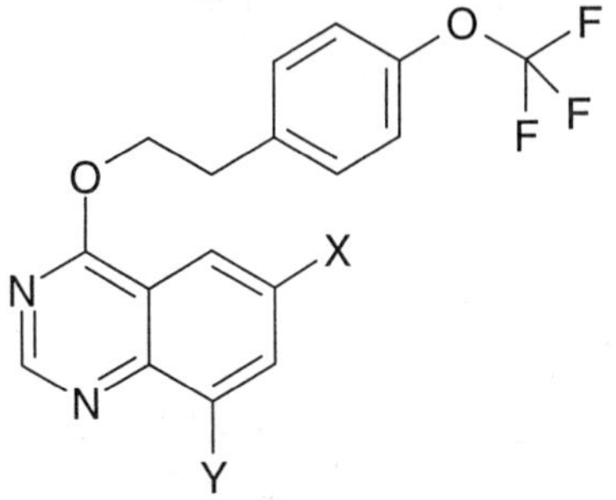

#	X	Y	LC_{90} / ppm
172	NO_2	F	89
173	NH_2	F	<1
174	Br	Me	>500
175	Me	Br	91
176	Me	Cl	34
177	Me	F	<1
178	Et	F	<1
179	F	Me	411
180	F	MeO	37
181	Me	MeO	10
182	MeO	Me	61
183	MeO	MeO	34
184	MeO	F	10
185	NMe_2	F	261
186	NHOH	F	>500
187	F	F	21

This study provided three very promising compounds. The 6-amino-8-fluoroquinazoline (Compound 173) was not tested further as it was photo-unstable.

We were not surprised by the close similarity between the activities of compounds 177 and 178. Earlier studies demonstrated little differentiation between the activities of the 6-methyl and 6-ethylquinazolines. The methyl analog was selected for more extensive tests as it was expected to be much less expensive to manufacture.

The results from the field trials of Compound 152 and the "safened" analog (Compound 177) are shown in Table XIV. The data clearly indicate the latter to be considerably safer across the four varieties of wheat tested.

Table XIV - Comparison of Crop Safety And Field Efficacy of Compounds 152 and 177 against PMW

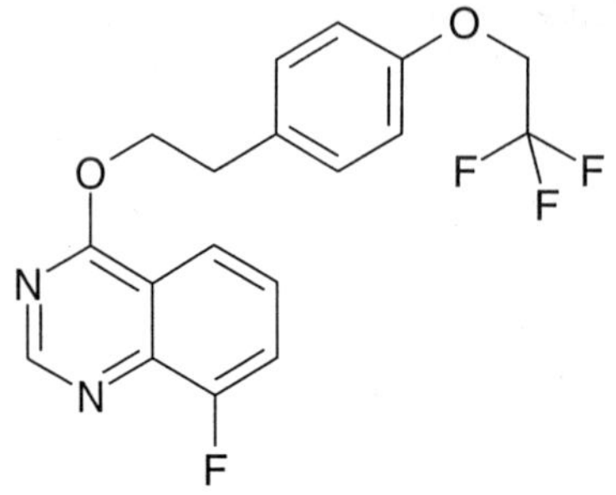

Compound 152

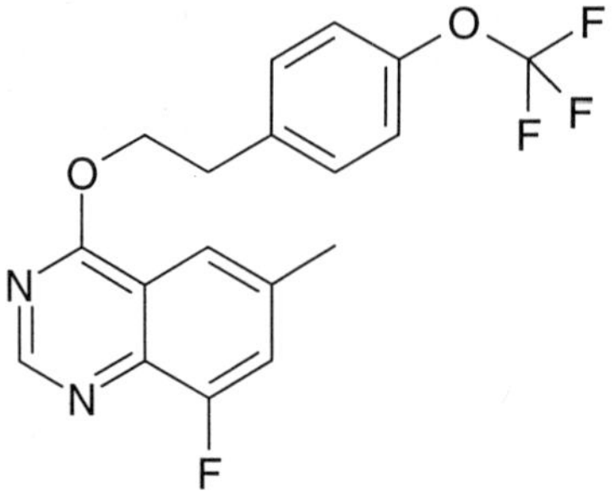

Compound 177

Wheat Variety	% Phytotoxicity 152	% Phytotoxicity 177	% Phytotoxicity Blank	Rates g ha^{-1}	% Infection in Field 152	% Infection in Field 177	% Infection in Field Blank
Hussar	25	10	10	500	17	3	24
Genesis	10	0	0	250	16	7	24
Galahad	5	0	0	125	12	9	24
Avalon	20	1	2				

In order to confirm that the safening of Compound 152 did not inadvertently affect the fungicidal activity, both analogs were tested side by side under field conditions. The results in Table XIV clearly demonstrates Compound 177 to be more efficient at suppressing infection by wheat powdery mildew.

At this stage we appear to have attained our goal of preparing a highly-active crop-safe fungicide that controlled powdery mildew in wheat under field conditions. In order to further improve market competitiveness, Compound 177 was required to have a broad biological spectrum of activity. An unfortunate side effect of the safening afforded by the introduction of a methyl group at position 6 was a narrowed spectrum of activity. Working in collaboration with molecular modellers, Compound 177 was analysed to ascertain if alterations could be made to broaden biological spectrum.

Replacement of the 'Tail' Benzene with 'Aza-benzene'

It was noticed that some of the earlier compounds which possessed broad spectrum activity had calculated lipophilicity values (cLog P) much lower than Compound 177.

The possibility that replacement of the benzene moiety in the "tail" by pyridine may reduce lipophilicity without adversely affecting PMW activity was investigated.

Calculations have shown that the cLog P values of quinazolines with "diaza-benzenes" in the tails had cLog P values which bore even closer resemblance to previous broadly active compounds. The pyrimidine, pyrazine and pyridazine were prepared and the PMW activity is shown in Table XV.

Table XV - PMW Activity of Quinazoline with "Aza-Benzene Tails"

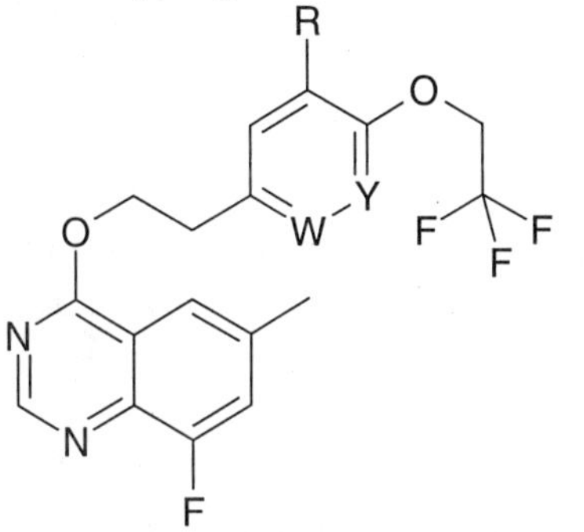

#	R	W	X	Y	Grams / ha			
					250	125	62.5	31.25
185	H	C	C	N	100	97	82	48
186	H	N	C	CH	85	48	13	13
187	H	N	C	N	4	0	0	22
188		CH	N	N	77	64	27	5
189		N	N	CH	100	98	97	82
190	F	CH	C	CH	97	89	56	19
191	F	CH	C	N	70	52	41	26
177					100	100	97	82

None of the compounds tested had wheat powdery mildew activity surpassing that of Compound 177. The molecular modeling predictions did not appear to be borne out by the inconsistent activities shown by some of the mono- and diaza-analogs. The pyridine 185 had activity comparable to 177 but the other pyridine 186 was less active. The pyrazine 189 had activity identical to 177 whereas the pyrimidine188 had only moderate activity and the pyridazine 187 was essentially inactive. Using fluorine as a pseudo nitrogen did not afford any advantages either.

The two most active compounds from Table XV (the pyridine 185 and the pyrazine 189) were further tested to determine the breadth of their biological spectrum with respect to Compound 177. Both compounds did have superior activity against leaf rust, glume blotch and leaf spot in wheat but the improvements were only slight.

Conclusion

Utilisation of structure-activity relationships has resulted in the transformation of a simple phenethylaminoquinoline with moderate activity against downy mildew in vines and squash into a phenethoxy quinazoline that was highly active across a broad biological spectrum of fungal diseases. The optimization of greenhouse activity gave a compound with unacceptably high levels of phytotoxicity in the field. Re-analysis of the available structure-activity data allowed the design of a crop-safe molecule but, unfortunately, this resulted in narrowing the biological spectrum. Once again, collaboration with molecular modelers and using the available structure-activity relationships data, the biological spectrum was re-broadened a little through increasing the overall hydrophilicity of the compound by replacing sp^2 carbons with nitrogens. Unfortunately, the improvement did not allow the new compounds to attain commercial levels of activity.

Acknowledgments

We are grateful for the biological contributions from K. Hellwig, M. N. Riches, W. R. Arnold, D. Johnson, C. Carroll, T. Mathieson, G. Davis and C. Flake. We also like to thank P. Lobb for field trial studies and J Liebeschuetz for molecular modeling.

Literature Cited

1. Dreikorn B. A. ; Jourdan G. P. and Suhr R. G. U S Patent 5 114 939 (1992).
2. Dreikorn B. A., Jourdan G. P., Davis L. N., Suhr R. G., Hall H. R. and Arnold W. R. in Synthesis and Chemistry of Agrochemicals II . ACS Symposium Series No 443 (1990). Baker, D. R.; Fenyes, J. G. and Moberg, W. K., Eds.;
3. Dreikorn B. A.; Jourdan G. P, and Suhr R. G. U S Patent 5 411 963 (1995).

Chapter 27

4-Phenethoxyquinazolines: The Development of a Predictive QSAR Analysis Against Wheat Powdery Mildew

B. A. Dreikhorn, Paul R. Graupner[1], J. W. Liebeschuetz[2], and M.Yap

Discovery Research Center, DowElanco, 9330 Zionsville Road, Indianapolis, IN 46268-1053

A predictive QSAR model has been developed that correlates *in vivo* activity of a series of quinazoline compounds against powdery mildew of wheat, to electronic and molecular size based properties. The model was tested by the synthesis of a number of new chemistries with a range of predicted activities. The predictions were successful for all these new compounds except in a series of 8-alkoxy substituted quinazolines, where actual activity was less than predicted. Experiments indicated that the lower activity may be due to metabolism of the alkoxy substituent.

A family of phenethoxy quinolines and quinazolines (QQ project) was discovered to have significant fungicidal activity against important plant pathogens (*1-4*) and an extensive synthesis program was undertaken to discover the extent of the activity. It has been demonstrated that these quinolines and quinazolines act as inhibitors of the mitochondrial electron transport (MET) pathway. Inhibition of the MET pathway is an important mode of action for insecticides and fungicides, and much research is underway to discover novel active chemistries. There are a number of different sites in the MET pathway in which inhibition may take place, one of the more important ones being the oxidation of NADH by NADH:ubiquinone oxidoreductase or Complex I (*5-6*). Activity at Complex I has been demonstrated for a range of chemistries, and there are a number of naturally occurring inhibitors. There are also a small number of chemistries for which a quantitative structure/activity relationship (QSAR) has been developed, for example the acridones (*7*), and a set of fluorinated arylamino-1-arylmethane phosphonic acid esters(*8*). It has been shown that the quinoline/quinazoline class of molecules investigated as insecticides and fungicides also act at Complex I (*9*) and, with a large number of molecules being synthesized in

[1]**Corresponding author.**

[2]**Current address: Proteus Molecular Design Ltd., Lyme Green Business Park, Macclesfield SK11 0JL, United Kingdom**

this area, may well be amenable for QSAR analysis. In the acridones, activity against submitochondrial particles was correlated against the sterimol parameters *L* and *L2* and by the addition of π (substituent lipophilicity). We were able to correlate activity in a plant assay with calculated atomic charges, orbital energies, and substituent molar refractivity (MR) in such a way as to enable facile prediction of activity of novel compounds. Thus we were able to test the model produced by synthesizing a number of new compounds and ensuring that the activity of the class had been optimized.
A principal goal of the QQ project was to develop qualitative and quantitative SARs that could be used in a predictive fashion to influence the direction of further research. Any attempt to derive SARs from *in vivo* data must, of necessity, deal with the separate subclasses of chemistry individually. Only at a later stage may it be possible to combine SARs from different areas to provide a more unified picture. The class of chemistry of most interest in terms of disease control for temperate cereals and top fruit - in particular powdery mildew species - is the phenethoxy quinoline/quinazoline class (Figure 1).

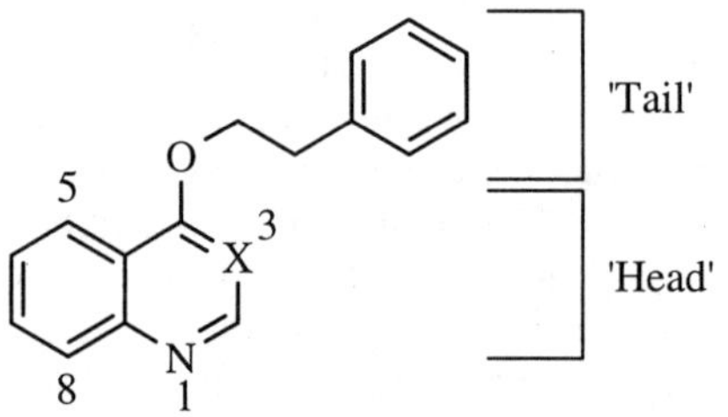

Figure 1: Basic Phenethoxy QQ Carbon Skeleton; X = N or C

This report details the studies carried out to examine how differential substitution on the benzo-ring of the head portion influence activity. In particular it was decided to look critically at quinazolines only. This was for two reasons: first, a fairly well exemplified data set of quinazolines existed in which the head was monosubstituted at the 5, 6, 7 or 8 position, and the phenethoxy tail was restricted to one of four different 4-substitutions including the substitution associated with highest mildew activity, namely OCH_2CF_3. Second, with synthesis in this area still ongoing and the chemistry required to make each differently substituted quinazoline requiring, in general, five or more separate steps, the existence of a predictive tool that could help prioritize possible synthetic targets would be of great benefit.

Numerous compounds have been prepared in this series and some highly active compounds have been discovered. Accordingly, it was decided to attempt to develop QSARs using *in vivo* data against powdery mildew of wheat (*Erysiphe graminis triticii (ERYSGT))*, the primary economically damaging pathogen of temperate cereals. Studies were initiated to look at the 'head' (quinazoline) and 'tail' (phenethoxy) portions of the molecule separately, in order to test the hypothesis that the 'head' portion of the molecule was the primary site for binding, whereas the 'tail' segment fine-tuned size and lipophilicity properties.

A useful QSAR was previously derived linking the properties of the head portion to activity against *Pyricularia oryzae* MET (C.Manly, N.Kirby; Unpublished results). This study did not confine itself to a subclass of the QQ series; instead, a wide cross section of head structures was analyzed. The major contribution to this SAR came from the electronic properties of the head calculated using semi-empirical MOPAC quantum chemical methods. In particular, the Highest Occupied Molecular Orbital (HOMO) energy and the dipole were found to be important in determining *in vitro* activity. The fact that electronic factors are found important raises interesting questions bearing in mind that these compounds inhibit electron transport a process that depends on factors such as redox potential.

Observations within the phenethoxy quinoline/quinazoline series suggested that electronic factors might also be important in describing activity against powdery mildew of wheat. For instance, substituting the 8-position with fluorine generally enhances activity significantly over the unsubstituted compound. Substitution at the 5 and 7 positions with fluorine significantly decreases activity. It was also felt that the magnitude of charge on the heteroatoms, in particular N1, might have an important role to play.

In vivo Data Analysis

The only comprehensive data on *ERYSGT* activity available for these compounds at the start of this study were those from a two hour protectant assay in which wheat was sprayed with the compound in solution and, two hours later, the wheat was challenged with the fungus. The results, read seven days later, are quoted on the Barratt Horsfall scale (*10*) which is non-linear, and is most differentiated around the 90% control level. Therefore the activity measure we used was the logarithm of the concentration (in ppm) at which 90% control of disease is achieved [log(LC90)]. There were several problems associated with using the two-hour protectant data to supply LC90 data. First, the data were non-comparative, non-replicated, and non-standardized, therefore a significant standard error had to be assumed. Second, concentration was measured in ppm rather than more correctly using molar concentration. This could be corrected for, but the error eliminated in so doing was considered much smaller than that already inherent in the data. Third, and most importantly, many of the compounds in the dataset were tested at only one rate (400 ppm) and were not active enough to be tested at lower rates. It was felt necessary to be able to use these compounds in the analysis in order to extend the range of activity under study. Accordingly a method was sought that would be able to estimate an LC90 from a single data point: the method used was the so called 'logit transform' (*11*) which assumes parallel dose response curves and follows a sigmoidal form.

It must be stressed that this method gives only a rough estimate for the log(LC90) and, like all extrapolation methods, must be used with great care. The logit transform was also used to calculate log(LC90)'s where more than one data point was available. In this case log(LC90)'s were calculated for individual data points and averaged over the whole dose response curve. This can be done to both interpolate and extrapolate log(LC90)'s. An advantage of this method of extrapolation is that unreasonably high or low LC90 values are never obtained, unlike with other extrapolation methods. This is because essentially the same shape curve is used each time to fit the data. Extrapolation standard errors in LC90's for heavily extrapolated points can be high, equivalent to 1 log C unit (*i.e.*, one order of magnitude). Nevertheless this error is comparable with that already inherent in the biological data used in this study. So long as a sufficiently large activity range was used (four log C units or four orders of magnitude in this study) and a sufficiently large dataset of compounds was selected, it was felt that these standard errors might be accommodated.

Choice of dataset: In order to obtain a data-set for the development of a QSAR model, the DowElanco discovery databases were searched for sets of quinazolines with the same 'tail' groups. The chosen list of 57 compounds consisted of simple substituents on the quinazoline for which there were at least two examples (in most cases) within the group of tails chosen (Figure 2). In this way, a form of replication was introduced, and any biasing effect due to outliers was reduced.

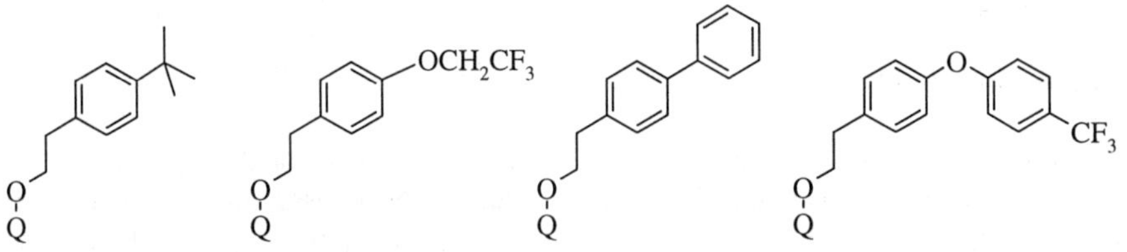

Tail A Tail B Tail C Tail D

Figure 2; Tail group substituents (Q=quinazoline)

Calculation Methods: All calculations were performed using the SYBYL (*12*) software suite of programs. The 'head-group' quinazoline molecules were built up and minimized using the SYBYL MAXIMIN2 force-field, and then submitted for semi-empirical calculations using MOPAC 5.0 (*13*). Calculations were initially performed using four methods with no appreciable difference being noted. All subsequent calculations were performed using PM3. For simple group substituents (*e.g.*, -F, -Me, *etc.*), the calculations were undertaken with full geometry minimization. Molecules with larger substituent groups (*e.g.*, -OMe, $-OCF_3$, *etc.*) were submitted to a semi-empirical conformational search routine around the rotatable bond of the substituent with no minimization, whereby the lowest energy conformations for the groups were identified for later full geometry minimization. Substituent *pi* and MR values were compiled from the literature (*14*). QSAR generation was carried out using the partial least squares (PLS) analysis as implemented in the COMFA module of SYBYL (*15*). Initially, runs were performed using cross-validation and examined for a minimum in the value of PRESS and the highest value of Q^2. Once the optimum number of components had been identified, the run was repeated twice, once using 'bootstrapping' in order to obtain the standard errors on the coefficients of the models, and secondly as a single run to obtain the final model.

Biological data were measured in a 2 hour protectant test, and activity measured using the Barratt Horsfall nomenclature (*10*). As this scale is non-linear and mostly differentiated at the 90% disease control level, log(LC90) measurements were extracted from all data points using the logit transform and averaged over the whole dose response. Where no activity was observed, an arbitrary value for log(LC90) of 4.5 was given.

Discussion

A total of 57 molecules were initially chosen for analysis (summarized in Table I), however, early studies indicated that the arbitrary activity value for inactive compounds of 4.5 was adversely affecting the predicted *versus* actual statistics and therefore the final model was constructed from a total of 47 molecules.

Table I: Biological activity of Selected Quinazolines Log(LC90)

Quinazoline Substituent	Tail 'A'	Tail 'B'	Tail 'C'	Tail 'D'
H	2.58	0.66	2.59	2.04
5-F	3.83		4.50	2.90
6-F	1.50	1.30	2.19	1.44
7-F	1.60	2.00	3.56	4.50
8-F	1.11	0.06	1.63	0.34
6-F,7-F	2.90	2.66	4.50	3.25
5-Cl	4.50	4.50		4.50
6-Cl	3.56	1.00		
7-Cl	4.50	4.50	4.50	
8-Cl	2.90		2.54	
5-Me	4.50	2.10	2.84	2.51
6-Me	1.21	1.19	1.60	
7-Me	2.09	1.51		
8-Me	3.25	1.55	3.13	2.25
6-OMe	2.45	1.12		
6-Br	3.56	2.30	3.00	
6-Br,8-Br	4.50			
6-Cl,8-Cl	4.50			
6-Me,8-F		0.06		
6-Pr		0.65		
6-*i*-Pr		1.877		
8-CH_2Cl		3.10		

The QSAR variables to be examined are given in Table II, where N1 is the MOPAC calculated partial charge on the nitrogen atom at position 1, and LUMO and HOMO are the calculated frontier orbital energies.

Table II: QSAR Parameter Data Set

Substituent	*pi*	MR	N1	LUMO	HOMO
H	0.00	1.03	-0.131	-0.78	-9.296
5-F	0.14	0.92	-0.139	-0.997	-9.377
6-F	0.14	0.92	-0.127	-1.026	-9.429
7-F	0.14	0.92	-0.134	-1.025	-9.54
8-F	0.14	0.92	-0.11	-1.011	-9.359
6,7-F	0.28	0.81	-0.131	-1.268	-9.674
5-Cl	0.71	6.03	-0.135	-0.951	-9.146
6-Cl	0.71	6.03	-0.13	-0.945	-9.228
7-Cl	0.71	6.03	-0.131	-0.967	-9.407
8-Cl	0.71	6.03	-0.123	-0.946	-9.062
5-Me	0.56	5.65	-0.135	-0.759	-9.115
6-Me	0.56	5.65	-0.129	-0.739	-9.164
7-Me	0.56	5.65	-0.133	-0.756	-9.242
8-Me	0.56	5.65	-0.131	-0.756	-9.121
6-OMe	-0.02	7.87	-0.121	-0.788	-8.946
6-Br	0.86	8.88	-0.132	-0.993	-9.429
6,8-Br	1.72	16.76	-0.116	-1.168	-9.461
6,8-Cl	1.42	11.06	-0.122	-1.101	-9.095
6-Me,8-Br	1.42	13.53	-0.111	-0.939	-9.219
6,8-Me	1.12	10.3	-0.126	-0.729	-8.993
6-Et	1.02	10.3	-0.129	-0.746	-9.184
6-Et	1.02	10.3	-0.129	-0.746	-9.184
6-Me,8-F	0.70	5.6	-0.107	-0.972	-9.233
6-Pr	1.55	14.96	-0.128	-0.723	-9.152
6-^{i}Pr	1.53	14.96	-0.129	-0.741	-9.514
6-CH_2Cl	0.17	10.49	-0.139	-0.984	-9.331

The effect of the different tail substituents was accounted for by the use of indicator variables in the PLS calculations, with any effect for the *t*-butyl tail being explained by the value of the constant. The results of examples of these calculations are given in Table III. It was considered that a value for R^2 of 0.75 would be highly satisfactory given the inherent errors in the biological screening data on which these calculations were based. The final model produced is highlighted and summarized in Table III, analysis of the predicted *versus* actual plot for the selected model (Figure 3), and the residuals indicated that a satisfactory model had been developed with no obvious trends left to be explained.

Table III: Examples of QSAR Statistics

Variables describing model	Q^2 (Comps.)	R^2(SD) Bootstrapped	Probability $R^2= 0$	Standard Error (SD)
Indicator Variables only	0.133 (3)	0.353 (0.110)	0.003	0.803 (0.329)
MR	-0.096 (1)	-	-	-
N1	0.310 (1)	0.388 (0.072)	0.000	0.793 (0.289)
LUMO	-0.031 (1)	-	-	-
Ind + N1	0.540 (3)	0.607 (0.101)	0.000	0.599 (0.240)
Ind + MR	0.108 (4)	0.383 (0.129)	0.005	0.819 (0.309)
Ind + LUMO	0.148 (1)	0.314 (0.051)	0.000	0.809 (0.303)
Ind + N1 + MR	0.516 (5)	0.652 (0.061)	0.000	0.617 (0.126)
Ind + N1 + LUMO	0.606 (2)	0.645 (0.115)	0.000	0.571 (0.274)
Ind + N1 + MR + LUMO	**0.631 (3)**	**0.752 (0.037)**	**0.000**	**0.511 (0.181)**

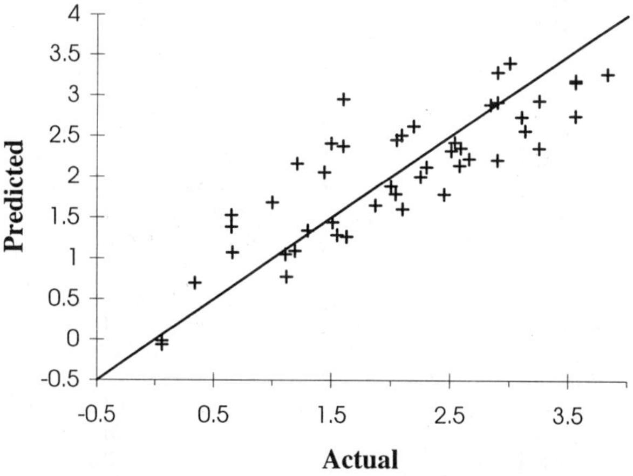

Figure 3: Predicted *vs* Actual (model)

$$\lg(lc90) = -9.7 - 1.0\times \mathbf{OCH_2CF_3} + 0.099\times \mathbf{Ph} - 0.51\times \mathbf{OPhCF_3} + 0.030\times \mathbf{MR} - 80.6\times \mathbf{N1} - 1.96\times \mathbf{LUMO}$$

$$n = 47;\ q^2 = 0.63;\ r^2=0.75;\ s = 0.51$$

The first four terms of this model indicate the significance of the tail substituent. As lower values of log(lc90) indicate higher activity, the most significant tail was the trifluoroethoxyphenyl tail, as had been shown in other work (*4*), the most significant property was the partial atomic charge on N1 thus confirming the hypothesis under test. From this model, it was apparent that the most beneficial substituents for improving activity on the quinazoline were those that were small (as determined by the small adverse effect on log(lc90) by the MR parameter), and those that reduced the value of the calculated atomic charge on the nitrogen, whilst not having a detrimental affect on the value of the LUMO energy, for example small electron donating groups on the 6 and 8 positions.

With a working model in hand, the activities of a large series of previously unscreened molecules were predicted, covering a diverse range of novel chemistries for the family (summarized in Table IV). A selection of these were selected and screened using the same screening procedures as before. The novel compounds were selected to cover all ranges of predicted activity, including non-active molecules.

Table IV: Predicted LC90's For Unscreened Materials

SUBST	LOG(LC90)	SUBST	LOG(LC90)
6,8-F	0.27	6-NH_2	1.39
5,6,7,8-F	2.26	6-NMe_2	1.69
6-Cl,8-F	0.52	6-SMe	2.80
6-Et,8-F	0.16	6-OEt-8-F	-0.07
6-OMe,8-F	-0.33	6-O-*i*-Pr-8-F	0.78
7-OMe	1.93	6-O-*t*-Bu-8-F	0.94
6,8-OMe	1.57	5-Me-6,8-OCF_3	1.90
8-Cl.	-1.17	6-OMe-6,8-OCF_3	2.45
5-OH	0.88	6-NHOH-8-F	0.69
6-OH	0.49	6-OH-8-F	-0.77
7-OH	2.32	6-OH-8-OCF_3	-0.41
8-OH	2.15	6-OMe-8-OCF_3	-0.19
5-CF_3	2.79	6-F-8-OCF_3	0.30
6-CF_3	2.94	6-F-8-OMe	-1.03
7-CF_3	2.20	6-NH_2CH_2-8-F	0.41
8-CF_3	2.83	6-NH_2CH_2-8-OCF_3	0.26
5-OCF_3	3.01	6-NH_2CH_2-8-OMe	-0.99
6-OCF_3	1.78	6-Cl-8-OCF_3	0.63
7-OCF_3	2.53	6-CH_2NMe_2-8-F	0.93
8-OCF_3	0.05	6-CH_2NMe_2-8-OCF_3	0.77
5,7-F	2.97	6-CH_2CN	2.18
5,6,7-F	3.24	8-OMe	-1.34
5,7-Cl	2.89	8-OEt	-1.09
6,8-Cl	1.77	8-OPr	-0.79
5-CN	2.99	8-OBu	-0.50
6-CN	2.79	5-OH-8-F	1.12
7-CN	2.55	5-OMe-8-F	1.28
8-CN	2.42	6-Br-8-Me	2.41
6-NH_2-8-F	0.40	6-Me-8-Br	0.93
6-NMe_2-8-F	0.54	6-CF_3-8-F	1.99
6-SMe-8-F	1.92	6,8-Br	1.13
6-Me-8-OCF_3	0.03	6-F-8-CF_3	3.14
6-Me-8-Cl	1.19	6-CF_3-8-OMe	2.04
5,6-OMe-8-F	0.99	5,6-methylendioxy	1.97
5-OMe-6,8-F	1.43	6,7-methylendioxy	1.99
5,6-Me-8-OCF_3	3.41	7,8-methylendioxy	0.39
6-F-8-Me	1.65	7,8-ethyloxy	-0.53

For all the new compounds, predicted *versus* actual activities were plotted and a very satisfactory correlation for most of the compounds was seen (Figure 4 - in this figure and figure 6, the line is not a least squares fit, but rather indicates where the ratio of predicted to actual activity is 1). However, there were some disappointing results, markedly the performance of the 8-alkoxy substituted compounds which are marked in the Figure. This was especially the case considering results from *in vitro* enzyme studies that were performed alongside the *in vivo* screens (Owen, W.J.; Unpublished data). In these tests, the compounds had inhibited activity at a very low rate, and within the values expected considering the predictions made. However, there was obviously some other factor in the *in vivo* screens that had not been taken into account.

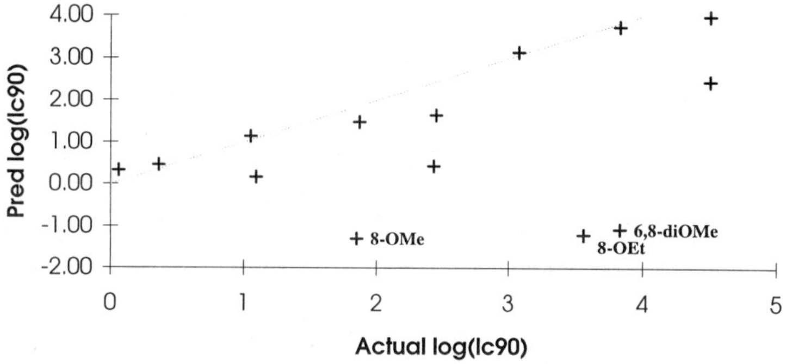

Figure 4:Predicted vs. Actual - (New Chemistries) 1

Three explanations were considered in order to explain the lack of *in vivo* activity for the alkoxy compounds:

1. Metabolism: O-demethylation is a known route for metabolism of alkoxy substituents *in planta*. Dealkyation of the 8-substituted compounds would lead to the much less (predicted) active 8-hydroxy compounds. In order to prevent this actuality, the 8-trifluoromethoxyquinazoline (2) was proposed for synthesis.
2. LUMO energy: For the model, the ranges of calculated LUMO energies had all fallen within a relatively narrow band. Thus calculation of the QSAR model had assumed linearity of the LUMO property. It is probable that the LUMO parameter is bi-linear in this study, with sampling only on one side of the maximum, and thus, if samples had been screened whose predicted LUMO energies were far removed from the other molecules in the study, the model may well not be accurate. The calculated LUMO energy for the alkoxy quinazolines fell well outside the range of LUMO energies sampled in the study. In order to remove this anomaly, 6-fluoro-8-methoxyquinazoline (3) was proposed for synthesis. Addition of the fluoro group had the effect of bringing the LUMO energy back into the sampled range for the modeled compounds, and yet still retained the excellent predicted activity.
3. Steric hindrance: The final scenario to be considered is that of steric hindrance of the alkoxy substituent. As the importance of the nitrogen at position 1 has been confirmed, it may be that this atom is involved in forming a hydrogen bond at the

receptor. The presence of a freely rotating group at the 8-position may well prevent such a bond from existing. In order to prevent this, the dihydrofurano derivative (4) was proposed for synthesis. In this molecule, the presence of the alkyl substituent was shown by the model to have little detrimental affect on the predicted activity, which would be enhanced by the presence of the 8-alkoxy substitution.

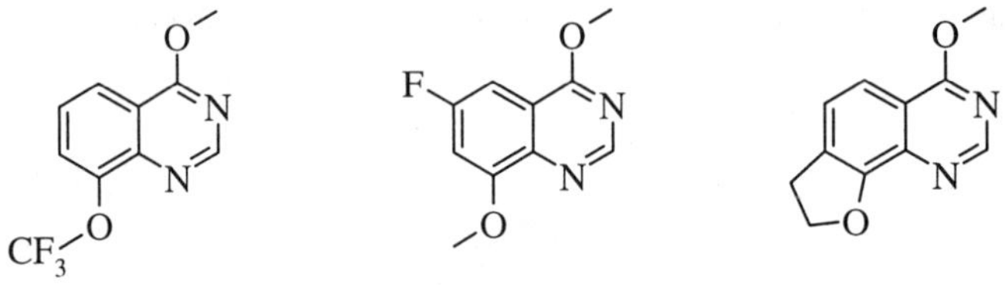

2 3 4

Figure 5

Two of the proposed compounds were synthesized, and their predicted *versus* actual activities are graphed in Figure 6. The addition of the 6-fluoro derivative had little effect on the poor performance of the alkoxy substitution, whereas the trifluoro methoxy substituent performed much better. Metabolism is therefore the primary reason for failure of the alkoxy substituted compounds, and needs to be considered in design of further molecules.

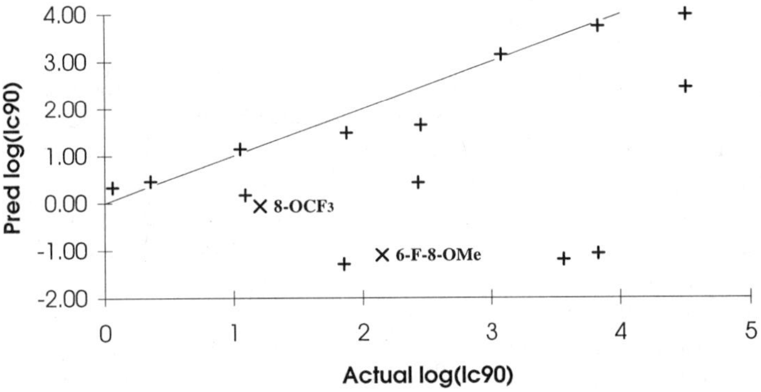

Figure 6 Predicted vs. Actual (New Chemistries) - 2

Conclusions

A model with good predictive activity has been developed for *in vivo Erysiphe* activity. The LUMO, partial charge and MR parameters take into account variation on the "head", and the indicator variables explain variation in the tail, very similar to the results in other studies of Complex I inhibitors (*7-8*), Design of future inhibitors based on the QQ chemistry can now concentrate on optimizing properties of both the head and tail portions, concentrating on electronic properties with quinazoline substituents, and on size/lipophilicity properties with tail group substitution.

Acknowledgments

The authors would like to acknowledge the assistance and encouragement of Greg Durst in this project.

Literature Cited

1: Dreikorn, Barry A.; Jourdan, Glen P.; Davis, L. Navelle; Suhr, Robert G.; Hall, Harold R.; and Arnold, Wendell R.; *Synthesis and Chemistry of Agrochemicals II*, **1991**, Chapter 44 ACS Symposium Series p443.
2: Dreikorn, Barry A.; Jourdan, Glen P.; and Suhr, Robert G.; **1992**, U.S. Patent 5,114,939.
3: Dreikorn, Barry A.; Jourdan, Glen P.; and Suhr, Robert G.; **1995**, U.S. Patent 5,411,963.
4: Yap, Maurice C.H.; Dreikorn, Barry A.; Davis, L. Navelle; Suhr, Robert G.; Kaster, S.V.; Kirby, N.V.; Paterson, Glen; Graupner, Paul R. and Erickson, W.R. *This publication.*
5: Weiss, H., Friedrich, T., Hofhaus, G., and Preis, D.; *Eur. J. Biochem.*, **1991**, *197*, p563.
6: Walker, J.E.; *Q. Rev. Biophy.*, **1992**, *25*, p253.
7: Oettmeier, W. Masson, K., and Soll, M.; *Biochem. Biophys. Acta* **1992,** *1099*, p262.
8: Dronia, H., Gruss, U., Hagele, G., Friedrich, T., and Weiss, H.; *JCAMD* **1996**, *10*, p100.
9: Burgos, J. and Redfearn, E.R.; *Biochem. Biophys. Acta,* **1965**, *110*, p475-483.
10: Horsfall, J. G., and Barratt, R. W.; *Phytopathology* **1945**, *35*, p655.
11: K-J Schaper, *QSAR in Drug Design and Toxicology*; D.Hadzi and B.Jerman-Blazic (Eds) Elsevier Sci.Publ., **1987,** p58-60.
12: Tripos Assoc., St. Louis MO.
13: QCPE-455, QCPE, Bloomington, IN.
14: Hansch, C., Leo, A.; *Substituent Constants for Correlation Analysis in Chemistry and Biology"*, J.Wiley and Sons, NY, **1979**.
15: Cramer, R., Patterson, D., Bunce, J.; *J. Amer. Chem. Soc.*, **1988**, *110*, p5959.

Chapter 28

Arylazo- and Arylazoxy-Oximes as Fungicides

William W. Wood, Andrew C. G. Gray, and Thomas W. Naisby

Cyanamid Agricultural Research Center, American Cyanamid Corporation, P.O. Box 400, Princeton, NJ 08543–0400

A number of azo- and azoxy-oximes have been synthesized and tested as potential agricultural fungicides. Several of these compounds showed impressive levels of activity, particularly against *Plasmopara viticola* and *Erisyphe graminis*. The most active compounds were those with a *p*-chlorine or *p*-methyl substituent on the aryl ring.

Arylazo- and arylazoxy-oximes have been known for many years, but have not been reported to show any useful biological function. The majority of literature references to these compounds deal with their coordination properties. One patent application has recently appeared in which the azoxime function appears as an adjunct to a strobilurin-type fungicide (*1*). As part of an extensive fungicide programme in the area, we have studied the fungicidal properties of azo- and azoxy-derivatives and report a summary of this work herein.

For the purposes of this study, azo- and azoxy-derivatives may be divided into four structural types according to the substitution at the C-terminus and oxidation level of the azo-double bond (Figure 1). Thus Class 1 azo- and Class 2 azoxy-compounds carry only a proton on the C-terminus, while Class 3 and Class 4 represent analogues which have alkyl or aryl groups attached to the oxime-bearing carbon.

Synthesis

Several different synthesis protocols were adopted or adapted from the literature to prepare the compounds detailed in this study. Class 1 azo-oximes could be prepared by addition of dipotassium malonate to an aryl diazonium salt as detailed in Scheme 1 (*2*). No intermediates were isolated in this procedure, but rather the malonate adduct was acidified with acetic acid and diazotized to form the nitroso moiety. Two successive decarboxylations then lead to a nitroso product, one of the tautomeric forms of the target.

As detailed in Table I, a number of compounds were prepared by this route, but yields were generally very poor, irrespective of the substitution pattern on the aryl ring. Extensive chromatography was also required to

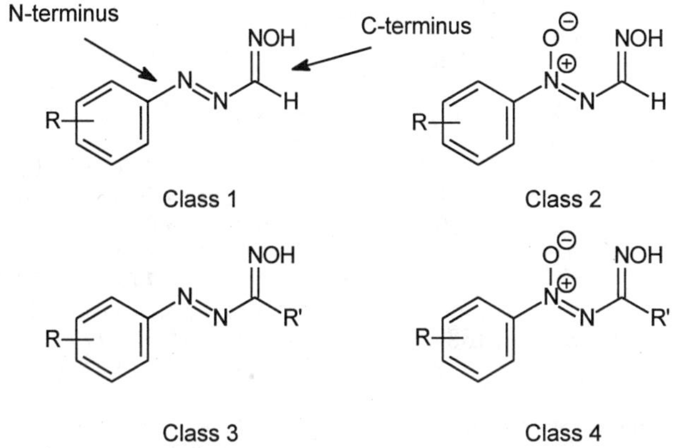

Figure 1: Structural Types of Azoxime

Scheme 1: Diazonium Route to Azo-oximes

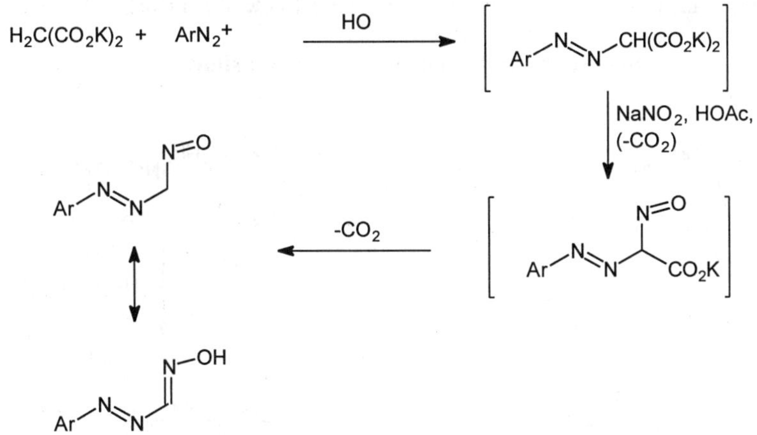

obtain samples of adequate purity. In the light of these problems, other synthetic approaches were investigated.

Table I: Examples of the Synthesis of Class 1 Azo-oximes by the Diazonium Route

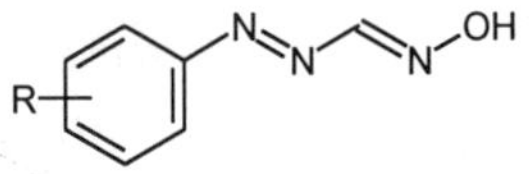

R	Yield (%)	M.p. (°C)	Analysis: Found (Theory) C	H	N
4-Cl	<1	140	47.2 (45.8)	4.0 (3.3)	21.7 (22.9)
4-Me	13	132	57.3 (58.9)	5.5 (5.5)	25.1 (25.8)
4-OEt	15	104	56.1 (56.0)	5.8 (5.7)	21.7 (21.8)
4-Br	20	145-7	37.1 (36.9)	2.8 (2.7)	18.4 (18.4)
4-*t*Bu	15	oil	65.1 (64.4)	7.7 (7.3)	18.9 (20.5)
2-OMe	7	153-5	53.9 (53.7)	5.2 (5.0)	23.5 (23.5)
4-OMe	8	138-140	54.3 (53.7)	5.3 (5.1)	22.6 (23.5)
4-$CONH_2$	3	222-4	50.2 (50.0)	4.3 (4.2)	29.1 (29.2)

A more successful strategy followed a very early synthesis of compounds in this class reported by Bamberger and Pemsel (*3*) and widely adapted by other later authors (*4*)-(*6*). In the more straightforward examples, this approach involved nitrosation of an appropriate hydrazone using an organic nitrite and base (Scheme 2). A number of different Class 3 azo-oximes were prepared using this route, the substitution at the N-terminus depending on the nature of the starting hydrazone. This chemistry could also be adapted to obtain Class 1 compounds by preparing a hydrazone from glyoxylic acid and nitrosating with sodium nitrite under aqueous conditions. This leads to concomitant decarboxylation and to the target Class 1 analogues.

Scheme 2: Hydrazone Route to Azo-oximes

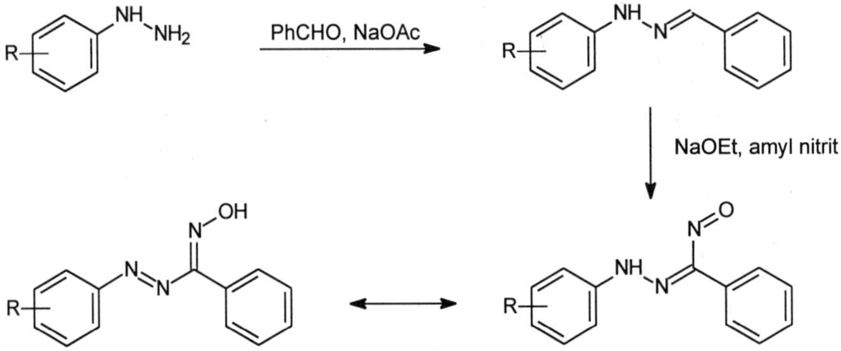

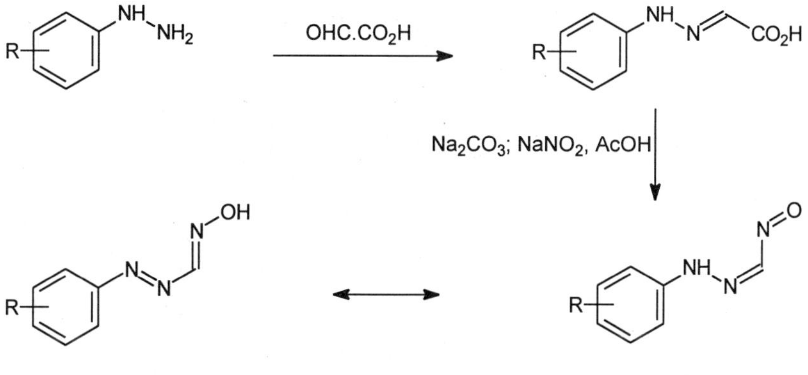

As shown in Table II, the yields from the hydrazone route to azo-oximes were better than those from the diazonium route. Using this approach, a range of analogues were prepared with different substituents at the N-terminus, including examples of aryl, alkyl and heteroaryl substitution. The compounds prepared in both Class 1 and Class 3 were usually stable, crystalline solids of relatively high melting point.

Table II: Examples of the Synthesis Class 1 and Class 3 Azo-oximes by the Hydrazone Route

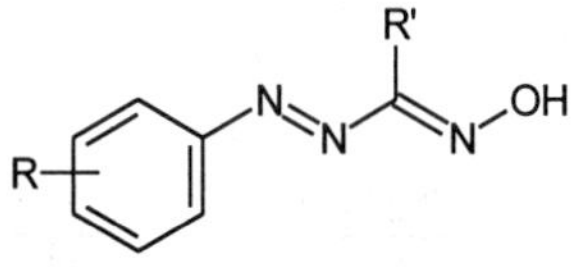

R	R'	Yield (%)	M.p. (°C)	Analysis: Found (Theory) C	H	N
3,4-di-Cl	Ph	13	133-5	52.6 (53.1)	3.2 (3.1)	13.8 (14.3)
4-Cl	Ph	52	123	60.4 (60.1)	4.3 (3.9)	16.5 (16.2)
H	Me	48	118-120	58.9 (58.9)	5.6 (5.5)	25.9 (25.8)
3,4-di-Cl	3-pyridyl	25	185-187	48.9 (48.8)	3.0 (2.7)	18.9 (19.0)
H	2-thienyl	28	99-100	56.6 (57.2)	4.0 (3.9)	17.8 (18.2)
4-Cl	Me	30	183-185	47.6 (48.6)	4.0 (4.1)	20.7 (21.3)
H	H	19	105-107	56.3 (56.4)	4.7 (4.7)	27.6 (28.2)
4-F	2-thienyl	28	136-138	53.7 (53.0)	3.6 (3.2)	16.9 (16.9)
2,4-di-Cl	H	10	142-145	38.4 (38.6)	2.7 (2.3)	19.2 (19.3)

A small number of heterocyclic azo-oximes were also prepared by a variation in the hydrazone route, as shown in Scheme 3. These compounds exhibited little biological activity and were not investigated further.

Scheme 3: Heterocyclic Azo-oximes

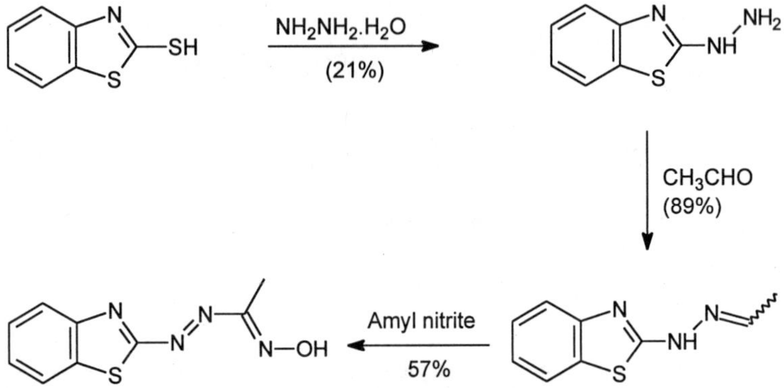

Oxidation of the azo-bond in both Class 1 and Class 3 azo-oximes was achieved using pertrifluoroacetic acid in tetrahydrofuran. The yields from this process were generally poor, but material suitable for biological evaluation was obtained and no attempts to optimize the process were made.

Scheme 4: Synthesis of Azoxyoximes

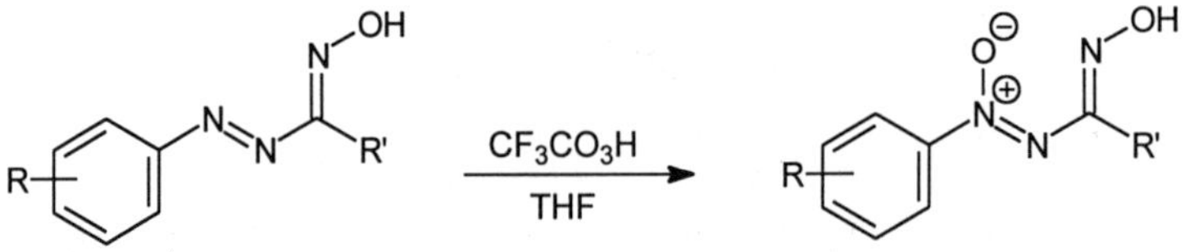

Table III: Examples of Class 2 Azoxyoximes

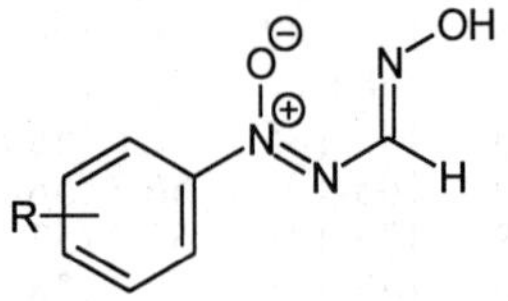

R	Yield (%)	M.p. (°C)	Analysis: Found (Theory) C	H	N
4-Cl	46	151-153	43.3 (42.1)	3.5 (3.0)	21.3 (21.1)
H	41	74-76	51.3 (50.9)	4.4 (4.3)	25.5 (25.5)
4-NO_2	25	163-165	41.4 (40.0)	3.2 (2.9)	27.7 (26.7)
4-Br	35	149-150	34.8 (34.5)	2.5 (2.5)	17.3 (17.2)
4-OEt	16	113-115	52.1 (51.7)	5.3 (5.3)	20.3 (20.1)
4-$CONH_2$	73	194-200	45.6 (46.1)	3.7 (3.9)	24.5 (26.9)
4-*t*Bu	11	124-125	59.1 (59.8)	6.7 (6.8)	18.6 (19.0)

Table IV: Examples of Class 4 Azoxyoximes

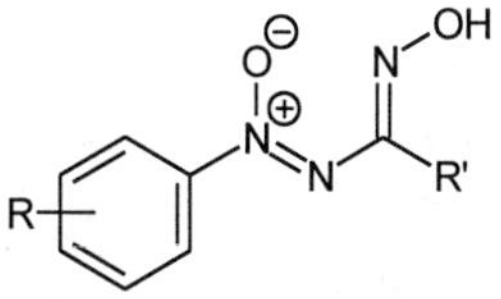

R	R'	Yield (%)	M.p. (°C)	Analysis: Found (Theory) C	H	N
H	Ph	23	122-124	65.7 (64.8)	4.7 (4.6)	17.4 (17.5)
H	Me	47	121	53.7 (53.7)	5.1 (5.0)	23.2 (23.5)
4-Cl	Ph	27	142-144	55.7 (56.6)	3.7 (3.7)	14.9 (15.3)
4-Cl	Me	49	150-153	45.4 (45.0)	3.1 (3.0)	15.7 (15.9)

Fungicidal Activity

Azo- and azoxy-oximes prepared in this study were screened against a number of commercially important diseases using a standard fungicide screen. Three screening modes were employed: prophylactic, where plants were sprayed with the test compound before innoculation with spores of the subject disease; therapeutic, where plants were innoculated with the subject disease before spraying with the test compound; and *in vitro,* where the test compound was incorporated into an agar medium whereon the isolated disease was grown. The diseases used in this assay are detailed in Table V.

Table V: Diseases in Primary Screen

Abbreviation	Disease	Common Name
Pv	*Plasmopara viticola*	Downy Mildew
Bc	*Botrytis cinerea*	Grey Mould
As	*Alternaria solani*	Tomato Early Blight
Ln	*Leptosphaeria nodorum*	Septoria Leaf Spot
Eg	*Erisyphe graminis*	Powdery Mildew
Pr	*Puccinia recondita*	Wheat Brown Rust
Po	*Pyricularia oryzae*	Rice Blast
Ph	*Pseudocercosporella herpotrichoides*	Agent of Wheat Eyespot
Fs	*Fusarium solani*	Agent of Fusarium Rot

As can be seen from Table VI, which details the screening results for 4-substituted Class 1 azo-oximes, the therapeutic and *in vitro* fungicidal activity of these compounds was much more significant than their prophylactic action. The most active compounds controlled a range of different diseases, but as will be seen below, showed their greatest depth of activity against *E. graminis* and *P. viticola.* Less active compounds in the series tended to show activity against only these diseases. In general, compounds with polar 4-substituents were less active than more lipophilic analogues. Thus, for example, the 4-nitro and 4-amido derivatives showed very little activity. This observation carried over into the *in vitro* screens, indicating a probable difference in intrinsic activity, rather than a transport effect. It was also noted that larger groups in the 4-position tended to decrease activity. For example, the 4-*tert*-butyl derivative showed less activity than the methyl substituted analogue.

Compounds with a 4-substituent were generally more active than the corresponding example substituted in the 2-position. Thus, for example, the 4-methoxy derivative showed a broad spectrum of activity giving complete control of five diseases at the primary screening dose (Table VII), whereas the isomeric 2-methoxy derivative was much less active. The 2- and 4-fluoro derivatives showed a similar effect. This general trend was observed for the other classes of azo- and azoxy-oximes when the appropriate compounds were available for screening. In this study, 3-substituted derivatives were not prepared.

The structure-activity relationships observed for Class 1 compounds were generally repeated in the oxidized series (Table VIII). As previously observed, polar groups and large substituents generally gave poor activity. Although the more active compounds in this series appeared at first to show some prophylactic action (4-Cl and 4-OEt derivatives, Table VIII), this activity did not extend to lower dose levels.

Table VI: Fungicidal Activity of Class 1 Azoximes: 4-Substituted Analogues (@ 1000 ppm)

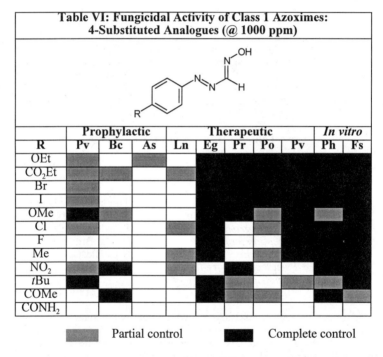

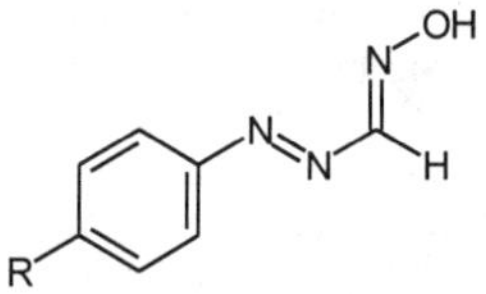

	Prophylactic			Therapeutic					*In vitro*	
R	Pv	Bc	As	Ln	Eg	Pr	Po	Pv	Ph	Fs
OEt										
CO_2Et										
Br										
I										
OMe										
Cl										
F										
Me										
NO_2										
*t*Bu										
COMe										
$CONH_2$										

Partial control Complete control

A number of examples of Class 3 azo-oximes were prepared, although a fully systematic study was not carried out. Results for two groups of compounds are given in Table IX, illustrating the general trends observed for this class. In both series, the C-phenyl and C-methyl compounds exhibited a broad spectrum of activity at the high dose rates, in most cases controlling more diseases than the Class 1 analogue (shown for comparison). As in previous cases, the apparent therapeutic action at the high primary screening dose rapidly disappeared at lower doses. Heterocyclic and many substituted aryl derivatives showed very little fungicidal activity (as illustrated for the 2-thienyl and *p*-anisyl derivatives).

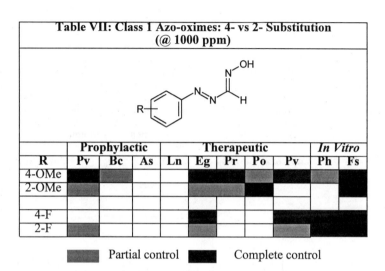

Table VII: Class 1 Azo-oximes: 4- vs 2- Substitution (@ 1000 ppm)

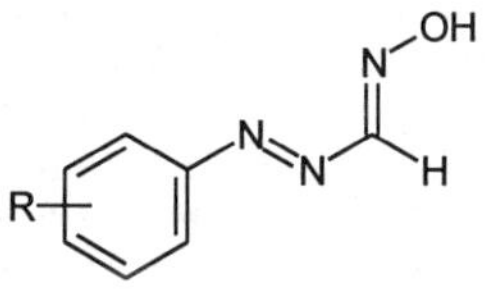

	Prophylactic			Therapeutic					*In Vitro*	
R	**Pv**	**Bc**	**As**	**Ln**	**Eg**	**Pr**	**Po**	**Pv**	**Ph**	**Fs**
4-OMe	Complete	Partial			Complete	Complete	Partial	Complete	Partial	Complete
2-OMe	Partial				Partial	Partial	Complete			Complete
4-F					Complete			Complete	Complete	Complete
2-F	Partial				Partial			Partial	Complete	Complete

Partial control (grey) Complete control (black)

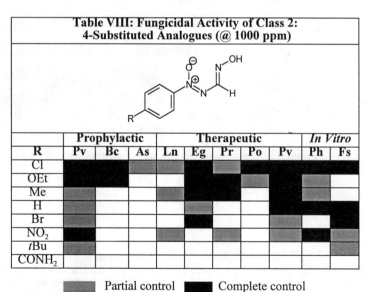

Table VIII: Fungicidal Activity of Class 2: 4-Substituted Analogues (@ 1000 ppm)

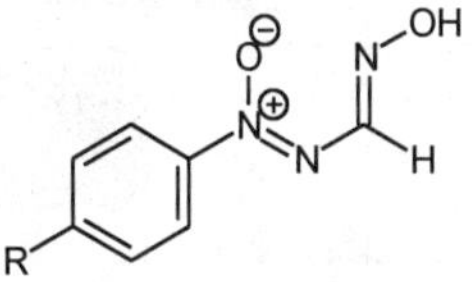

	Prophylactic			Therapeutic					*In Vitro*	
R	**Pv**	**Bc**	**As**	**Ln**	**Eg**	**Pr**	**Po**	**Pv**	**Ph**	**Fs**
Cl	Complete	Complete	Partial	Partial	Complete	Partial	Complete	Complete	Complete	Complete
OEt	Complete	Complete			Complete	Complete	Partial	Complete	Partial	
Me	Partial			Partial	Complete	Complete		Complete	Partial	
H	Partial				Partial			Complete	Complete	Complete
Br	Partial				Complete			Partial		Complete
NO_2	Complete			Partial		Partial		Partial	Complete	Partial
*t*Bu	Partial									Partial
$CONH_2$										

Partial control (grey) Complete control (black)

Relatively few Class 4 azoxyoximes were prepared in this study. These compounds were generally less active than compounds in the other three classes. The same general patterns found in the other series were also observed in this class. The 4-chloro-C-methyl derivative showed the broadest spectrum of activity, with the 4-chloro-C-phenyl compound showing a similar breadth of control.

Compounds in this study which showed interesting activity at the high dose rate were examined further in a simple dose response test. Although few of the azo- or azoxy- oximes prepared in the study showed significant levels of activity below 300 ppm, some analogues gave control at lower doses, particularly against *E. graminis* and *P. viticola.* In these studies, the generalizations which were useful for predicting spectrum of activity became less clear cut. Some illustrative examples of the activity of these compounds are given below. Activity was measured *in vivo* at 300, 100 and 30 ppm on a 0-9 scale, where 0 indicated no activity and 9 indicated complete control.

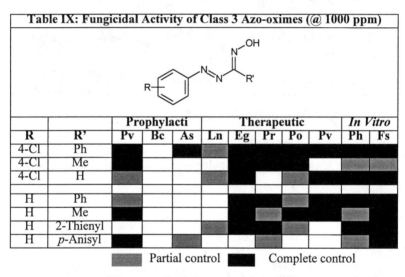

Table IX: Fungicidal Activity of Class 3 Azo-oximes (@ 1000 ppm)

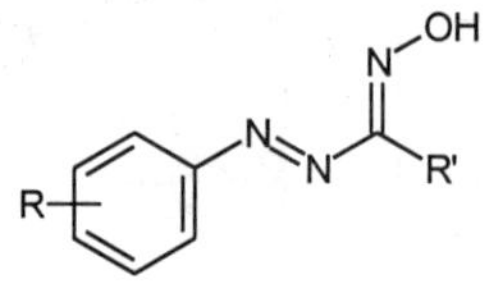

		Prophylacti			Therapeutic					*In Vitro*	
R	**R'**	**Pv**	**Bc**	**As**	**Ln**	**Eg**	**Pr**	**Po**	**Pv**	**Ph**	**Fs**
4-Cl	Ph	■		■	▒	■	■	■	■	■	■
4-Cl	Me	■				■	■	■		▒	▒
4-Cl	H	▒			▒	■		▒	■	■	■
H	Ph	▒				■	■	▒	■	■	■
H	Me	■				■	▒	■	■	▒	■
H	2-Thienyl				▒	■	■	▒		■	■
H	*p*-Anisyl	■		▒			▒			▒	■

▒ Partial control ■ Complete control

Table X: Fungicidal Activity of Class 4 Azoxy-oximes (@ 1000 ppm)

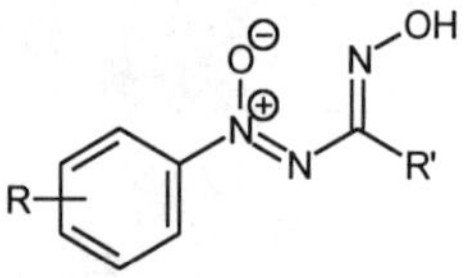

		Prophylactic			Therapeutic					*In Vitro*	
R	**R'**	**Pv**	**Bc**	**As**	**Ln**	**Eg**	**Pr**	**Po**	**Pv**	**Ph**	**Fs**
4-Cl	Me	▒	■			■	■		■	■	■
H	Ph	■		▒	▒	■				■	
4-Cl	Ph					■	■	▒	■		▒
4-F	2-Thienyl					▒	■				▒
H	2-Thienyl										▒

▒ Partial control ■ Complete control

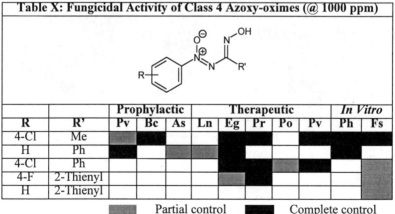

Tables XI and XII show comparisons of Class 1 and Class 2 compounds, differing only by oxidation level at the N-terminus. As indicated, the unoxidized analogue **(1)** showed good control of both *Eg* and *Pv*, with activity just beginning to break at the 30 ppm level for both diseases. The oxidized analogue **(2)**, on the other hand showed equivalent control on *Eg*, but a much more rapid fall off in activity against *Pv*, being essentially inactive at 30 ppm on this disease.

The 4-chloro analogues **(3)** and **(4)**, in contrast, gave rather different results. The unoxidized analogue **(3)** gave no control of *Eg* below 1000 ppm, despite its intense activity against *Pv*, where it gave complete control at the lowest dose rate tested. Oxidation to give **(4)**, in this case led to one of the most active compounds tested against *Eg*, giving almost complete control at 30 ppm. Against *Pv*, the activity of **(4)** broke between 100 and 30 ppm.

Table XI: Class 1 vs Class 2 Activity

Dose (ppm)	300	100	30
Eg (1)	8.8	8.8	6.8
Eg(2)	8.8	6.5	6.9
Pv(1)	9	8.9	5.1
Pv(2)	9	6.8	0.2

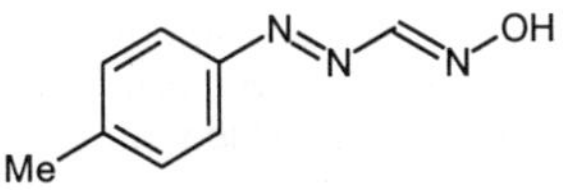

(1)

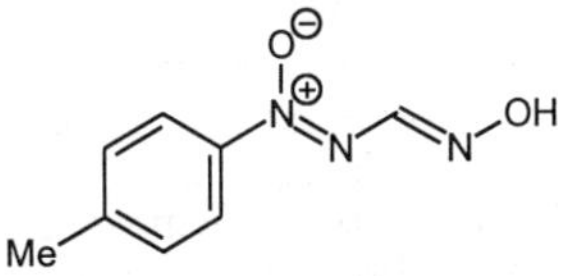

(2)

Table XII: Class 1 vs Class 2 Activity

Dose (ppm)	300	100	30
Eg (3)	2.6	1.1	0.4
Eg(4)	9	8.4	7.5
Pv(3)	9	9.0	8.9
Pv(4)	9	8.9	3.7

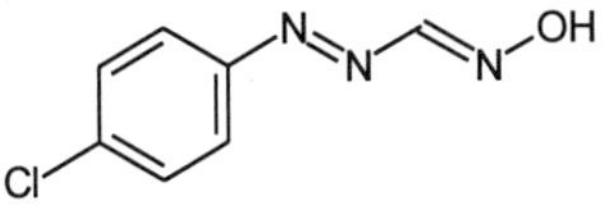

(3)

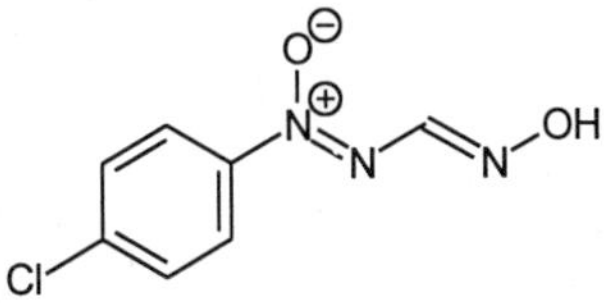

(4)

Compounds **(5)** and **(6)**, representing Classes 1 and 3, were also among the most active compounds tested at the lower dose rates (Table XIII). Both of these compounds gave good control of *Eg* even at the lowest dose rate of 30 ppm. They were, however, clearly differentiated against *Pv* where the unsubstituted compound **(5)** broke between 100 and 30 ppm, while the C-phenyl derivative **(6)** broke sharply at 100 ppm.

Table XIII: Class 1 vs Class 3 Activity

Dose (ppm)	300	100	30
Eg (5)	8.8	7.1	6.8
Eg(6)	8.8	6.5	6.9
Pv(5)	9	8.9	5.1
Pv(6)	9	6.8	0.2

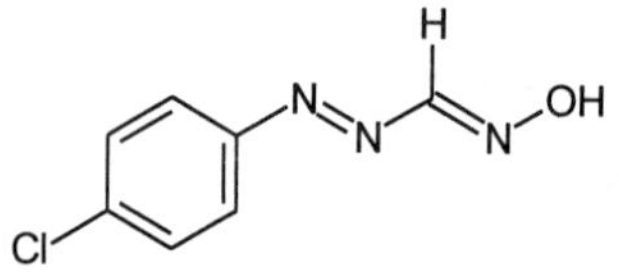

(5)

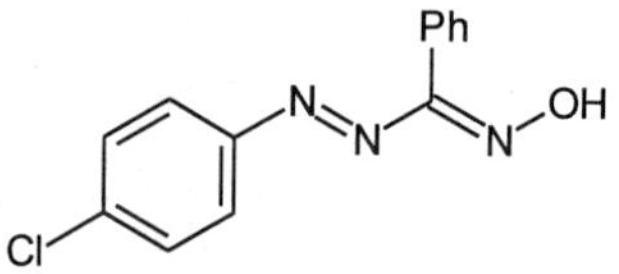

(6)

Conclusion

A number of azo- and azoxy- oximes have been synthesized and tested as potential agricultural fungicides. Several of these compounds showed impressive levels of activity, particularly against *Plasmopara viticola* and *Erisyphe graminis*. The most active compounds were those with a *p*-chlorine or *p*-methyl substituent on the aryl ring.

Literature Cited

1. Ziegler, H; Trah, S WO 9316986 (to Ciba-Geigy).
2. Shawali, A.S.; Altahou, B.M. *Tetrahedron,* **1977**, *33(13)*, 1625.
3. Bamberger, E.; Pemsel, W.; *Chem.Ber.,* **1903**, *36*, 85.
4. Kalia, K.C.; Chakravorty, A.; *J.Org.Chem.,* **1970**, *35(7)*, 2231.
5. Armand, J.; Furth, B; Kossanyi, J.; Morizur, J.P.; *Bull.Soc.Chim.Fr.,* **1968**, *(6)*, 2499.
6. Hunter, L.; Roberts, C.B.; *J.Chem.Soc., **1941**, 823.*

Chapter 29

Aminatophosphorous(1+) Salts: A New Class of Nonmobile Protectant Fungicides

John Miesel[1], Carl Denny[1], Leah Reimer[1], Greg Kemmitt[1], Todd Mathieson[1], Mary Barkham[2], and Sarah Cranstone[2]

[1]Discovery Research Center, DowElanco, 9330 Zionsville Road, Indianapolis, IN 46268-1053

[2]Letcombe Laboratory, DowElanco Europe, Letcombe Regis, Oxon 0W12 9JT, United Kingdom

It was found that members of a series of aryl and heterocyclic aminatophosphorus (1+) salts form a new class of non-mobile protectant fungicides especially efficacious against *Plasmopara viticola*, the causative organism of grape downy mildew. The synthesis and structure-activity relationships of this class of fungicides are discussed. Development of this series was discontinued due to phytotoxicity to vines and extreme ocular irritation found in the most active compounds.

Non-mobile protectant fungicides such as the dithiocarbamates have a long history of use as mixing partners for crop disease management (*1*). More recently the use of existing products, mancozeb for example, has come under scrutiny because of their very high rate of application and possible toxicological problems. Newer compounds are needed to fill this niche in the marketplace. Aminatophosphorus(1+) salts are a new class of non-mobile protectant fungicides which were examined at DowElanco. Members of this series possess efficacy against several fungal organisms of commercial interest, *Plasmopara viticola*, *Erysiphe graminis*, and *Puccinia recondata*. The activity against *P. viticola*, the causative agent of grape downy mildew, was the most significant and used to direct the study of the SAR of this series.

Plasmopara Screening Methods

Key points in the screening methods for *P. viticola* are outlined below. To be passed from initial to follow-up screening generally required a disease control rating of 7 to 9 at 100 ppm in the initial screening.

Initial *P. viticola* Screening:

- 4 week old grape seedlings - single replicate, 2 plants.
- Technical material applied at 400, 100, 25, 6.25 ppm.
- Inoculation 2 hrs (or 24 hrs) after application.
- Read test 4-5 days after inoculation.
- Disease control expressed on a modified Barratt-Horsfall scale (*2*), 9 = 100% control, 8 = 99-97%, 7 = 97-90%, 5 = 74-60%, 1 = 0-19%.

Follow-up *P. viticola* Screening:

- 4 week old grape seedlings - 3 replicates, 1 plant each.
- Technical material applied at 100, 25, and 6.25 ppm.
- Inoculation 24, 96, and 240 hrs after application.
- Disease control expressed as a percentage relative to untreated control.

Compound Structure and Synthesis

Aminatophosphorus(1+) salts are essentially N-alkylated phosphine imines. However the products are best described by the phosphorus to nitrogen single bond structures shown in Figure 1. This structure was confirmed by single crystal X-ray analysis of two members of the series.

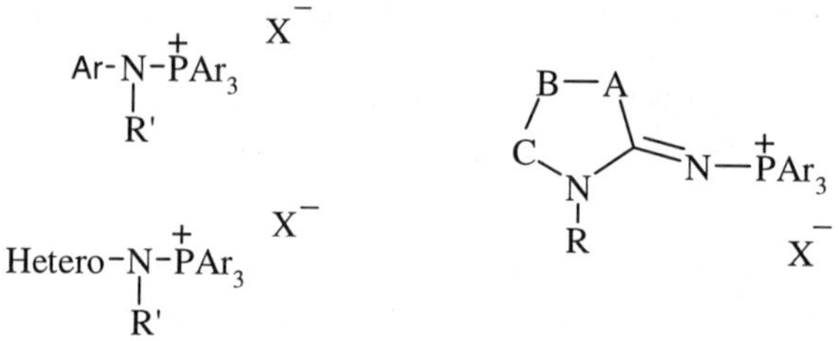

Figure 1 - Structure of Aminatophosphorus(1+) Salts

The phosphine imine intermediate is prepared in one of two ways as shown in Figure 2. The Staudinger reaction of an azide with a phosphine is most useful when examining a variety of phosphines but requires the synthesis of sometimes unstable azides (*3*). The other general synthesis uses *in situ* generation of a phosphine dibromide followed by reaction with an aniline or aminoheterocycle (*4*). This route is favored when all imine substituents are aromatic in nature. Alkylation of phosphine imines has been long known (*5*). The rate of alkylation of the phosphine imine is strongly dependent on its electronegativity, the ease of displacement of the leaving group, and steric factors. When the alkylation solvent is dichloromethane, pure product can be isolated by precipitation on addition of ether or a mixture of ether and ethyl acetate.

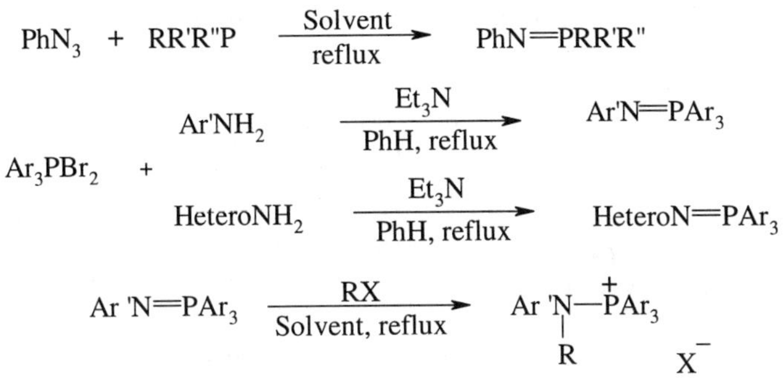

Figure 2 - Synthesis of Aminatophosphorus(1+) Salts

Structure-Activity Relationships

The compounds which are discussed below are only a portion of those synthesized and tested. They are representative of the series and accurately reflect the trends discussed. In some cases, the counter ions will vary in a particular series. We have found no significant difference in fungicidal efficacy due to the counter ion and will not present this data due to space limitations.

Effect of Phosphine Substitution on *P. viticola* Activity. As indicated in Table I, aryl substitution on phosphorus provided higher activity, especially beyond the 2 hour test.

Table I - Effect of Phosphine Substituent on *P. viticola* Activity

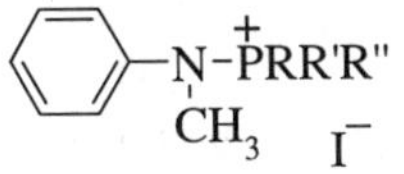

R	R'	R"	Activity 2 Hr	Activity 24 Hr
Ph	Ph	Ph	9, 6, 1, 1	
Cyclohexyl	Cyclohexyl	Cyclohexyl	1	
Ph	Ph	Bu	8, 8, 6, 1	53, 43, 5
4-CH3Ph	4-CH3Ph	4-CH3Ph	9, 9, 9, 3	100, 60, 22
2-CH3Ph	2-CH3Ph	2-CH3Ph	9, 8, 8, 5	97, 50, 6
4-CH3OPh	4-CH3OPh	4-CH3OPh	9, 8, 6, 6	98, 90, 73

Effect of N-Aryl Substitution on *P. viticola* Activity. The effect of N-aryl mono-substitution is given below in Table II. Monosubstitution was generally disappointing although compounds with electron-donating substituents were more active than those with electron-withdrawing substituents. The more lipophilic phenoxy group was an improvement especially in the ortho position.

Table II - Effect of Aryl Mono-Substituent on *P. viticola* Activity

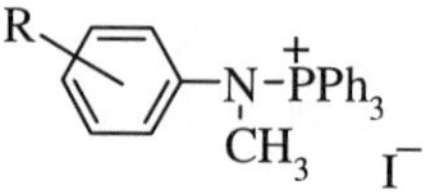

	Activity			Activity
R	2 Hr	24 Hr	R	2 Hr
H	9, 6, 1, 1		4-CN	5, 1, 1, 1
2-OCH_3	9, 7, 1, 1		3-CF_3	7, 3, 1, 1
2-OPh	9, 9, 9, 6	100, 93, 87	2-C_2H_5	9, 6, 6, 2
4-OPh	9, 9, 9, 9	93, 68, 44	2-SCH_3	8, 3, 1, 1
3-OPh	9, 7, 6, 6	84, 73, 8	2-$SOCH_3$	1, 1, 1, 1
4-NO_2	9, 9, 1, 1	28, 10, 5	2-SO_2CH_3	1, 1, 1, 1

However, the effect of multiple substitution with electron-donating substituents, especially with one of the substituents in the ortho position, was quite favorable as shown in Table III. Both the 2,4-dimethoxy analog and the dibenzofuran compound **1** provide substantial protection against *P. viticola* when grape seedlings were infected 10 days after treatment.

Table III - Effect of Aryl Substituent on *P. viticola* Activity

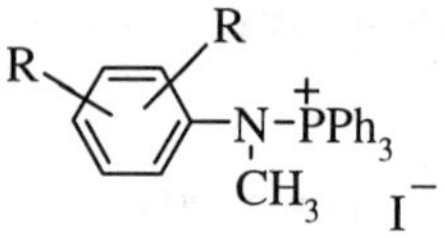

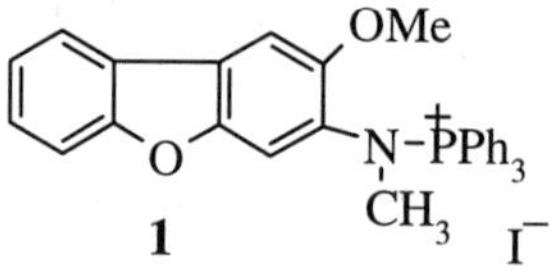

1

			Activity	
R	R'	2 Hr	96 Hr	240 Hr
2-OCH_3	4-OCH_3	9, 9, 8, 6	100, 93, 13	96, 87, 23
2-OCH_3	5-OCH_3	9, 9, 9, 9	92, 67, 20	
3-OCH_3	4-OCH_3	9, 1, 1, 1		
3-OCH_3	5-OCH_3	8, 1, 1, 1		
2-CH_3	4-CH_3	8, 8, 1, 1		
1		9, 9, 5, 5	93, 89, 50	87, 77, 57

Effect of N-alkyl substitution on *P. viticola* activity. The effects of the N-alkyl substituent on *P. viticola* activity are outlined in Table IV. Simple alkyl groups larger than methyl offered no activity advantage and the more electron-poor cyanomethyl group decreased activity. Benzyl substitution most often improved activity, especially residual activity, but the optimum benzyl substituent would vary with substitution on the N-aryl ring. Throughout the entire series, the *in vivo* activity of any specific compound appeared to depend on whole molecule properties which could not be predicted by simply combining the individual substituents that displayed the most activity.

Table IV - Effect of Alkyl Substituent on *P. viticola* Activity

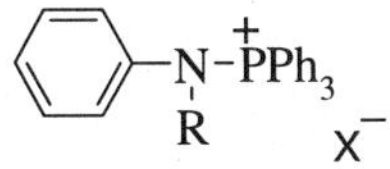

		Activity	
R	24 Hr	96 Hr	240 Hr
CH_3	9, 6, 1, 1		
n-Bu	9, 7, 6, 1		
i-Bu	9, 9, 3, 1		
$PhCH_2$	9, 9, 9, 9	100, 85, 61	99, 92, 73
4-CH_3OPhCH_2	9, 9, 8, 8		69, 0, 0
3,4-Cl_2PhCH_2	9, 8, 3, 1		62, 31, 5

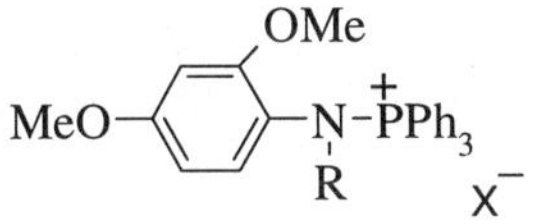

		Activity	
R	24 Hr	96 Hr	240 Hr
CH_3	9, 9, 8, 6	100, 93, 13	96, 87, 23
$NCCH_2$	9, 6, 1, 1		
CH_2=$CHCH_2$	9, 8, 3, 1		93, 71, 61
$PhCH_2$	9, 9, 7, 3	100, 61, 2	
4-CH_3OPhCH_2	9, 9, 7, 3	98, 93, 40	100, 99, 82
3,4-Cl_2PhCH_2	9, 8, 8, 6	93, 87, 9	99, 90, 42

Effect of N-heterocyclic substitution on *P. viticola* Activity. Synthesis of N-heterocyclic aminatophosphorus(1+) salts provides an additional complication. Alkylation can occur endo at the ring nitrogen, as in the thiazole below, or at the exocyclic nitrogen as in the very similar isoxazole. Endo addition occurs in about 80 - 90% of the examples, and mixtures are obtained in a few cases. Proton NMR is diagnostic for the position of alkylation since groups adding exo show vicinal coupling between the α protons on the alkyl group and phosphorous. Those adding endo have the α protons unsplit.

Endo Alkylation

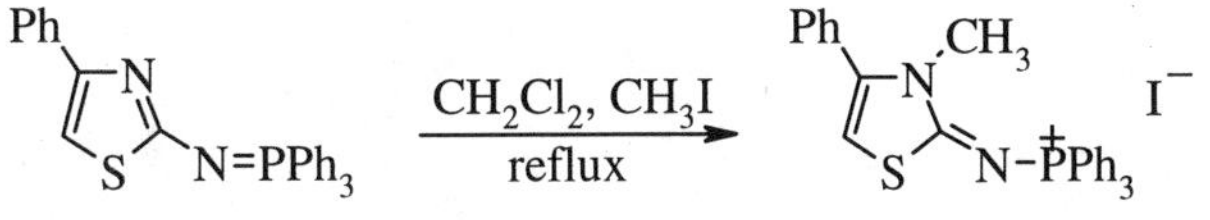

Exo Alkylation

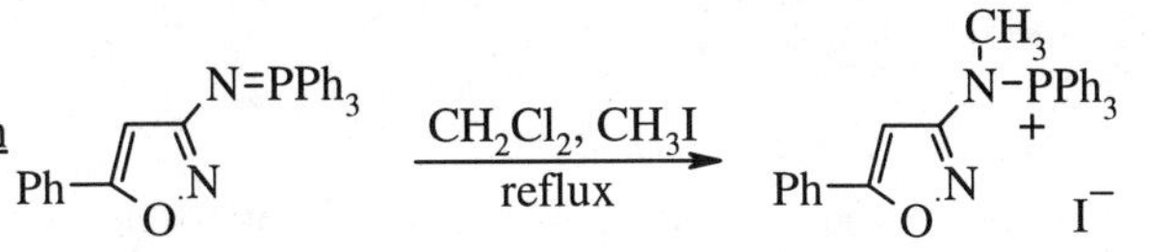

The electronegativity of the N-heterocyclic substituent has a strong effect on biological efficacy. The relationship of ring electron density with activity mirrors that seen with the N-aryl members of the series. The more electron-rich the ring, the greater is the *P. viticola* control. This is shown in Table V for heterocycles with either phenyl or *t*-butyl substituents.

Table V - Effect of Heterocycle on *P. viticola* Activity

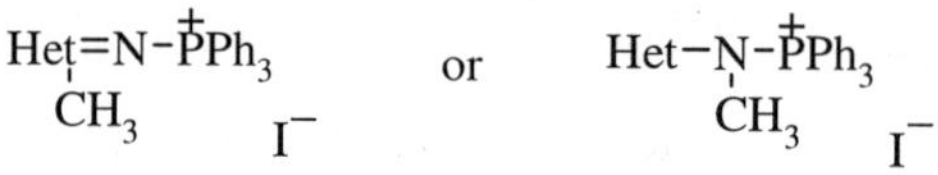

endo — or — exo

Hetero	Position	Activity 24 Hr	96 Hr	240 Hr
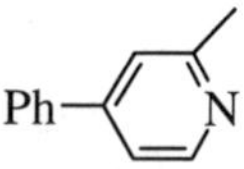	endo	9, 9, 9, 6	86, 67, 33	85, 25, 20
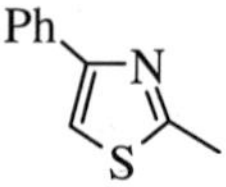	endo	9, 9, 9, 9	98, 97, 77	93, 93, 73
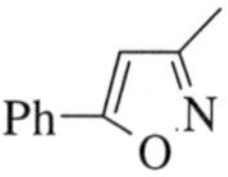	exo	7,-,-,-		
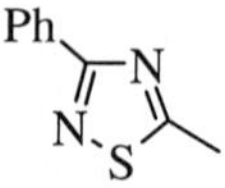	endo	6, 5, 3, 1		
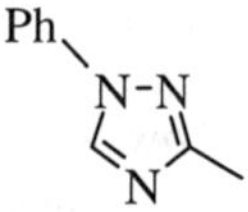	endo	9, 9, 5, 1		
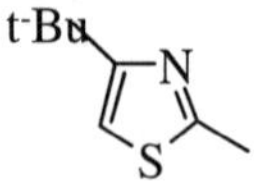	endo	9, 9, 9, 7		96, 43, 43
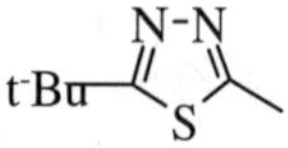	endo	9, 7, 1, 1		
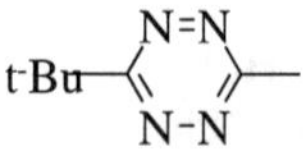	endo	6, 1, 1, 1		

The same trend is obtained by varying the electronegativity of a substituent on the heterocyclic ring. This is exemplified in Table VI by a group of pyridine compounds. The benzosubstituent, in this case the quinoline ring, is often an analog equivalent to a phenyl substituent.

Table VI - Effect of Heterocycle Substituent on *P. viticola* Activity

$$\text{Het}_{|}^{=}\text{N}-\overset{+}{\text{P}}\text{Ph}_3 \quad \text{or} \quad \text{Het}-\text{N}-\overset{+}{\text{P}}\text{Ph}_3$$
$$\text{CH}_3 \quad \text{I}^- \qquad\qquad \text{CH}_3 \quad \text{I}^-$$
$$\text{endo} \qquad\qquad\qquad \text{exo}$$

Hetero	Position	24 Hr	96 Hr	240 Hr
pyridin-2-yl	endo	8, 7, 2, 1		
5-Cl-pyridin-2-yl	endo	9, 3, 1, 1		
5-NC-pyridin-2-yl	endo	3, 1, 1, 1		
6-CH₃-pyridin-2-yl	endo	9, 7, 4, 3		
5-CH₃-pyridin-2-yl	endo	8, 6, 2, 1		
4-CH₃-pyridin-2-yl	exo	9, 9, 9, 6	99, 80, 19	
quinolin-2-yl	endo	9, 9, 9, 9	100, 85, 70	80, 52, 3

The non-predictability of the effect of the N-alkyl substituent on efficacy, especially residual activity, is demonstrated in Table VII. The thiazole compound **2** has excellent 240 hour residual activity with the N-methyl substituent, but much poorer with the N-benzyl. The reverse is true with the quinoline analogs **3**.

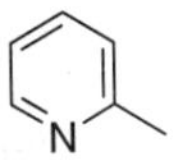

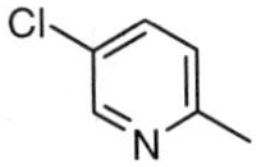

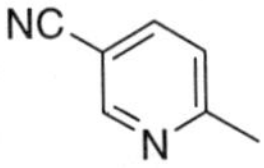

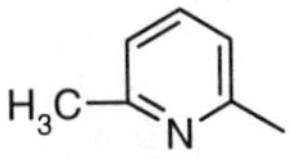

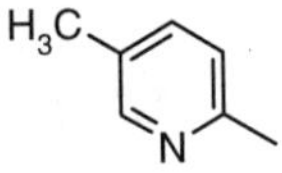

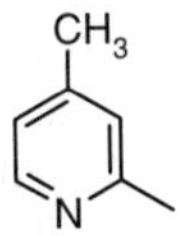

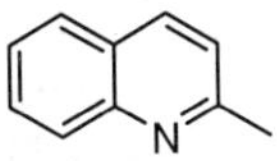

Table VII - Effect of Alkyl Substituent on *P. viticola* Activity

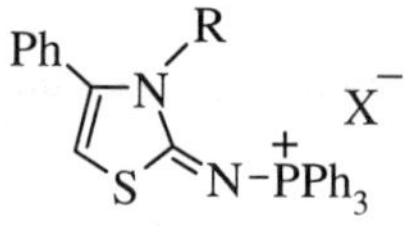

2

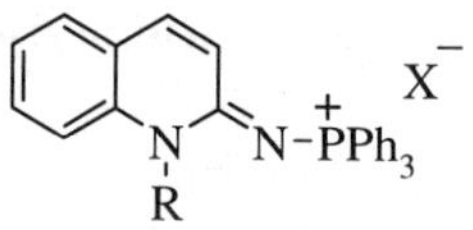

3

Compound	R	24 Hr	Activity 96 Hr	240 Hr
2	CH_3	9, 9, 9, 9	98, 97, 77	93, 93, 73
2	$PhCH_2$	8, 7, 1, 1	100, 99, 68	25, 16, 0
3	CH_3	9, 9, 9, 9	100, 85, 70	80, 52, 3
3	$PhCH_2$	9, 9, 9, 9	99, 99, 90	97, 97, 8

Pot Trials

At DowElanco the next stage in structure-activity relationship testing for *P. viticola* is outdoor pot tests. Pot tests allow us to examine some of the features of a field test using less compound. Large vines that are initially 2.5 - 3 feet tall and that are hardened to the outdoors are used. The test compound in a more commercial formulation is examined after multiple applications at 7 day intervals. Compounds **1**, **4**, and **5** were tested in this fashion.

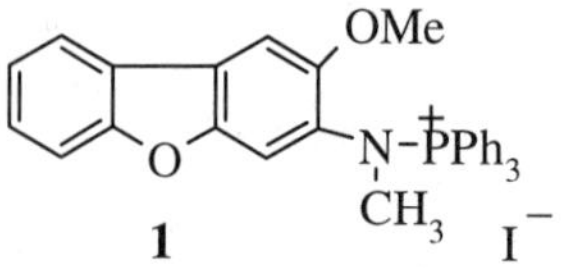

1

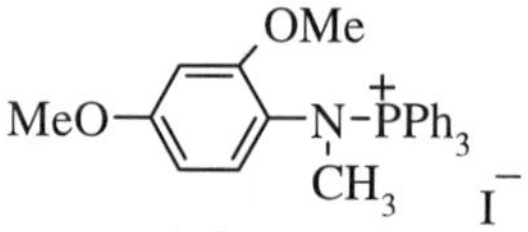

4

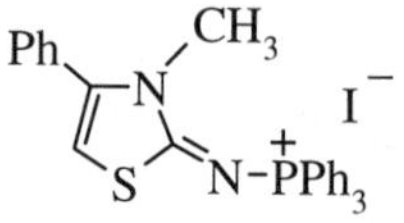

5

Initial results were disappointing. All three salts were superior to Dithane in the greenhouse. However, in the pot test only compound **5** had an LC90 below Dithane. More troubling was the unexpected appearance of phytotoxicity in the test. The more efficacious **5** was also more phytotoxic. This phytotoxicity was quite characteristic at low rates where leaf cupping and a leathery surface on the affected leaves were observed. This damage progressed to partial leaf necrosis at higher rates. The affected leaves were newly emerged at the time of compound application and the damage appeared as the leaves expanded. Mature leaves were unaffected by compounds in this series.

Special Phytotoxicity Testing

To continue our examination of this series, it was necessary to reproduce this phytotoxicity in a greenhouse test. Some simple changes in the screen were sufficient. The presence of newly emerged leaves was ensured by using larger vines. The plants were held at 100% humidity for 24 hours after compound application which markedly increased phytotoxicity. Symptom development was encouraged by holding the plants for 10 days. Damage similar to that in the pot test was expressed using this protocol. Based on the data from the special phytotoxicity test for the three pot trial compounds as well as the two other active analogs shown in Table VIII, compound **7** was an obvious choice for further examination.

Table VIII - Special Greenhouse Vine Phytotoxicity Test

	Percent total leaf area expressing symptoms		
	Rate (ppm)		
Compound	100	500	1000
1	0	5	0
4	1	13	22
5	27	15	43
6	10	15	13
7	0	0	1.7

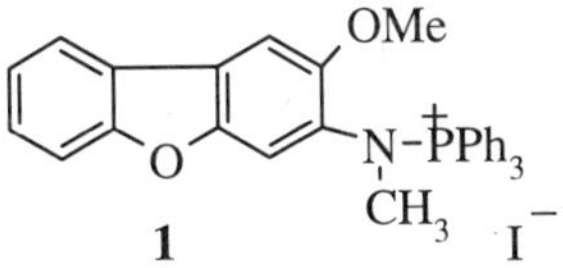

1

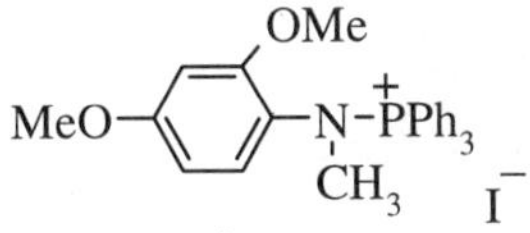

4

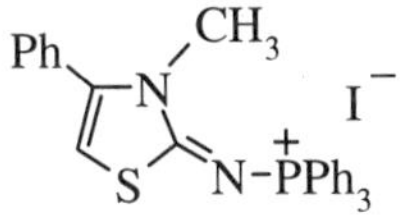

5

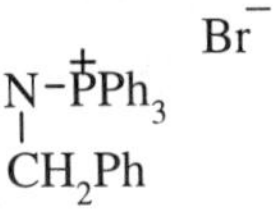

6

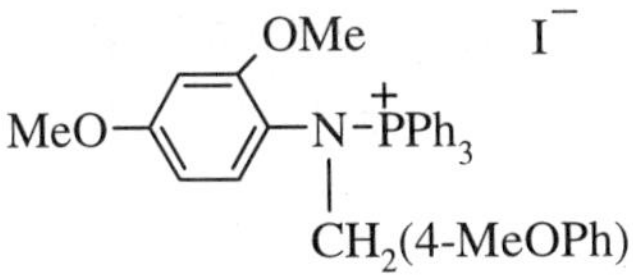

7

Further examination of compound 7

The pot trial of compound **7** was quite favorable. The efficacy of this compound against *P. viticola* was equivalent to Dithane at 0.25 the rate. As predicted by the special greenhouse testing, there was no phytotoxicity to vines after 4 applications of **7** at 1000 ppm. Before going to a full field trial, a mammalian hazard evaluation on **7** was conducted. Although non-toxic by oral or dermal routes, the compound was found to be an extremely dangerous ocular irritant. The ocular damage was so severe that the animals were euthanized after 24 hours. No further work with **7** was conducted.

***Caution*:** All aminatophosphorus(1+) salts should be considered as possible severe occular irritants. Full eye protection should be used when preparing or using compounds in this class of chemistry.

Summary

A series of aminatophosphorus(1+) salts was examined as non-mobile protectant fungicides for the control of *P. viticola*. Structure-activity relationship studies were successful in developing a number of compounds with excellent residual control in the greenhouse. Special trials indicated that many of the most active compounds caused unacceptable phytotoxicity to vines with repeat applications at efficacious rates. Extreme ocular irritation found with the most promising compound in this series ended further work on the series.

Acknowledgments

The authors would like to thank the other scientists in DowElanco Discovery Research who provided helpful assistance through the course of these studies. Special acknowledgment must be given to Prof. Phil Fanwick of Purdue University for determining the X-ray structures mentioned in this report.

Literature Cited

1. *Encyclopedia of Chemical Technology*, Grayson, M., Eckworth, D. Eds, John Wiley and Sons: New York, NY, 1980; *Vol. 11*, pp 490-98.
2. Horsfall, J.G.; Barratt, R.W.; *Phytopath.*, **1945**, *35*, 655.
3. Staudinger, H.; Meyer, J.; *Helv. Chim. Acta*, **1919**, *2*, 635.
4. Horner, L.; Oediger, H.; *Liebigs Ann.*, **1959**, *627*, 142.
5. Staudinger, H.; Hauser, E.; *Helv. Chim. Acta*, **1921**, *4*, 861.

Chapter 30

Taxane Biofungicides from Ornamental Yews: A Potential New Agrochemical

Mary Jane Incorvia Mattina[1], Gerri MacEachern Keith[1], and Wade Elmer[2]

[1]Departments of Analytical Chemistry and [2]Plant Pathology and Ecology, The Connecticut Agricultural Experimental Station, 123 Huntington Street, New Haven, CT 06511

Research on the natural product paclitaxel, better known as taxol, has spanned four decades beginning in the 1960s. Paclitaxel is a member of the taxane class of compounds, diterpenes in which the C_{20} skeleton is arranged in a fused, three ring structure. Originally isolated in 1971 from the bark of the Pacific yew tree, *Taxus brevifolia*, paclitaxel has been found in a variety of tissues from both native and cultivated *Taxus* spp. The unique cytokinetic mode of action of certain taxanes in mammalian cells has resulted in their exploitation as effective antineoplastic agents. Research into the possible cytokinetic activity of taxanes in non-mammalian systems is far more limited. The work reported here is an investigation of the inhibition of radial growth of twelve plant pathogenic fungi on potato-dextrose agar by the amendment of taxanes into the solid medium. Fungal growth inhibition produced by a mixed taxane fraction, extracted from the needles of cultivated *Taxus* spp., followed by its partial purification using solid phase extraction techniques, was compared with that produced by single authentic taxanes. Fungal growth inhibition produced by the mixed taxane fraction was also compared with the inhibition produced by a combination of three authentic taxanes. The data indicated that plant pathogens in the Ascomycetes and Deuteromycetes can be classified as taxane insensitive. Although taxane sensitivity varied across the five Oomycetes examined, our data suggest that the taxane class of compounds may provide an alternate to current methods of control of Oomycetes plant pathogens.

Considerable coverage of the clinical successes of Taxol® as a chemotherapeutic agent has appeared in the popular media. Excellent technical reviews of the discovery and development of paclitaxel are available as well (*1*, *2*). Suffice it to say here that in 1964 an alcoholic extract of *Taxus brevifolia* Nutt. bark was shown to be cytotoxic against several tumor cell lines. In 1971 the active principal in the crude extract was identified as the compound paclitaxel (*3*) and its structure elucidated. This structure and that of several other common taxanes are illustrated in Figure 1. Interest in the compound increased considerably when it was shown in

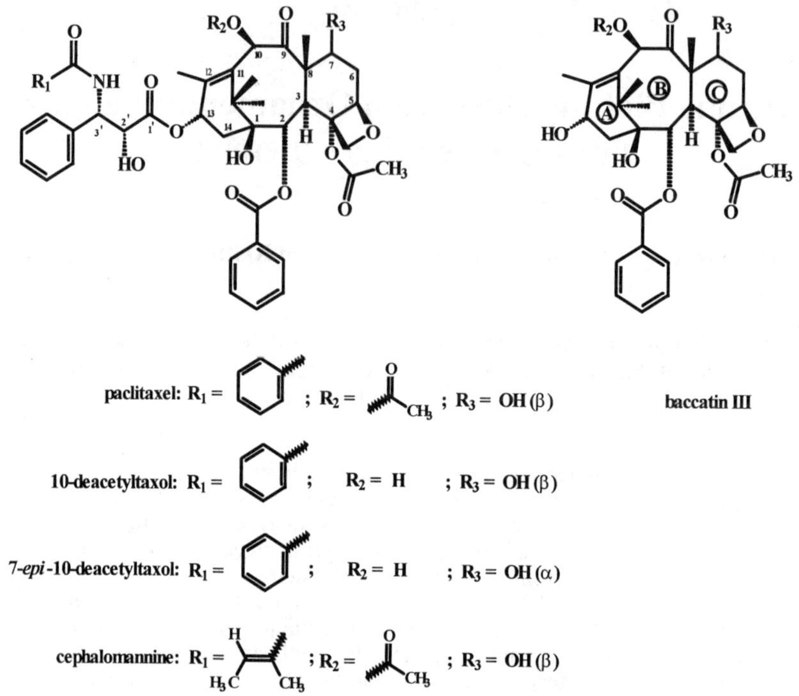

Figure 1. Structures of several common taxanes.

1979 that paclitaxel is a mitotic spindle poison with a unique mode of action (*4*). Today the clinical formulation of paclitaxel, Taxol®, is approved for the treatment of ovarian and breast cancer.

Limited studies of the effects of mitotic disrupters, such as paclitaxel, on non-mammalian eukaryotic cells appeared in the 1980s (*5, 6*). Vaughn and Vaughn were apparently the first to suggest the use of paclitaxel as an agrochemical, specifically as an herbicide (*7*).

Studies of the non-clinical aspects of taxanes are most appropriate at our institution, a state agricultural experiment station. Accordingly, since 1991 we have been investigating cultivated *Taxus* spp. as a source of taxanes. Phase I of our work consisted of a survey of the paclitaxel content in the needles of ornamental yew shrubs grown extensively in the Northeast. Unlike the bark of the Pacific yew tree, *Taxus* needles from cultivated species might provide a sustainable and renewable taxane resource. In Phase II we investigated what influence, if any, cultivation practices have on taxane biosynthesis. Phase III is an examination of the cytokinetic effects of taxanes on fungal plant pathogens. In contrast to the considerable expense incurred in the extraction and purification of a single taxane from its botanical source for use as a pharmaceutical, a partially purified, but well-characterized taxane extract of *Taxus* needles, intended as an agrochemical, can be achieved more economically. We focused our attention on the efficacy of such a mixed taxane extract on fungal cells. The results of our work indicate possible economic success of "farmaceuticals", biomass cultivated not for its food, fiber, or landscape potential, but for its chemical content as potential agrochemicals and medicinals. Results from the three phases of the taxane project at The Connecticut Agricultural Experiment Station will be discussed below, with the emphasis placed on Phase III, the potential development of taxanes as agrochemicals.

Taxane Study: Phase I

Serious supply issues became apparent early in the clinical development of paclitaxel. The original source of paclitaxel required the destructive harvest of Pacific yew trees, which form part of the habitat of an endangered species, the Northern spotted owl. As opposed to the bark of the Pacific yew, needles from cultivated *Taxus* species might serve as an environmentally non-destructive and renewable source of the natural product. Connecticut, Rhode Island, and Massachusetts are major commercial growers of *Taxus* shrubs for the landscape industry. For these reasons we chose to determine the concentration of paclitaxel found in the needles of common yew cultivars grown in the Northeast (*8*). Biomass was obtained from three nurseries in Connecticut and one in Rhode Island, extracted, and analyzed using high performance liquid chromatography (HPLC). The graph in Figure 2 shows some of the data from this Phase I study. Since the average concentration of paclitaxel in the bark of the Pacific yew tree is 400 ppm on a dry weight basis, the data in Figure 2 suggest that the needles of some cultivars, for example *T.* x. *media* 'Nigra' and 'Hicksii', may be a viable alternative to the original source of paclitaxel. This is a valid conclusion even though we subsequently determined that these data are an overestimation of paclitaxel concentration by approximately 15%. This overestimation is due to the coelution of paclitaxel with cinnamyl taxanes on the C_{18} reverse phase column used in these early HPLC analyses (*9*).

Taxane Study: Phase II

Phase II consisted of both a field and a greenhouse component. In the field component rooted cuttings of three cultivars, 'Greenwave', 'Tauntonii', and 'Hicksii', were established at Lockwood Farm, The Station's experimental facility. Several beds of each cultivar were planted, each bed measuring 4 feet on a side and containing 81 cuttings of the particular cultivar on 6 inch centers. The beds were spaced three feet apart. During the first summer, the beds were exposed to direct sunlight and watered periodically. In subsequent summers, the beds were shaded and watered regularly with an installed overhead irrigation system. No herbicides were applied; hand weeding was used within each bed and a hay mulch was laid down between beds. After three years of growth, all three cultivars were well established with the growth of 'Hicksii' and 'Greenwave' cultivars particularly impressive. Paclitaxel concentrations in the needles of all three cultivars were similar to the values shown in Figure 2, but reduced by 15%. The close planting density and the lack of periodic fertilization did not appear to diminish plant growth or taxane needle content during the three year observation period.

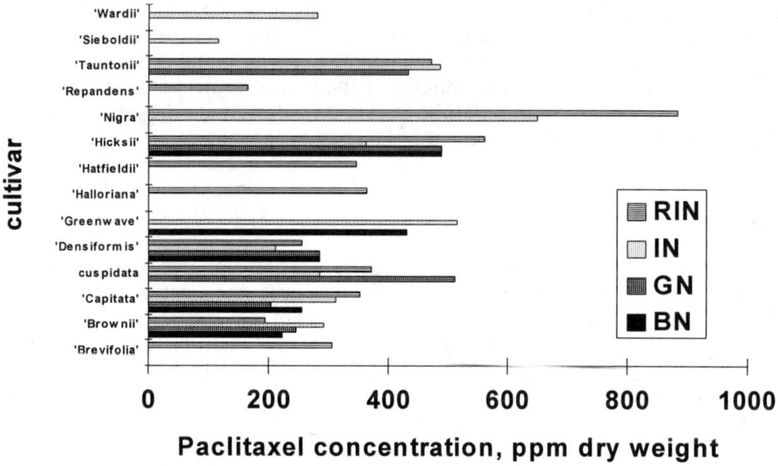

Figure 2. Paclitaxel content in the needles of various ornamental *Taxus* cultivars obtained from four nurseries as denoted in the legend. (Adapted from ref. 8)

In the greenhouse, rooted cuttings of several different cultivars were established in random blocks. We examined the effects of a variety of nutrient regimes and applications of exogenous chemicals on taxane biosynthesis. The applications consisted of various amounts of Hoagland's plant nutrient solution (*10*); the herbicides, acifluorfen and oxyfluorfen; the growth regulators, daminozide and chlormequat chloride; and sodium chloride and sodium salicylate. We observed no increase or decrease of taxane biosynthesis as a result of any of the nutrient regimes or applied chemicals (*11*).

Taxane Study: Phase III

Having decided to study the cytokinetic effects of a mixed taxane fraction on fungal systems, we undertook to improve our analytical methodologies for the extraction, clean-up, and HPLC analysis of taxanes from *Taxus* needles. The steps in the process are presented in the flow chart in Figure 3.

Several points regarding the method require comment. Our Phase III study necessitated the preparation of substantial amounts of the mixed taxane fraction. Our initial use of commercially available C_{18} SPE cartridges for the partial purification of the reconstituted crude extract was far too costly and time consuming for preparing large amounts of the mixed taxane fraction. Accordingly, we investigated using C_{18} SPE cartridges packed by us with large amounts of the C_{18} adsorbent. Since we did not achieve the desired improvement in cost and time savings, we turned to C_{18} SPE membranes for scaling up this purification step. Table I presents a comparison of the costs involved in the SPE clean-up using commercially available C_{18} cartridges, ultra-high capacity cartridges prepared in our laboratory, and 47mm and 90mm Empore™ disks (3M Corporation) packed with

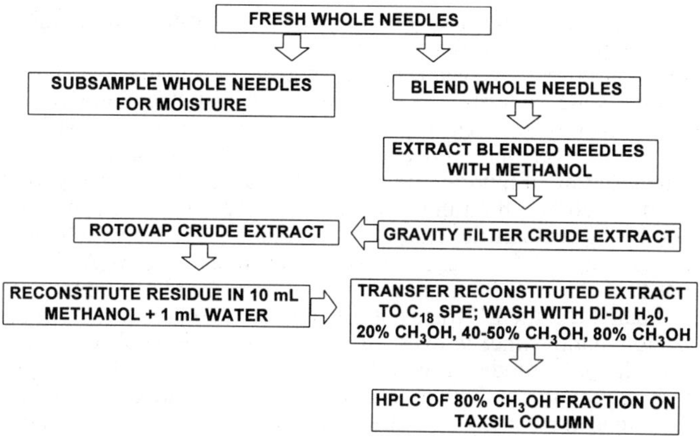

Figure 3. Flow chart for the analysis of taxane content in *Taxus* biomass.

C_{18} (*12*). The fraction eluted off the C_{18} adsorbent using 80% methanol/20% water (see Figure 4A), which we refer to as the "partially purified mixed taxane fraction," contained quantitative amounts of paclitaxel and other non-polar taxanes. It was equivalent regardless of whether the C_{18} was packed in a cartridge or in a teflon membrane. The cost saving in replacing the cartridge with the disk format is apparent from the information in Table I. The saving in time was similar.

Table I. Comparison of C_{18} SPE of Crude Extract from Taxus Needles

	High Capacity Cartridge	Ultra High Capacity Cartridge	47mm disk	90 mm disk
vol. crude	0.5mL	3.5---3.9mL	7.0mL	25mL
vol. H_2O	5mL	20mL	15mL	25mL
vol. 20%	10mL	40mL	15mL	25mL
vol. 40%	-0-	-0-	-0-	25mL
vol. 45%	-0-	-0-	15mL	-0-
vol. 50%	10mL	40mL	-0-	-0-
vol. 80%	10mL	40mL	20mL	50mL
paclitaxel recovery[a]	104.0%±4.0 (n=4)	103.5% ±13.0 (n=2)	115.8%±5.3 (n=2)	103.2%±5.0 (n=3)
paclitaxel recovery[b]			89.5%±0.1 (n=2)	89.8%±4.1 (n=3)
est. cost/mL of crude extract	$5.70	$2.80	$1.00	$1.00

[a] Authentic paclitaxel spiked directly into crude extract prior to partitioning by means of SPE.

[b] Recovery based on paclitaxel concentration calculated from partitioning on J.T. Baker 7020-07 high capacity cartridge.

SOURCE: Adapted from ref. 12.

The coelution of paclitaxel on the C_{18} HPLC column with cinnamyl taxanes present in the partially purified extract from *Taxus* needles was not optimal for quantitation purposes. Several manufacturers eventually developed speciality columns for taxane analysis. We chose the Taxsil™ column from MetaChem Technologies (Torrance, CA). Equipped with the appropriate guard cartridge, this column is robust over many injections and provides the separation necessary for reliable quantitation. Figure 4 shows the HPLC analysis of the partially purified mixed taxane fraction run on this column under the following conditions: reservoir A contains 100% acetonitrile, reservoir B contains 70% water/30% methanol; 0 to 15 minutes isocratic at 41% A; 15 to 20 minutes linear to 65% A; 20 to 30 minutes isocratic at 65% A; 30 to 35 minutes linear to 41% A; 35 to 40 minutes isocratic at 41% A. In panel (A) the LC trace collected on the 230 nm channel of the UV diode array detector (darker line) is overlaid with the LC trace collected on the 280 nm channel of the UV diode array detector (lighter line). Paclitaxel has a UV absorbance maximum at 230nm and a small absorbance at 280nm, while the cinnamyl taxanes have a UV absorbance maximum at 280nm. Comparison of the two traces shows that baseline resolution of paclitaxel (retention time ≅ 10 min) from the cinnamyl taxanes has been achieved. However, if we overlay the 230nm UV channel trace from the LC of the partially purified mixed taxane fraction with the 230nm UV channel trace from a five taxane standard run under the same conditions, as shown in panel (B), we note that paclitaxel coelutes with 7-*epi*-10-deacetyltaxol under these conditions. Varying the HPLC mobile phase conditions slightly from the above permits us to resolve paclitaxel from 7-*epi*-10-deacetyltaxol in the partially purified mixed taxane fraction. In so doing we have shown that the *Taxus* needle extracts which we are examining contain very small quantities of 7-*epi*-10-deacetyltaxol. Hence, the use of the HPLC conditions enumerated above provide us with reliable quantitation results for paclitaxel.

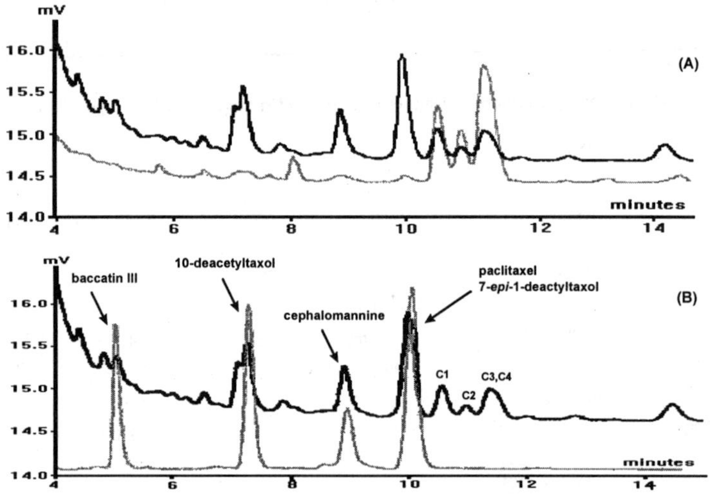

Figure 4. HPLC traces of taxane needle extract and authentic standards.

***In vitro* Cytokinetic Activity of the Taxane Fraction from Taxus Needles.** With these two improvements to our analytical methodology in place, we were ready to examine whether the partially purified mixed taxane extract had any *in vitro* cytokinetic effect on several plant pathogenic fungi. The fraction eluted from the solid phase extraction membrane with 80%methanol/20% water was evaporated to dryness using rotary evaporation, and was reconstituted in 95% ethanol. The reconstituted fraction was amended into cooled (50 °C) molten potato-dextrose agar (PDA), which was then transferred to petri plates. The final concentration of paclitaxel in the PDA ranged from 0.05 μg/mL (0.058 μM) to 4.0 μg/mL (4.68 μM), with the amount of ethanol in the amended and control agar equal to 1.3%. Additional controls were prepared which contained no 95% ethanol. In all Tables and Figures the agar amended with the reconstituted mixed taxane fraction is characterized by the final paclitaxel concentration in the agar. The plates were seeded on the perimeter with a 4 mm diameter agar plug removed from an actively growing agar plate colonized with the particular fungus. The plates were incubated in the dark at 24-26 °C. Radial growth was measured to the nearest mm twice daily, daily, or every three days, depending on the growth rate of the particular fungus. The EC_{50} values summarized in Table II were calculated by linear regression from plots of radial growth vs. concentration. From the EC_{50} data it is obvious that species in the Deuteromycetes and Ascomycetes can be considered taxane-insensitive, although there is a limited response to taxanes as evinced by the Inhibition Time data in Table II. Based on the EC_{50} data, species in the Oomycetes

Table II. Inhibition of Fungal Growth on Taxane-amended PDA.

Fungus	Taxonomic group	EC 50 [a] (μg paclitaxel/ml)	Inhibition Time [b] (days)
Botrytis cinerea	Deuteromycetes	>4.00 [c]	1.0
Rhizoctonia solani	Deuteromycetes	>4.00	1.3
Fusarium oxysporum	Deuteromycetes	>4.00	1.0
Fusarium proliferatum	Deuteromycetes	>4.00	1.5
Verticillium dahliae	Deuteromycetes	>4.00	ND [d]
Venturia inaequalis	Ascomycetes	>4.00	ND [d]
Monilinia fruticola	Ascomycetes	>4.00	0
Phytophthora cactorum	Oomycetes	0.50	ND [d]
Phytophthora citricola	Oomycetes	0.50	16.0
Pythium aphanidermatum	Oomycetes	0.05	ND [d]
Pythium myriotylum	Oomycetes	0.82	ND [d]
Pythium irregulare	Oomycetes	1.37	5.5

[a] Effective Concentration for inhibition of radial growth by 50%.

[b] The difference between the time required for fungus to grow half-way across the ethanol control plate vs. 4 μg/ml plate.

[c] Values represent the mean of six replicate plates.

[d] ND = not determined; the radial growth on the 4.0 μg/ml at 26 days was 1.45 cm for *V. dahliae* and *V. inaequalis*, 0.9 cm for *P. cactorum*, 0.4 cm for *P. aphanidermatum*.

SOURCE: Adapted from ref. 14.

are taxane-sensitive. These observations are in agreement with earlier reports by Lataste *et al.* (*5*) and Young *et al.* (*13*) in spite of the many variations among the assays conducted in the three laboratories. Considerable variation in taxane sensitivity is exhibited by the radial growth data for the Oomycetes, with *P.*

aphanidermatum the most taxane sensitive of the *Pythium* species examined. It was observed, however, that although slow radial growth was noted for all *Pythium* spp. at low concentrations of paclitaxel in the mixed taxane fraction, the fungi lacked the robust mycelial growth observed for fungi on the control plates. This may be seen in the photos provided in Figure 5.

Comparison of Cytokinetic Activity of Taxane Fraction with Authentic Taxane Standards. As is evident from the darker traces in both panels of Figure 4, the partially purified extract from *Taxus* needles contains a mixture of taxanes such as paclitaxel, cephalomannine, 10-deacetyltaxol, and baccatin III. It was of interest to determine how the cytotoxity of the mixed taxane fraction compared with fungal toxicity of single authentic taxanes. Accordingly, we prepared individual standards of paclitaxel, cephalomannine, and baccatin III, and amended these into different batches of PDA. Fungal radial growth was then measured as before and compared with fungal growth on media amended with the mixed taxane fraction. The data are presented in Figure 6. Once again the mixed taxane fraction is characterized by the

a

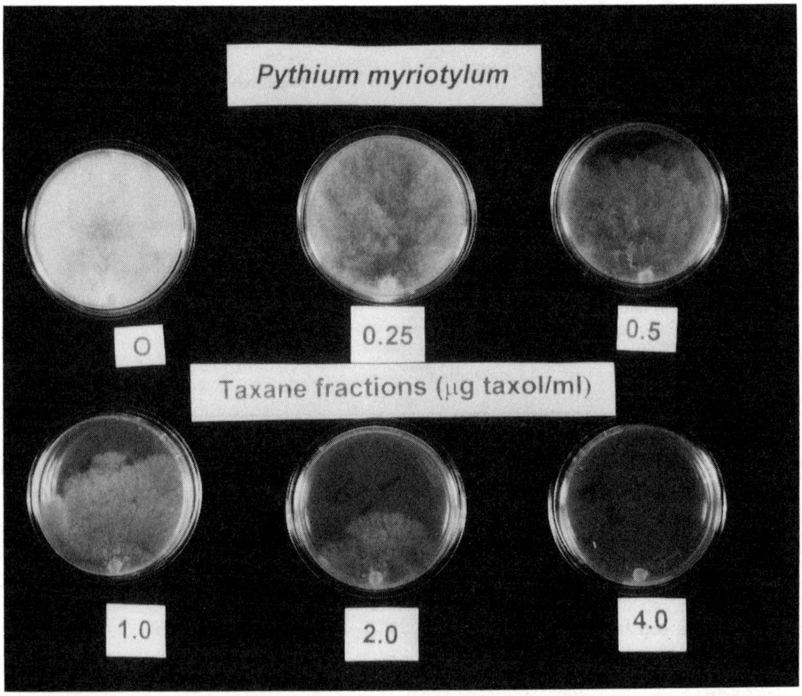

Figure 5. Radial growth of *Pythium* spp. on PDA amended with partially purified taxane fraction (Adapted from ref. 14).

b

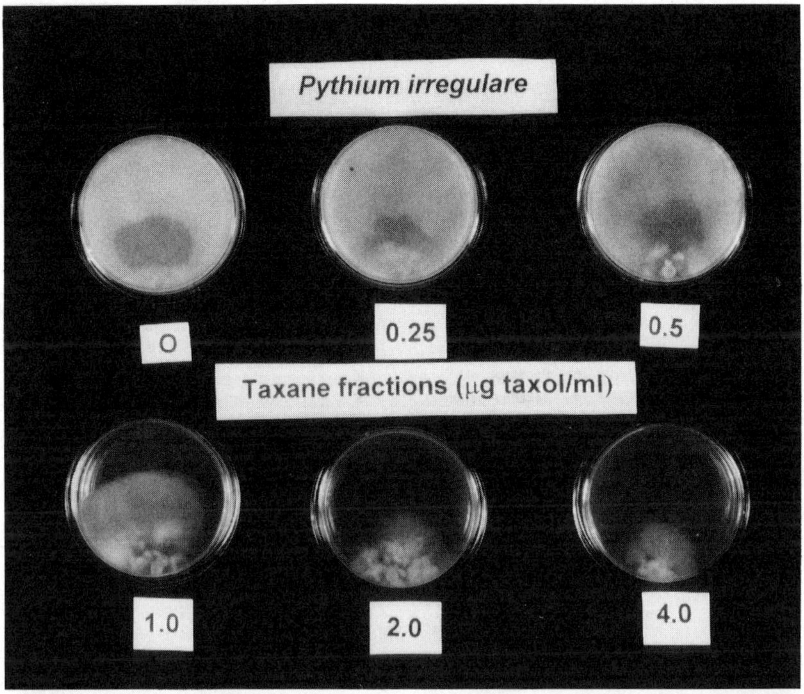

c

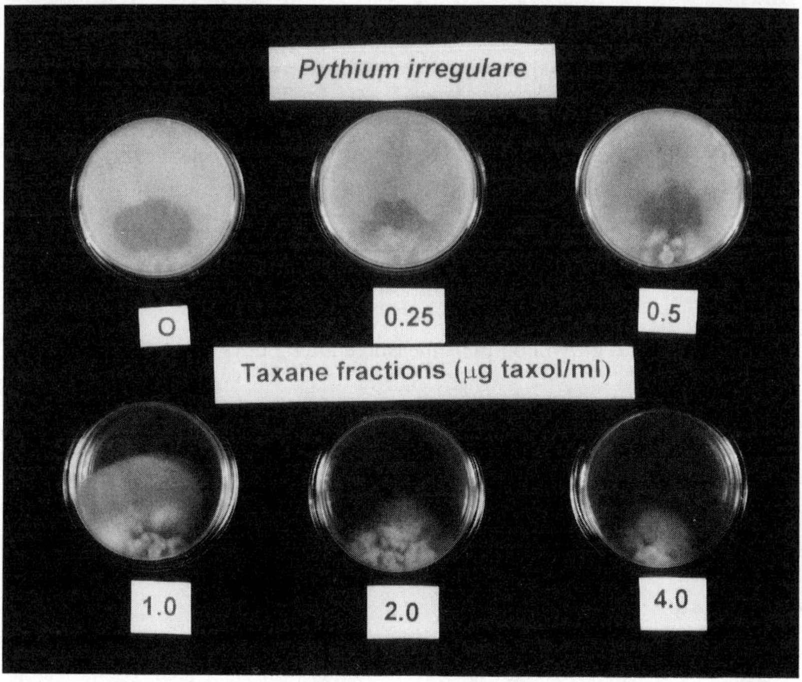

Figure 5. *Continued.*

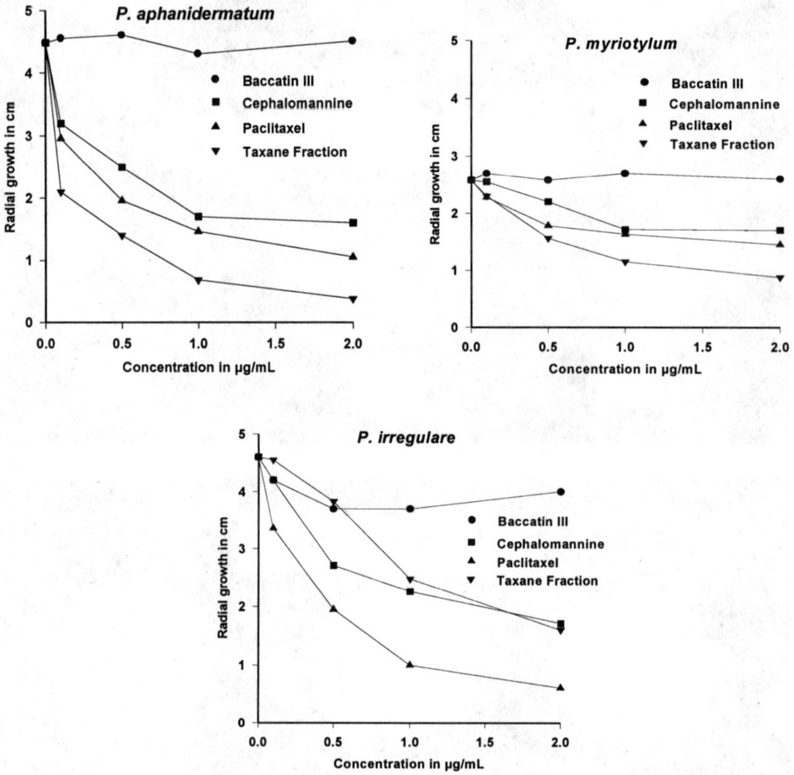

Figure 6. Radial growth on PDA measured after 3 days (Adapted from ref. 14).

paclitaxel concentration in the PDA, but it must be noted that this fraction also contains significant amounts of other taxanes. Baccatin III, which lacks the side chain at C13 of the taxane A-ring, exhibited no inhibition of fungal growth. Given the absence of a cytokinetic effect of baccatin III on mammalian systems, a similar absence on non-mammalian systems is not surprising. Cephalomannine and paclitaxel both inhibit fungal growth. However, the mixed taxane fraction is more effective than either paclitaxel or cephalomannine individually in inhibiting the growth of both *P. aphanidermatum* and *P. myriotylum*. For *P. irregulare* the opposite is true, that is, both cephalomannine and paclitaxel appear more effective individually in inhibiting fungal growth than the mixed taxane fraction. In all cases paclitaxel is more effective in inhibiting fungal growth when compared directly to cephalomannine.

Comparison of Cytokinetic Activity of Taxane Fraction with a Mixture of Authentic Taxane Standards. We also compared the cytokinetic activity of the mixed taxane fraction with a mixture of three authentic taxanes. Individual standards of authentic paclitaxel, cephalomannine, and baccatin III were prepared and combined in the same proportions as typically found in the partially purified taxane extract from *Taxus* needles, namely 10:5:1. This combination was then amended into PDA and the fungal radial growth measured as before. The results are shown in the graphs of Figure 7. In this Figure both the partially purified taxane fraction and the standard mixture prepared from authentic standards are characterized by its paclitaxel content. *P. aphanidermatum* and *P. myriotylum* again respond similarly, with both of these sensitive fungal species exhibiting more sensitivity to the partially purified taxane fraction than to the prepared standard mixture. For *P. irregulare*, however, the prepared standard mixture is more effective in limiting fungal growth than the *Taxus* extract.

Conclusions

Substantial quantities of a non-selective alcoholic extract of *Taxus* needles can be partially purified rapidly and cost effectively using solid phase extraction membranes. The resulting fraction eluted from the SPE membrane with 80% methanol/20% water contains a mixture of taxanes, including paclitaxel, cephalomannine, and 10-deacetyltaxol. This fraction was reconstituted in 95% ethanol, amended into potato dextrose agar, and the radial growth of several fungal plant pathogens on the modified agar was examined. Response of Deuteromycetes and Ascomycetes was substantially less than the response of Oomycetes. The effect on the Oomycetes was considered to be fungistatic since hyphal tips of *P. irregulare* and *P. myriotylum* transferred from PDA amended with the partially purified taxane fraction (4.00 μg paclitaxel/mL PDA) to non-amended PDA grew at rates equivalent to hyphal growth on control plates. *P. aphanidermatum* similarly subcultured grew at lower rates than hyphal growth on control plates.

Although Oomycetes response varied across the species examined, the mixed taxane fraction was generally more effective in reducing fungal growth than single authentic taxanes and competitive with a mixture of authentic taxanes. The development of a single taxane such as paclitaxel as an agrochemical is highly unrealistic at the present time both because of its demand in chemotherapy and the substantial expense of production. A partially purified fraction from the needles of ornamental yews with a consistent taxane profile can be more economically produced. As our research indicates, its potential as an agrochemical warrants further research.

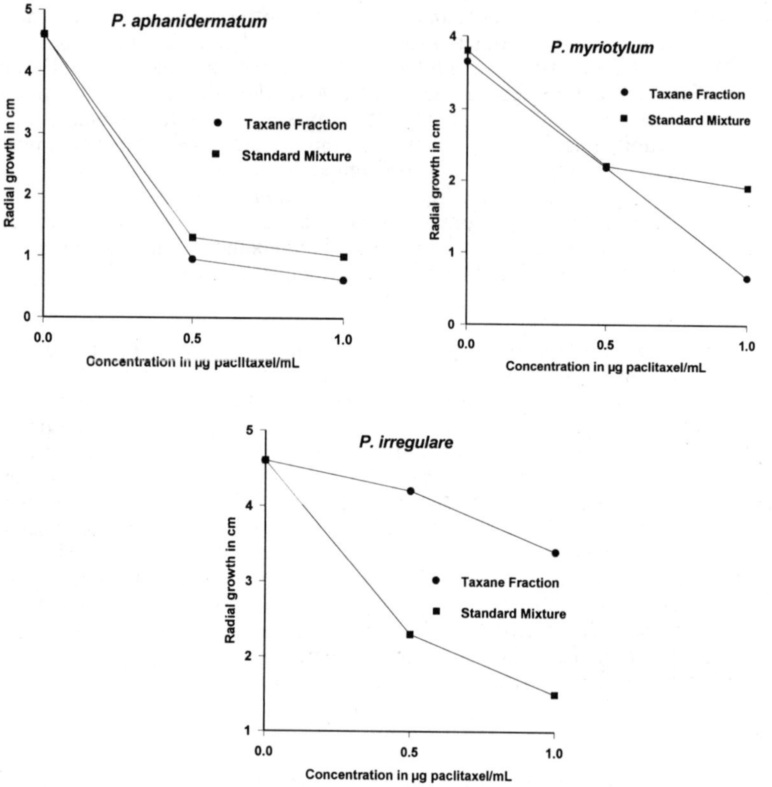

Figure 7. Radial growth on PDA measured after 3 days (Adapted from ref. 14).

Literature Cited

1. Borman, S. *Chem. Engineering News*; Am. Chem. Soc., Washington, D.C. **1991**, *69,* 11.
2. Huizing, M. T.; Sewberath Misser, V. H.; Pieters, R. C.; ten Bokkel Huinink, W. W.; Veenhof, C. H. N.; Vermorken, J. B.; Pinedo, H. M.; Beijnen, J. H. *Cancer Invest.* **1995**, *13*, 381-404.
3. Wani, M.; Taylor, H. L.; Wall, M. E. *J. Am. Chem. Soc.* **1971**, *93*, 2325-2327.
4. Schiff, P. B.; Fant, J.; Horwitz, S. B. *Nature* **1979**, *277*, 665-667.
5. Lataste, H.; Senilh, V.; Wright, M.; Guenard, D.; Potier, P. *Proc. Natl. Acad. Sci. USA* **1984**, *81*, 4090-4094.

6. Baum, S. G.; Wittmer, M.; Nadler, J. P.; Horwitz, S. B.; Denner, J. E.; Schiff, P. B.; Tanowitz, H. B. *Proc. Natl. Acad. Sci. USA* **1981**, *79*, 6569-6573.
7. Vaughn, K. C. and Vaughn, M. A. In *Biologically Active Natural Products: Potential Use in Agriculture*; Cutler, H. G., Ed.; Symposium Series 380; Am. Chem. Soc.: Washington, D. C., 1988, 273-293.
8. Mattina, M. J. I.; Paiva, A. A. *J. Environ. Hortic.* **1992**, *10*, 187-191.
9. Castor, T. P. ; Tyler, T. A. *J. Chromatogr.* **1993**, *16*, 723-731.
10. Hoagland, D. R. ; Arnon, D. I. University of California Agricultural Experiment Station Circular 347, **1950**.
11. Mattina, M. J. I.; MacEachern, G. J.; Elmer, W.; Gent, M. *Abstracts of Papers*; 207th National Meeting; Am. Chem. Soc.: Washington, D. C., 1994.
12. Mattina, M. J. I. ; MacEachern, G. J. *J. Chromatogr.*, **1994**, *679*, 269-275.
13. Young, D. H.; Michelotti, E. L.; Swindell, C. S.; Krauss, N. E., *Experientia* **1992**, *48*, 882-885.
14. Elmer, W. H.; Mattina, M. J. Incorvia; MacEachern, G. J. *Phytopathology* **1994**, *84,* 1179-1185.

Chapter 31

Biological Activity and Synthesis of a Kenaf Phytoalexin Highly Active Against Fungal Wilt Pathogens

R. D. Stipanovic, L. S. Puckhaber, and A. A. Bell

Agricultural Research Center, Southern Crops Research Laboratory, U.S. Department of Agriculture, College Station, TX 77845

A new potent phytoalexin has been isolated from kenaf (*Hibiscus cannabinus*) and identified as 3,8-dimethyl-1,2-naphthoquinone, which we call *o*-hibiscanone. *o*-Hibiscanone had ED_{50} values of 0.5 μg/ml and 1.1 μg/ml against conidia of *Verticillium dahliae* and *Fusarium oxysporum* f. sp. *vasinfectum*, respectively. In comparison, desoxyhemigossypol, the most potent phytoalexin in cotton xylem tissue, had ED_{50} values of 5.2 μg/ml and 8.8 μg/ml, respectively. *V. dahliae* detoxifies *o*-hibiscanone by converting it to the benign hydroquinone. Mansanone C, a phytoalexin from elm which differs from *o*-hibiscanone in that the latter lacks the isopropyl group of the former, was less than one-half as toxic to *V. dahliae*. *o*-Hibiscanone is believed to be derived from δ-cadinene [2,3,4,6,7,8-hexahydro-1,6-dimethyl-4-(1-methylethyl)naphthalene]. A biosynthetic pathway to account for the loss of the isopropyl group is proposed.

Losses to cotton production in the U.S. from Fusarium and Verticillium wilt were estimated to be in excess of $120M in 1996 (*1*) despite extensive efforts to increase resistance to these pathogens using traditional breeding techniques. New sources of resistance to these diseases are therefore needed. In 1966, Idessis tested the resistance of fifteen plant species to twenty strains of *Verticillium* isolated from 17 plant species (*2*). Results of this test showed kenaf, camomile, alfalfa, and snapdragon were highly resistant to most isolates tested. Kenaf (*Hibiscus cannabinus*) is of particular interest because it resides in the same family as cotton (i.e. Malvaceae). In our own study, we found four varieties of kenaf had fresh leaf weights two weeks after inoculation with the *Verticillium dahliae* strain V-76 (a strain virulent to cotton) which averaged 89% (median 87%) of the uninoculated control. Under similar conditions, susceptible Acala 44 and resistant Acala Prema cultivars had fresh leaf weights two weeks after inoculation that averaged 8% and 59%, respectively, of the uninoculated controls (Table I).

Table I. Effect of *Verticillium dahliae* Infection on Kenaf and Cotton Leaf Weight.

Plant	Cultivar	Leaf Weight[a,b]
Kenaf	Everglade 71	95 (13)
	Guatamala 4	79 (7)
	45-9	88 (4)
	Everglade 41	95 (22)
Cotton	Acala 44	8 (8)
	Acala Prema	59 (22)

[a]Relative mean leaf weight at 2 weeks after inoculation at 27°C: infected leaf weight divided by average uninfected leaf weight times 100.
[b]Means and standard deviations for Everglade 71, Guatamala 4 and 45-9 are based on four leaf samples, for Everglade 41 on three leaf samples, and for Acala 44 and Acala Prema on eight leaf samples.

The mechanism of resistance in kenaf is not known, although, in the case of cotton we have shown the phytoalexins play a critical role in the defense response (*3*). In order to examine the role of phytoalexins in the disease response in kenaf, we analyzed extracts from kenaf stems inoculated with *V. dahliae*. These extracts were spotted on TLC plates, developed in two dimensions, then over-sprayed with

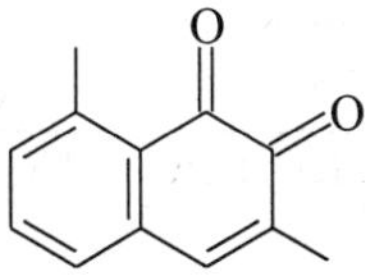

HBQ
o-Hibiscanone

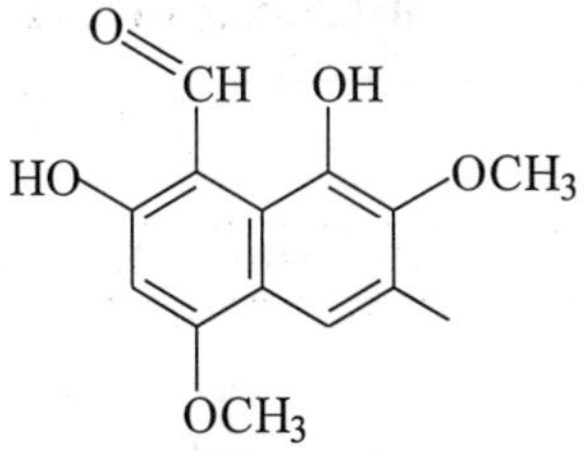

Hibiscanal

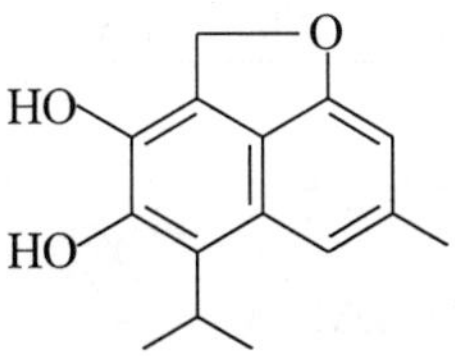

Desoxyhemigossypol
dHG

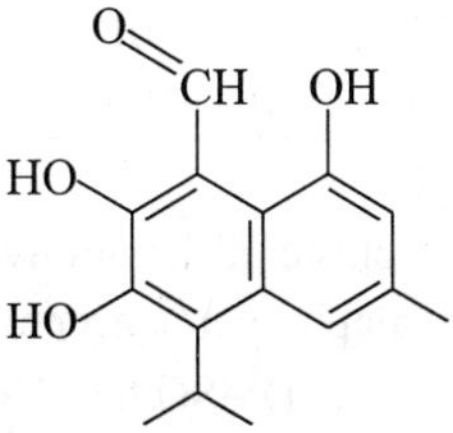

Hemigossypol
HG

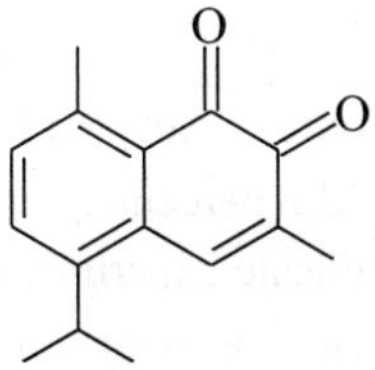

Mansanone C

agar and with a spore suspension of *V. dahliae*. The spores were allowed to germinate over 2 days, and the plates were examined for clear zones which indicated antibiotic activity. Extracts from uninoculated plants were used as controls. Two compounds exhibiting antibiotic activity were isolated from the inoculated stems and identified as *o*-hibiscanone and hibiscanal. The characterization of these compounds is reported elsewhere (*4*). These compounds have certain structural characteristics in common with the phytoalexins from cotton [i.e. desoxyhemigossypol (dHG) and hemigossypol (HG)].

Fungitoxicity

The fungitoxicity of *o*-hibiscanone and hibiscanal to both *V. dahliae* and *Fusarium oxysporum* f. sp. *vasinfectum* have been determined using a direct turbimetric bioassay (*5*). Hibiscanal exhibit ED_{50} values of 25.8 μg/ml and 36.5 μg/ml to *V. dahliae* and *F.o.v.*, respectively. *o*-Hibiscanone was significantly more toxic to these pathogens exhibiting ED_{50} values of 0.5 μg/ml and 1.1 μg/ml, to *V. dahliae* and *F.o.v.*, respectively. dHG, the most toxic cotton phytoalexin (*6*), exhibited in the same bioassay ED_{50} values of 5.2 μg/ml and 8.8 μg/ml (*5*) to *V. dahliae* and *F.o.v.*, respectively.

The toxicity of *o*-hibiscanone to mycelia of four isolates of *V. dahliae* has also been determined. The results of this experiment are shown in Table II and are compared to the toxicity of dHG as previously reported (*6*). These bioassays show *o*-hibiscanone is significantly more toxic to both *V. dahliae* and *F.o.v.* than the most effective phytoalexin produced by cotton.

Table II. Toxicity of desoxyhemigosssypol (dHG)[a] and *o*-hibiscanone (HBQ) to mycelia of 4 isolates of *Verticillium* dahliae (cotton nondefoliating isolates PH and TS-2 and defoliating isolates V-44 and V-76).

	Percentage Live mycelia[b]							
Concentration	PH		TS-2		V-44		V-76	
(µg/ml)[c]	dHG	HBQ	dHG	HBQ	dHG	HBQ	dHG	HBQ
20	0	-	0	-	0	-	0	-
15	64	3	56	0	22	3	42	0
10	100	6	100	3	100	6	100	6
5	-	36	-	61	-	25	-	33
0	100	100	100	100	100	100	100	100

[a]Source: Adapted from ref. 6.

[b]Mean percentage of wells in cluster dish with live mycelia; three replicate experiments of 12 samples per concentration.

[c]pH6.3 nutrient solutions with 1% DMSO for dHG and 2% DMSO for HBQ.

Figure 1 compares the toxicity of *o*-hibiscanone in the turbimetric bioassay to that of mansanone C. The latter is a phytoalexin isolated from elm (*7,8*). Mansanone C differs from *o*-hibiscanone in that the latter lacks the isopropyl group of the former. Loss of this isopropyl group increases the toxicity by more than five fold (ED_{50} mansanone C = 2.29 μg/ml; ED_{50} o-hibiscanone = 0.44 μg/ml).

Biotransformation of *o*-Hibiscanone by *V. dahliae*

o-Hibiscanone was synthesized as shown in Figure 2. By utilizing ^{13}C-labeled methyl magnesium iodide, it was possible to prepare ^{13}C-labeled *o*-hibiscanone. When *o*-hibiscanone was allowed to incubate with *V. dahliae*, the solution which was originally bright yellow turned colorless. A ^{13}C-NMR study showed that the ^{13}C-labeled methyl group which occurs at 23.6 ppm is gradually replaced over 140 minutes by a peak at 24.75 ppm. On mixing with air, the solution returns to its original color. The colorless conversion product was shown to be the hydroquinone by synthesis (i.e. chemical reduction of *o*-hibiscanone by sodium dithionite).

o-Hibiscanone also is reduced slowly by glutathione. On standing overnight with glutathione (132 μg/ml), *o*-hibiscanone (40 μg/ml) was completely reduced to its hydroquinone. Under anaerobic conditions, this hydroquinone solution or a fresh solution of the quinone (40 μg/ml) and glutathione (132 μg/ml) were mixed with *V. dahliae* (V-76) conidia (10^6 conidia/ml) and allowed to stand for 0.2, 15, 30 and 60 minutes. After the indicated time, aliquots were diluted and pipetted onto individual PDA plates, and smeared to distribute the conidia evenly over the plate. After 5 days the number of colonies on each plate were counted. The bioassay was repeated three times. As shown in Table III, the hydroquinone was essentially non-toxic to the pathogen under conditions in which the quinone was lethal.

Table III. Comparison of the fungitoxicity of *o*-hibiscanone (HBQ) and its hydroquinone (HBQ-HQ) to conidia of *Verticillium dahliae* (V-76) exposed for the indicated time under anaerobic conditions.

Time[a] minutes	Precentage of Live Conidia HBQ[b]	HBQ-HQ
0.2[c]	10	100
15	1	100
30	0	100
60	0	100

[a]Time of exposure of conidia ($1x10^6$) to compounds.
[b]Concentration of HBQ and HBQ-HQ were 40 μg/ml; media contained 132 μg/ml glutathione and 1.8% DMSO.
[c]Approximate time required to mix chemicals with conidia and immediately dilute solution in preparation for transfer to PDA plates.

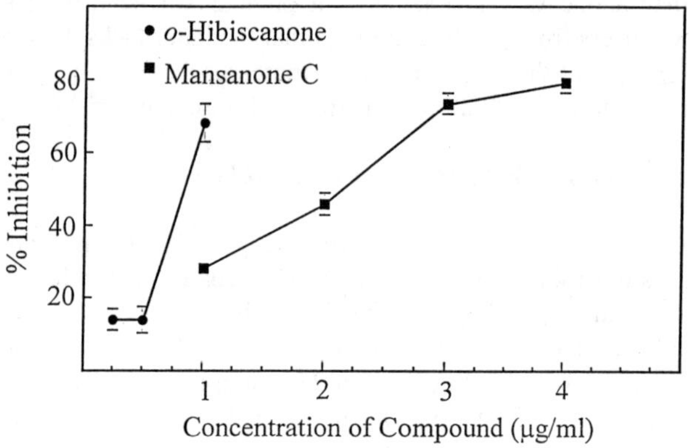

Figure 1. Toxicity of *o*-hibiscanone and Mansanone C to *Verticillium dahliae* (V-76)

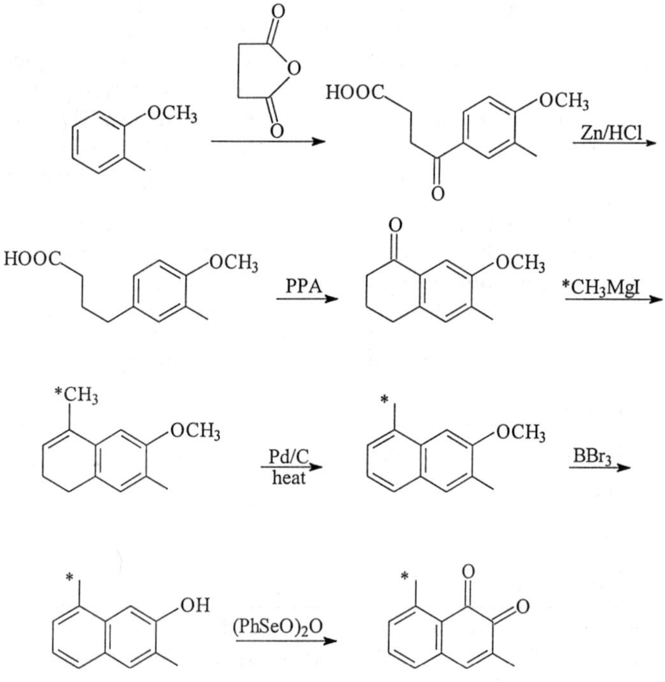

Figure 2. Synthesis of *o*-hibiscanone (* indicates location of ^{13}C-label).

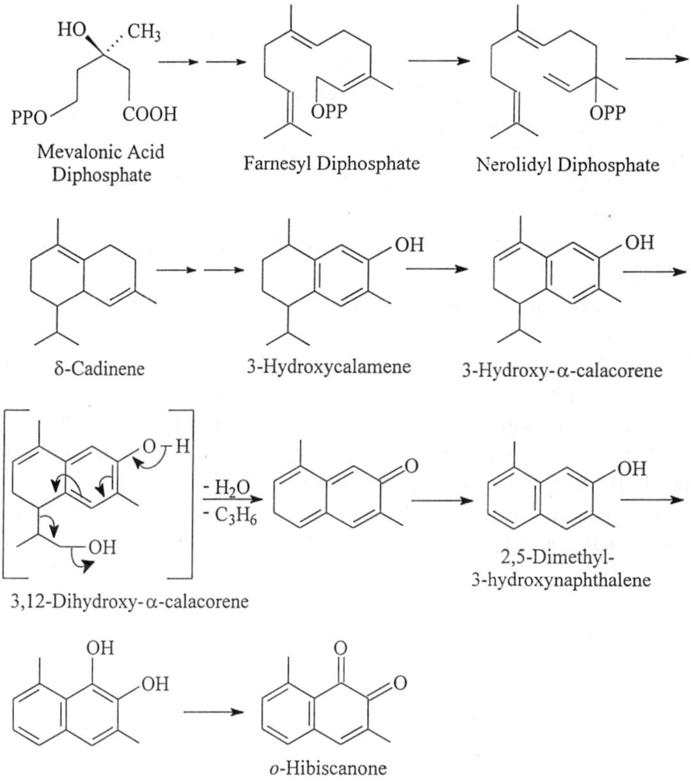

Figure 3. Proposed biosynthesis of *o*-hibiscanone in kenaf.

Proposed Biosynthesis of *o*-Hibiscanone

o-Hibiscanone is thought to be derived from δ-cadinene via the isoprenoid pathway. A proposed biosynthetic pathway is shown in Figure 3. Benedict and coworkers (*9*) have shown nerolidyl diphosphate is the precursor to δ-cadinene in cotton and Essenberg and coworkers (*10*) have shown the cadalenes are derived from δ-cadinene. Using GC/MS comparisons with synthetic compounds, we have identified 3-hydroxycalamene, 3-hydroxy-α-calacorene, and 2,5-dimethyl-3-hydroxynaphthalene in extracts of cold shocked kenaf seedlings. 3,12-Dihydroxy-α-calacorene is proposed as the intermediate between 3-hydroxy-α-calacorene and 2,5-dimethyl-3-hydroxynaphthalene. 3,12-Dihydroxy-α-calacorene is expected to spontaneously dehydrate with the concomitant loss of propene.

Literature Cited

1. Blasingame, D. In *Beltwide Cotton Production Research Conference*; P. Dugger; D. Richter, Eds.; National Cotton Council, Memphis, TN, 1996, p 227.

2. Idessis, V. F. In *Cotton Wilt*; M.V. Mukhamedzhanov, Ed.; Academy of Sciences of the Uzbek, USSR:1966; pp 61-64.
3. Bell, A. A.; Stipanovic, R. D.; Mace, M. E.; Kohel, R. J. In *Recent Advances in Phytochemistry Vol. 28*; B. F. Ellis; G. W. Kuroki; H. A. Stafford, Eds.; Plenum Press, New York:1994, pp 231-249.
4. Bell, A. A.; Stipanovic, R. D.; Zhang, J.; Reibenspies, J.; Mace, M. E. *Phytochemistry* (submitted).
5. Zhang, J.; Mace, M. E.; Stipanovic, R. D.; Bell, A. A. *J. Phytopathology* **1993,** *139*, 247-252.
6. Mace, M. E.; Stipanovic, R. D.; Bell, A. A. *Pest. Biochem. Physiol.* **1990,** *36*, 79-82.
7. Dumas, M. T.; Strunz, G. M.; Hubbes, M.; Jeng, R. S., *Experientia* **1983**, *39*, 1089-1090.
8. Burden, R. S.; Kemp, M. S. *Phytochemistry* **1984**, *23*, 383-385.
9. Alchanati, I.; Acreman-Patel, J. A.; Liu, J.; Benedict, C. R.; Stipanovic, R. D.; Bell, A. A. *Phytochemistry* (in press).
10. Davis, E. M.; Tsuji, J.; Davis, G. D.; Piera, M. L.; Essenberg, M. *Phytochemistry* **1996,** *41*, 1047-1055.

INDEXES

Author Index

Subject Index

D

E

F

G

H

I

N

O

P

U

V